NETWORK MODELS IN OPTIMIZATION AND THEIR APPLICATIONS IN PRACTICE

NETWORK MODELS IN OPTIMIZATION AND THEIR APPLICATIONS IN PRACTICE

FRED GLOVER

DARWIN KLINGMAN

NANCY V. PHILLIPS

A WILEY–INTERSCIENCE PUBLICATION

JOHN WILEY & SONS, INC.

NEW YORK • CHICHESTER • BRISBANE • TORONTO • SINGAPORE

Library of Congress Cataloging in Publication Data:

Glover, Fred

Network models in optimization and their applications in practice /by Fred Glover, Darwin Klingman, Nancy V. Phillips

p. cm.

"A Wiley-Interscience publication."

Includes bibliographical references and index.

ISBN 0-471-57138-5 (cloth : alk. paper)

1. Mathematical optimization. 2. Operations research. 3. System design. I. Klingman, Darwin, II. Phillips, Nancy V. III. Title.

QA402.5.P47 1992

003--dc20 91-41110

CIP

Printed and bound in the United States of America by Braun-Brumfield, Inc.

10 9 8 7 6 5 4 3 2 1

To Darwin (February 5, 1944–October 27, 1989),
whose aspirations for excellence inspired us all.

Nancy Phillips
Fred Glover

PREFACE

INTRODUCTION

In a world where terms like "ribosomal RNA," "quasars," and (around April 15) "accelerated cost recovery system" are commonplace, the term "network flow optimization" is perhaps not unduly forbidding. In spite of the growing influence of the network realm on our lives, however, it is not often publicized by the popular press, and comments about it are not to be heard rolling off the tongue either of talk show hosts or of the public at large. What then is network flow optimization, and what is its connection to the concerns of the modern world?

To answer this question, it is helpful to go back to the early days of World War II, when the eminent US economist and logistics analyst Tjalling Koopmans was faced with the problem of moving personnel, supplies, and equipment from various US bases to their foreign counterparts. The goal was to do this in a way that would optimize one or more objectives: minimize total transportation cost, minimize total transit time, and/or maximize defensive effectiveness.

Although Koopmans was the first to analyze maritime distribution problems in the network model format, he was not in fact the first to discover this approach. A few years earlier, but unknown to the western world and, ironically, largely disregarded by his own countrymen for two decades, a distinguished Russian mathematician/economist, L. V. Kantorovich, was studying important problems associated with the Soviet economy. One problem that concerned him was the allocation of production levels to factories and the distribution of the resulting products to markets. He visualized his problem in a form highly analogous to that of Koopmans' approach, and thus developed one of the earliest network models, within a framework relevant to practical concerns.

The significance of these models lies not in the particular applications that first inspired them, but rather in the fact that they provide the seeds of a general methodology for structuring and analyzing complex decision problems. Koopmans and Kantorovich could not have foreseen the consequences of the special frameworks they pioneered. We now live in a world where network flow optimization directly and indirectly influences high-level decision making around the globe—not only in economics, but in all aspects of industry and government planning. Whether determining the most profitable levels of production, the best cycles of harvesting and planting, the most effective uses of energy, the optimum allocation of scarce resources, or the best decisions in financial planning, network flow optimization is consistently relied on to improve our analysis capabilities and has been documented with saving millions of dollars. Even so, relative to its full potential, we are witnessing only the beginning.

Nobel Prize committees are often behind the times in recognizing the importance of innovations that have changed our lives. In the case of networks, however, they may justifiably be credited with being ahead of the times. They bestowed the honor of the Nobel Prize on Tjalling Koopmans in 1974 and L. V. Kantorovich in 1975. This gave recognition of the importance of network models as a decision framework—a foresight vindicated by the explosive development of the field and its applications since the early 1970s.

Three properties are responsible for the widespread use of network flow optimization: (1) visual content, (2) model flexibility and comprehensiveness, and (3) solvability.

1. *Visual Content.* Network flow optimization allows a problem to be visualized by means of diagrams, thereby making it possible to capture important interrelationships in an easily understood "pictorial" framework. This property has allowed nontechnical and technical individuals alike to gain valuable insights into their problems, confirming the old saying that "a picture is worth a thousand words."

Due to recent discoveries, the visual aspect of networks may have interesting consequences from a learning and intelligence perspective. Psychologists have found that systems which represent concepts in a variety of visual and auditory ways improve learning and memory and thereby enhance our cognitive processes. Further, studies of information-processing functions of the brain disclose that different hemispheres of the brain are specialized to perform different mental functions, where typically the right hemisphere takes a dominant role in functions related to spatial imagery; and the left hemisphere is devoted more fully to processing serial, analytical, or linguistic information. By dual coding—putting information in both verbal and visual languages—researchers conjecture that the words, sounds, and images naturally lead the separate parts of the brain into a highly coordinated effort. With this type of learning device, one does not just accumulate knowledge; one actually acquires greater facility in problem solving. If this is true, then we conjecture that network modeling and analysis is, in addition to its other virtues, an ideal tool for increasing the ability to grasp critical problem features and to reason more effectively about the domains to which it is applied. Conceivably, this integration of visual and symbolic elements may stimulate "dual coding" in other parts of our thought processes as well.

2. *Model Flexibility and Comprehensiveness.* A major characteristic of the network framework is the diversity of problems that can be modeled and solved by its application. Today it is difficult to name a field which has not benefited from network modeling or in which network flow optimization is not an important research component. The algorithmic aspect of the field has attracted mathematicians for many years. In the realm of the physical sciences and engineering, networks have found application to such things as electrical circuit board design, telecommunications, water management, design of transportation systems, metalworking, chemical processes, aircraft design, fluid dynamic analysis, and computer job processing.

Networks also are used in the arts to analyze group communication phenomena in sociology, to determine ancestral orderings in archaeology, to probe the effect of government fiscal and regulatory policies, and to allocate library budgets to determine what types of publications to acquire.

Businesses are using network flow optimization in practically every sector. A wide array of problems in production, marketing, distribution, financial planning, project selection, facility location, and accounting—to mention only a few—falls naturally into the network domain.

3. *Solvability.* An early network problem (although not formulated as a network model at the time) dates back to the French Academy of Sciences in 1781, which proposed an award for developing a solution procedure for the civil engineering problem of "cutting and filling"—the problem of creating a level plot of land by cutting down hills and filling in valleys. The award went unclaimed and the problem remained unsolved for more than 150 years. Kantorovich, Koopmans, and others similarly attempted to devise solution procedures for the problems they faced, but none were successful. The first solution method with a theoretical guarantee of optimality appeared in 1947 when US Science Medalist G. B. Dantzig developed the ***Simplex Method***. Even so, the ability to solve many network problems of real-world complexity remained undemonstrated, because the theoretical solution capability did not at once translate into computer procedures that could solve such networks within a reasonable length of time.

The 1950s and early 1960s saw the further development of solution algorithms and the identification of new problems which could be formulated and solved as networks. By the late 1960s, the state of the art was still beset by notable practical limitations, however. The largest network problems to be solved contained about 600 nodes and 2000 arcs. Government and business applications, however, demanded the solution of much larger problems.

Research in the 1970s and 1980s resulted in computer codes 100–150 times faster than the best codes of the 1960s. The computer software provided by this technology is currently being used by more than 100 government agencies and

200 companies, and has been credited with saving over $500 million in practical applications. Perhaps not surprisingly, these developments have brought about a shift of research focus on a worldwide scale during the past decade, causing increased numbers of researchers to study networks and closely related problems. Commenting on these events, Nobel laureate Herbert Simon notes that these contributions have "brought computational mathematics back into the mainstream of mathematics, as a source of fundamental new problems and theory."

In reflecting on the resulting increase in our ability to deal with a variety of practical problems, it is important to keep in mind that advances in modeling can be as important as advances in solution methods. By the mid-1970s, it became apparent that expanded problem representations were needed to handle features beyond the scope of existing formulations. Accordingly, an approach called netform modeling emerged that allows go/no-go problems (often called 0-1 discrete optimization problems) to be formulated in a network-related format. A bonus of netform modeling is that it fosters insights into problem structure that often lead to efficient specialized solution methods. This close link between models and solution methods has led to the term "netform approach" to describe the combined result of using a netform model and a specialized solution procedure based upon it.

For the past several years, the netform approach has been used by several government agencies and private corporations to solve problems which were heretofore unsolvable. For example, one company credits this approach with solving a problem in less than 30 min of computer time that previously had not yielded successfully to efforts spanning 25 person years. In another application of the netform approach, an agency was able to obtain a solution to a problem which was $10 million better than the best solution identified by other means.

The ability to economically solve large-scale problems on many fronts opens the door to new perspectives and options, inviting a redefinition (and re-evaluation) of many older but still prevalent approaches to problems that range from economics to ecology. In many respects the research challenges are just beginning! New opportunities for practical application are bringing with them new demands on the underlying theory—by developing further elaborations and extensions of the models and solution approaches that have proved successful. Although we may not expect very soon to hear late-night talk show hosts exchanging remarks with their guests about the latest in "network flow optimization," this is not likely to slow the continued dramatic progress of the field.

PREREQUISITES

This book is written primarily for college students who have completed an introductory course covering linear programming. No mathematical sophistication beyond college algebra is required. The text can be used in a one-semester course for juniors, seniors, or graduates in business or engineering curricula.

COVERAGE AND ORGANIZATION

The book covers modeling techniques for pure, generalized, and integer networks, equivalent formulations, and successful applications of network models. At the end of each chapter is (in order) a synopsis of a real-world application, one or more case studies, exercises, and references. The references, cited throughout the text by [author, year], guide the reader to more information about the particular subject being discussed. (A list of selected readings not cited specifically in the text appears at the end of the book in Appendix C.) The cases and exercises are often scaled-down versions of real problems we have encountered. More difficult exercises are marked by an asterisk (*). Important terms appear throughout in boldface italic type. In teaching this course, we have found in-class discussion of the cases and exercises to be very helpful. A solutions manual is available from the publisher.

ACKNOWLEDGMENTS

We gratefully acknowledge the people who assisted in preparing this book. Special thanks go to Dr. Peter Mevert for reviewing the manuscript and making many valuable suggestions, and to Connie Pechal for her tireless efforts in typing the manuscript and preparing figures.

FRED GLOVER
NANCY PHILLIPS

Boulder, Colorado
Greer, South Carolina
June, 1992

CONTENTS

NETWORK MODELS IN OPTIMIZATION AND THEIR APPLICATIONS IN PRACTICE

1

NETFORM ORIGINS AND USES: WHY MODELING AND NETFORMS ARE IMPORTANT

1.1 BACKGROUND

The term netform stands for network flow-based (or network-related) formulation. Netforms are characterized by the use of diagrams that have emerged, by progressive elaboration, from those used traditionally in network flow and graph theory. A major discovery of the last decade is that computer and mathematical models once believed to have little or nothing in common with network flow problems can be treated productively in the netform framework. The range of important decision support problems that netforms can model, for example, embraces many of the traditional problems of management science and mathematical optimization, including those of the widely known domains of linear and integer programming.

The visual component of netform models makes a wide array of problems from industry and government far easier to comprehend than by classical formulation techniques, and constitutes one of the most conspicuous features of these models. Yet there are other features by which netforms are proving useful as a practical modeling tool. Among them is the ability to write down a mathematical version of a netform model directly "from the picture," providing a better understanding of the connection between alternative forms of problem representation.

The focus on netform models in this book derives from the conviction that effective modeling is the most important and fundamental skill any potential "problem solver" can learn. The crucial first order-of-business is to know which problem to solve, and as a corollary to know how to represent the problem so that its features can be reviewed and analyzed. Unfortunately, many do not have the modeling skill to be sure whether the problem they think they are solving is the real one. The most refined mathematical solution techniques and the most powerful computers cannot make up for a poor problem representation. Practitioners are well acquainted with the frustration of taking great pains to make a problem formulation or model accurate, only to find the result so opaque that no one can crosscheck it to confirm or deny its validity. When conditions change (as they inevitably do), such models quickly turn into dinosaurs. It happens also that some models prove very difficult to solve, chewing up incredible chunks of computer time (and money) before they yield answers. Knowing good frameworks for representing problems can help greatly to avoid such pitfalls.

In terms of a popular analogy, a modeler may be viewed as a "map maker" and a problem solver as a "navigator." A poorly prepared map (model) may distort the features of the terrain sufficiently that even the best navigators will be unable to find a way from one point to another—or, having found such a way, may arrive at the wrong destination after all. Clearly, a navigator must know quite a bit about maps to be effective. This points up once again why modeling is among the most fundamental of pursuits.

Yet some may suggest that today's informal decision-maker need not possess an understanding of models. After all, it is always possible to hire a map maker to lay out the terrain and a guide to take over after that. But the decision-maker on the scene knows more about the terrain than anyone else, and unless he has some skill at sketching it—or, at the very least, has some facility to comprehend a sketch by someone else—he is like a passenger who hopes to reach a destination blindfolded (with a blindfold provided for the guide as well). All too often decision-makers have been "led down the path" as a result of this sort of arrangement. Additionally, many good models are never used because the decision-maker does not understand them. In many cases, managers would rather live with a simple-minded solution to a model they understand than accept a sophisticated solution to a model beyond their comprehension. The emergence of netform models, with their intuitive visual component and the wealth of settings to which they apply, has contributed significantly to mitigating this situation.

1.2 NETFORM MODELING IN THE CONTEXT OF MANAGEMENT SCIENCE

Because netform modeling, as a decision support tool, has evolved through the years from the practice of management science, it is appropriate to provide a bit of the background of management science, and to indicate some of its major perspectives.

Historically, the fields of management science (MS) and operations research (OR)—now viewed as the same domain—began with primitive mathematical economic models developed in the eighteenth and nineteenth centuries by Quesnay and Walras and extended in the early twentieth century by Von

Neumann and Kantorovich. The active application of OR emerged as a highly interdisciplinary undertaking during World War II when engineers and physical scientists, five of whom later won Nobel Prizes, were called upon to research the operational problem of using radar to defend Britain against air attack. Applications of OR grew rapidly in the years following and spread to diverse fields, including the public and private sectors. Today, societies such as The Institute of Management Sciences (TIMS), the Operations Research Society of America (ORSA), the Decision Sciences Institute (DSI), the Canadian Operations Research Society (CORS), the International Federation of Operations Research (IFORS), the European Congress on Operations Research (EURO), and the Mathematical Programming Society (a list that is by no means exhaustive) hold national and international meetings that are typically attended by thousands of people.

In the early days, MS/OR teams, assembled to study particular problems of the military or government—and later, increasingly, of industry—were typically composed of individuals of diverse specialities, each possessing his or her own expertise. Gradually, a body of special models and problem-solving methods were developed, with the specialists called management scientists. This is the setting in which netform models emerged, and were imparted with a flexibility that made them applicable to a broad range of practical settings. As early as 1956, it was estimated that 70% of all linear programming applications, for example, were in fact problems that could be classed as "network flow problems."

The cross-fertilization between MS and other fields, and the effect of this on developing versatile models, is very evident today. The extent to which MS practitioners have made their way into, and left their mark upon, a diverse array of disciplines is little short of staggering. Management scientists have established "home bases" in most segments of industry and government, and also in significant areas of social science, psychology, biological science, and physical science. Academic departments devoted to teaching MS are typically found in business and engineering schools. Frequently, MS practitioners investigate problems that transcend traditional boundaries.

Most of us are well aware that problems arising in large organizations and complex systems are not all susceptible to solution by common sense and practical experience. Many disciplines within the domains of business and organizational study, however, rely heavily on these problem-solving "techniques". It is sometimes tempting to suppose we can do better on our own than with an assist from technology. In the framework of management science, on the other hand, the goal is to provide analytical models and methods for dealing with problems that cannot adequately be treated by rules of thumb and follow-the-dotted-line analyses. Management science promotes the view that there is such a thing as a right tool for the right application, where "tools" include such things as model representations, mathematical procedures, and computer programs, as well as cookbooks of generally accepted principles.

OR model-based systems are especially useful to help organizations analyze and gain insight from the multitude of data which many collect and maintain today with help from readily available cheap computer power. As stated in an article by the Committee on the Next Decade in Operations Research [CONDOR 1988], "Data availability, once the Achilles' heel of OR, is now a driving force." Yet some would argue that computers, having done much to promote the growth of management science, are now rendering a good deal of management science superfluous. To be more specific, the power of present-day computers now typically reduces the task of evaluating complex alternatives to a simple exercise. For example, one might argue that a manager who seeks the best assignment of 70 employees to 70 tasks can simply have the computer itemize all alternatives and pick the best. In such a case, so the argument goes, no sophisticated analysis technique is necessary.

There are of course a variety of possible responses to the foregoing perspective. One of the more intriguing is due to George Dantzig, "the father of linear programming," who has posed the preceding employee-to-task assignment example as a rather ingenious way of showing how reliance on computer power can be misplaced even in an extraordinarily simple setting. With only 70 employees to be assigned to 70 tasks, the number of alternatives is $70 \times 69 \times 68 \ldots$, since 70 employees are available to be assigned to the first task, and for each of these, 69 remain for the second, etc. This number is just a bit larger than 10^{100}. Suppose we have a reasonably powerful computer capable of evaluating a million assignments/second. If we started the computer when the earth began, it still would not have generated all the assignments! So suppose instead we have a "nanosecond computer" able to generate a billion assignments/second. Starting this computer when the earth began also would not generate all assignments by today. If we started this super computer at the beginning of the Big Bang (some 10–15 billion years ago) and let it run until the sun grew cold . . . still its work would not be done. If we filled the earth with these super computers, started them all at the Big Bang, let them run till the sun grew cold, again not all assignments would be generated. We would succeed in generating all assignments by filling the solar system with 10^{50} earths (or 10^{44} suns) each filled with these nanosecond computers, and letting them all run from the moment of the Big Bang until the sun grows cold!

It is interesting to note that such an assignment problem is indeed one of the simplest kinds of network models, and can be solved with present methods in seconds on some personal computers. Real-world applications involving the assignment of hundreds of thousands of personnel to jobs are routinely solved in minutes on mainframe computers.

Many people today, however, are pretty knowledgeable about the limitations and potentials of computers, and about the important roles played by appropriately selected models and methods. Quite probably, a lot of this is due directly or indirectly to the expanding applications of management science. Moreover, many people in fields outside of MS are picking up bits and pieces—in some cases, large chunks—of management science technology as evidenced by the applications discussed later and listed in the references. Such individuals, often holding important managerial positions in industry and government, are acquiring knowledge of models and methods at a level that makes them increasingly competent to recognize the settings where MS can make a valuable contribution, and pass judgment on the relative merits of alternative proposals. These individuals are also increasingly found working together with MS and decision support specialists in solving problems of common concern. This healthy process reflects a rising awareness about the kinds of approaches that are becoming available for solving practical problems, as shaped by—and also, in turn, shaping—the "computer revolution."

Still, even though practitioners of other disciplines are beginning to adopt some of the orientation and assimilate some of the techniques of management scientists, there remain some general, fairly concrete distinctions between the primary emphasis of management science and other areas.

To illustrate, consider the domain of inventory management, as approached from an accounting orientation and an MS orientation. The two kinds of approaches might roughly be differentiated as follows:

Accounting Approach to Inventory Management, Primary Functions

1. Prepare and keep records of purchases and sales of inventory items (thus, for example, enabling vendors to be paid and purchasers to be billed in an appropriate fashion).
2. Keep track of the processes and products in which different inventory items are used (as for specifying costs of operations and goods sold).
3. Identify appropriate inventory valuations (such as original cost, market cost, cost based on different usage sequences, cost based on tax considerations, etc.).
4. Link inventory records with other records (for preparing balance sheets, profit and loss statements, etc.).

In summary, the primary role of accounting in inventory management is to record, organize, and report transactions to document the progress of the enterprise, and to provide meaningful data that managers can use to make decisions.

Management Science Approach to Inventory Management, Primary Functions

Develop models and methods for answering questions such as:

1. How can inventories be controlled to maximize profits?
2. What are the optimum amounts of each type of inventory item to be purchased? How much of various inventories should be stored and carried over from one period to another?
3. How much of each type of raw material or intermediate goods inventory should be allocated to various production processes? (This ideally in-

volves integrating inventory and production, by coordinating inventory decisions with the determination of best amounts to be produced by alternative processes.)

4. What are the optimum quantities of finished goods to be delivered to specific customers in given periods? (How should backorders be managed?)
5. Which warehouses should store and ship different components of the finished goods inventory?
6. To which customers, and by what distribution channels, should the shipments from each warehouse be made?

In summary, the primary focus of MS in inventory management is to identify those decisions, within a set of numerous complex and interacting alternatives, that help to optimize the company's profits. That is, MS may be used to analyze the impact of a decision on the total organization rather than just the functional area which implements the decision.

There is, of course, some distortion inevitably in any differentiation based on "primary emphasis." In the practical world, MS strongly overlaps not only accounting but many of the traditional disciplines.

1.2.1 The Specific Role and Background of Netforms

Within this setting, the netform modeling techniques make it possible to represent many MS problems in a very natural way. The inventory management example just described is a good illustration. Netforms are ideally suited to represent the ramifications of coordinating inventories with production and distribution activities as illustrated by the applications discussed later and in the references.

But where precisely in this scheme of things did netform models get the impetus that brought them into prominence? It might seem their versatility would provide impetus enough, but this was not the case. Many of the features we now see as important benefits of netforms were initially discounted or not anticipated. Although networks were appreciated by specialists as an important problem area from the start, a good deal of this appreciation was wedded to theoretical considerations, and the full practical potential of the first models was beyond the dreams of most.

The diagrammatic representations used in network models owe a considerable debt to those of graph theory, whose origins date back to the 1700s with the work of the Swiss mathematician Euler. One of the first practical network problems (not classed as such at the time) was a "cut and fill" problem (creating a level plot of land by cutting down hills and filling in valleys) posed by the French Academy of Sciences in 1781. A prize was offered to anyone who could provide a solution, but was never claimed. This problem was a variant of the now famous bipartite (two-piece) "transportation" problem formulated by the Soviet economist Kantorovich [1939]. Closely related network problems were studied by two other mathematical economists, Hitchcock [1941] and

Koopmans [1947]. (The transportation model is also known as the Hitchcock–Koopmans model. Kantorovich, Hitchcock and Koopmans all subsequently received Nobel Prizes, in part because of their studies of such models.) However, the real beginnings of the general awareness of network models did not come about until after World War II. The first general solution capability for these problems was provided by George Dantzig with the development of the simplex method for solving linear programs in 1947, which he specialized to provide the first "network method" for bipartite problems [Dantzig, 1951]. From the mid to late 1950s, with the contributions of several researchers, including Orden [1956], Charnes and Cooper [1961] and, especially, Ford and Fulkerson [1962], the foundations of pure network models and methods were laid.

If any period can be picked as a watershed in network modeling and solution, however, it was probably the late 1960s. At that point, computational experiments in ways of refining solution procedures for networks, from the standpoint of practical computer implementation, led to advances in computational efficiency that caught theoreticians and practitioners alike by surprise [see, e.g., Golden and Magnanti 1977 and Grigoriadis 1980 for surveys]. With each new gain came an increased awareness of the savings in computer time and dollar savings by formulating problems as networks. This was soon joined by the discovery that a variety of problems customarily viewed from a "mathematical" orientation could be viewed alternately—and more conveniently—as networks.

By the early 1970s network applications were springing up in abundance. Management scientists were also beginning to notice that many additional problems were "almost" networks. These problems, called *generalized networks*, were found likewise to be susceptible to extremely efficient solution. Then came the discovery that a significant portion of more complex problems—including some of the infamously difficult discrete (integer programming) variety—could be treated advantageously by isolating components which were networks or "network-related" structures. The class of what we now call netform models, bringing these various network-related models together, came into existence. Continuing research in this area is documented in thousands of articles and books [see Ahuja et al. 1989 for examples]. This includes research on modeling innovations, algorithm development, and analysis of algorithm performance (both empirically and in terms of worst or average case theoretical bounds). This book focuses on modeling. Other texts, such as Ahuja et al. [1992], Jensen and Barnes [1980], Kennington and Helgason [1980], Murty [1992], and Rockafellar [1984], provide detailed discussion of algorithms and their performance.

What are the consequences of these advances in solution and modeling technology? From the standpoint of methods, two stand as paramount:

1. Problems whose vast size put them beyond the reach of earlier methodology can now be solved routinely. Some of these involve hundreds of thousands or even millions of variables.
2. Small to moderate sized problems, formerly requiring many minutes or hours of computer time to solve, can now be solved in a small fraction of that time using personal computers. This enables the analyst to deal with

such problems interactively, thus allowing management to explore the consequences of alternative assumptions and scenarios.

The benefits that have accrued as a result of the "visual content" of netform models are no less impressive. Netforms illuminate underlying problem relationships in a way not possible with more abstract formulations, whose attendant symbols and mathematical expressions can often be constructed and deciphered only with great difficulty. At the same time netform models are completely rigorous, and can be translated directly into mathematical language by means of a few simple rules.

From the practitioner's standpoint, some of the more conspicuous attributes of the "visual content" of netform models include:

1. Improved communication between managers and management science practitioners (facilitating feedback and review through the pictorial element);
2. Improved model fidelity (reducing the chances of leaving out important relationships and of inadvertently creating bogus ones, through the increased visibility of problem interconnections);
3. Improved sensitivity analyses (encouraging and more readily providing answers to "what if" questions);
4. Improved implementation of results (enabling solution prescriptions to be better understood, and hence more readily and effectively applied).

1.3 A PREVIEW OF NETFORM APPLICATIONS

Throughout subsequent chapters many different types of netform models will be illustrated in detail. As a preliminary, to provide a glimpse of the variety of applications for these models, we provide the following examples taken from real-world settings.

Example 1: The marketing department of a major airline company develops forecasts of the number of passengers available to take different flights in each of several fare classes (first class, coach, economy, etc.). The marketing vice president, in reviewing the forecasts, observed that the company's profitability could be significantly affected by the number of available seats the airline allocated to different fare classes on different flights. He also recognized that, although standard techniques from marketing were inapplicable, the problem form involved optimizing over a network-related structure of flight and fare class possibilities. A team experienced in such applications was called in to work closely with the VP and his associates, yielding a netform model and method that are now used routinely by the airline to improve the profitability of its marketing operation.

Example 2: Large grocery stores (in common with certain fast food stores and hotels) face the problem of determining how to schedule their employees. For example, in dealing with check-out clerks, the store manager must determine, each week, which (and how many) employees should be assigned to check-out stations at each hour of the day, taking into account varying volumes of business at different hours and on different days. These employees, many of whom are part-timers (such as students), have individual restrictions on the hours and days when they can be available. Their scheduling must also account for seniority considerations, and allow for lunch periods and breaks. The goal is to provide a weekly schedule of working hours that best meets the requirements generated by the expected volume of business. Employing netform techniques, a computer-based system has been designed that provides schedules significantly better and faster than those produced by an experienced manager, reducing labor hours and improving customer service.

Example 3: Determining the best use of energy resources is a vital concern of economists, engineers, statisticians, geologists, ecologists and businessmen. A major issue is to identify the ideal levels and types of use of solar, biomass-based, oil-shale, crude oil, natural gas, and other energy sources, given the enormous array of industries that rely on energy resources and other products that can be generated from them. Experts have argued at length on what is best and why. After years of contradictory assertions and faulty estimates, recognition has emerged that intelligent answers require an analytical approach that can handle vastly more interacting considerations than a human is capable of entertaining. A computer-based netform model and solution procedure has been implemented to address these energy issues.

Example 4: Changing interest rates and opportunites for investment have underscored the need for financial institutions to develop ways to manage their funds more effectively. Issues of liquidity and rate of return must be balanced carefully to achieve proper relationships between inflows and outflows. A netform model has been developed and implemented to help financial officers find the best composition and timing of investments.

In the following, a number of additional (no less important) netform applications are described in abbreviated form. The types of model structures represented by these applications range from the simplest to the most advanced. "Classical" names for various types of network models, such as shortest path, assignment, pure and generalized networks, used in these examples will be defined specifically in later chapters.

1. The US Department of Transportation determines the best routes for proposed highways and other transportation channels by shortest path and assignment network methods.
2. A national bakery products company uses a pure network model to

determine the optimal number of each type of its cookies to make on successive production runs.

3. The US Treasury Department, in conjunction with the Internal Revenue Service, uses a bipartite ("two-piece") network to determine the composition and characteristics of various population segments in order to determine appropriate welfare and tax classifications [Gilliam and Turner 1974].
4. A clothes manufacturer determines the integration of its cloth inventories with production (sew line) operations using pure network techniques.
5. The State of Texas Water Board, Poland and California use pure networks to determine optimal levels and optimal water allocations to various industrial and agricultural users in different time periods [Bhaumik 1973].
6. The US Army, Navy and Marines use large-scale network methods to determine which personnel to assign to various jobs, in consideration of multiple objectives [Klingman and Phillips 1984, Bausch et al. 1991].
7. The Transportation and Road Research Laboratory of England uses trans-shipment network methods to solve large-scale aggregate distribution problems.
8. A large car manufacturer devotes many computer hours each week to a network model that determines how many of each car type to produce at various plants and then to ship to various cities [Glover and Klingman 1977].
9. NASA uses netform techniques to determine characteristics of optimal space satellite orbits.
10. Chemical products companies use a tailored netform package to integrate production, inventory, and distribution operations. The package is also used to analyze plant location and sizing problems [Glover et al. 1979, Klingman et al. 1988].
11. An international pharmaceutical company uses netform models and software to determine which and how much of various drugs and medications it should produce.
12. The US Bureau of Land Management employs a netform model and solution method to determine allocations of land (and vegetation) to animal use, in order to assure proper nutrition for various animal populations [Glover et al. 1984].
13. A lumber company uses a netform model for scheduling logs to plants, then to products, and finally to local markets.
14. The US Air Force determines optimal flight training schedules for its pilots, while satisfying restrictions on timing and classroom availabilities, by means of netform procedures [Glover and Klingman 1977].
15. A car manufacturer uses a netform package to determine optimal lot sizes for different plastic components of automotive assembly.

16. A major oil company employs a discrete netform model and solution software to determine the optimal timing of oil use and distribution from its leased and owned wells over a multi-year horizon.
17. A major US chemical company uses netforms to determine customer zones to be served by its warehouses, for different alternative commodity and freight mode combinations.
18. A major oil company uses a network optimization-based decision support system for supply, distribution, and marketing planning. The multi-period model spans a short-term time horizon which incorporates both manufacturing and distribution time lags along with inventory planning [Klingman et al. 1986, 1987].
19. The Tennessee Valley Authority's problem of determining the minimum cost refueling schedule for nuclear reactors yielded greatly improved solutions through the use of a discrete netform model and solution strategy [Glover, Klingman, and Phillips 1989].
20. The US army equipment procurement and allocation decisions with an objective to maximize readiness are addressed by a discrete netform model [Glover, Klingman, and Mote 1989].
21. In option trading, the calculation of the minimum deposit or margin that the broker must require of the investor can be formulated as a netform problem [Rudd and Schroeder 1982].
22. The state of Texas chose the location of out-of-state tax audit offices (for collecting sales taxes and other taxes due to Texas) and assigned states and auditors to them using a discrete netform model [Fitzsimmons and Allen 1983].
23. Some statistics problems, such as controlled rounding and least-absolute-value estimators for two-way tables, and least-absolute-value and Tchebycheff linear regression analyses, can be formulated and solved as netform models [Armstrong and Frome 1979, Causey et al. 1985, Klingman and Mote 1982].

The foregoing applications are, of course, not exhaustive. They have been selected to give a preview of the broad spectrum of undertakings to which netform models and methods usefully apply. The fundamental modeling ideas and techniques underlying these and many other applications will be presented in the following chapters.

APPLICATION: THE OIL INDUSTRY

The petroleum industry has been a major innovator of management science applications [Bodington and Baker 1990]. In fact, one of the first industrial applications of linear programming was refinery planning models [Adams and Griffin 1972, Charnes and Cooper 1961, Manne 1958, Symonds 1955, Wagner

1969]. In addition, management science has been used to develop decision-making aids in such areas as oil and gas operations [Aronofsky and Lee 1957, Aronofsky and Williams 1962, Charnes 1954, Charnes and Cooper 1961, Durrer and Slater 1977, Garvin et al. 1957, Manne 1958], crude oil acquisition [Aronofsky and Lee 1957, Durrer and Slater 1977, Jackson 1980], unit process control [Baker 1981, Charnes and Cooper 1961], refinery scheduling [Baker 1981, Dantzig 1963], blending [Charnes 1954, Charnes and Cooper 1961, Dantzig 1963], and distribution planning [Klingman et al. 1986, 1987, Zierer et al. 1976]. Major oil companies typically employ a large number of management scientists to analyze these and other aspects of the business on a regular basis using either separate or comprehensive models. Application of linear optimization to support both strategic and operational planning has saved the petroleum industry millions of dollars.

CASE: AN OIL COMPANY PROBLEM

Presented on the following pages is a simplified linear programming formulation for an oil production, refinement, and distribution operation. Study the formulation carefully and then answer the questions that follow. Initially, ignore variables x_{17} and x_{19}. The coefficients to be placed in these columns are specified later.

Consider the situation confronting the management of a small integrated oil company that operates two refineries and five oil wells for the production and marketing of three finished products and two crudes. Oil well production yields both heavy crude and light crude; three of the wells produce heavy crude exclusively and the remaining two produce light crude exclusively. In addition to selling part of this raw crude production, the company processes some of it into gasoline, kerosene, and fuel oil for sale at each of its refineries. Management is interested in determining the best production and marketing strategy for scheduling the process from the well to the customer.

After an analysis of the process it was evident that the interrelations between elements could be characterized as linear. Therefore, the company decided to build a linear programming model. Consequently, they identified the relevant variables and the mathematical expressions which describe the interrelations between them.

For convenience, the activities involved can be divided into raw crude production, finished product manufacturing (refining), and distribution of products to customers (both crude oil and finished goods). The model formulation and variable definitions are given below (see tables).

The first three groups of constraints express limitations on the physical facilities in terms of well capacity, pipeline capacity, and refinery capacity, respectively. Similarly, the last two groups reflect the market demand limitations for products distributed by the company, including both the raw crude sales and finished product sales. The remaining constraints express internal interrelationships which must be discussed in more depth.

Constraint Number	Production					Transportation		Inputs to Refinery				Crude Sold		Finished Product Sold									
	Heavy Crude			Light Crude		Tot. Hvy. Cd.	Tot. Lt. Cd.							Gasoline		Kerosene				Fuel Oil			
								To_A	To_B	To_A	To_B	Hvy.	Lt.	At_A	At_B	At_A	At_A	At_B	At_B	At_A	At_B		
	x_1	x_2	x_3	x_4	x_5	x_6	x_7	x_8	x_9	x_{10}	x_{11}	x_{12}	x_{13}	x_{14}	x_{15}	x_{16}	x_{17}	x_{18}	x_{19}	x_{20}	x_{21}		
1	1																					≤10	1000 barrels/year Crude Production Capacity
2		1																				≤ 5	
3			1																			≤12	
4				1																		≤13	
5					1																	≤ 8	
6												1	1									≤ 6	Pipeline Capacity on Crude Sales
7								1		1												≤21	Refinery Input Capacity
8									1		1											≤27	
9	1	1	1			-1																= 0	Crude Flow Summary
10				1	1		-1															= 0	
11						1		-1	-1			-1										= 0	
12							1			-1	-1		-1									= 0	
13								.1		.5				-1								= 0	Refinery Process Yields
14								.25		.3						-1						= 0	
15								.6		.15										-1		= 0	
16									.15		.45				-1							= 0	
17									.27		.3							-1				= 0	
18									.53		.2										-1	= 0	
19												1										≤ 5	Crude Sales Demand
20													1									≤ 5	
21														1	1							≤12	Finished Product Sales Demand
22																1		1				≤ 9	
23																				1	1	≤ 8	
	-1.2	-1.5	-1.3	-1.6	-1.7	-.2	-.1	-.6	-.5	-.55	-.45	2.8	3.0	6.0	5.8	4.1	0	3.8	0	2.6	2.4	Objective (Max)	
	Costs $/barrel											Sales Prices $/barrel											

Constraints 9–12 are included merely to define the summary variables x_6 and x_7. The first two derive from the input side of the process (i.e., well production). The remaining two derive from the distribution of crude to the refineries and direct sales. In this way, these four constraints together define a set of variables which permit us to trace the flow of crude through the system.

Constraints 13–18 define the relationships between crude allocations to the refineries and the resulting mixes of finished products. In other words, the cracking process yields a particular amount of gasoline, kerosene, and fuel oil for each barrel of crude oil refined. The actual quantity produced depends upon which refinery is used and whether the crude is heavy or light. For example, consider constraints 13–15 which represent the production of gasoline, kerosene, and fuel oil, respectively, at refinery A. Constraint 13 indicates that each barrel of heavy crude (x_8) yields 0.1 barrel of gasoline; each barrel of light crude (x_{10}) yields 0.5 barrel of gasoline. The total amount of gasoline produced at refinery A (x_{14}) must, therefore, equal $0.1x_8 + 0.5x_{10}$. The same analysis can be performed for constraints 14 and 15 when considering the total production of kerosene and fuel oil at refinery A. The objective function indicates the cost or revenue associated with each variable.

Variable Definitions

Variable	Meaning	Activity
x_1	Heavy crude from well 1	
x_2	Heavy crude from well 2	Crude production
x_3	Heavy crude from well 3	
x_4	Light crude from well 4	
x_5	Light crude from well 5	
x_6	Total heavy crude production	Summary transportation
x_7	Total light crude production	variables
x_8	Heavy crude processed at refinery A	
x_9	Heavy crude processed at refinery B	Refinery inputs
x_{10}	Light crude processed at refinery A	
x_{11}	Light crude processed at refinery B	
x_{12}	Unprocessed heavy crude sold directly	Crude sales
x_{13}	Unprocessed light crude sold directly	
x_{14}	Gasoline produced and sold at refinery A	Finished product
x_{15}	Gasoline produced and sold at refinery B	manufacture and sale
x_{16}	Kerosene produced and sold at refinery A	
x_{17}		
x_{18}	Kerosene produced and sold at refinery B	
x_{19}		
x_{20}	Fuel oil produced and sold at refinery A	
x_{21}	Fuel oil produced and sold at refinery B	

Case Questions

1. Create a graphical representation of the specific problem situation modeled by this formulation which depicts the flows of products and interrelationships of these flows. (In other words, draw a picture of the problem situation which would help you to understand it and to communicate it to someone who was unfamiliar with it.)
2. According to this formulation, can heavy crude and light crude be transported in the same pipeline?
3. **a.** How much total crude (heavy and light) can be sold in its unrefined form per year?
   b. What is the limiting factor?
4. What is the sales price of kerosene refined at A?
5. How many barrels of fuel oil are produced from one barrel of heavy crude at A? At B? Light crude at A? At B?
6. According to this formulation, what is the maximum amount of kerosene that will be produced by the company? What is the limiting factor?
7. How many total barrels of refined products are produced from each barrel of heavy crude at A? At B? Each barrel of light crude at A? At B?

***8.** Formulate the dual of this problem.

9. In the primal formulation given place a -1 in the row 14, column 17 position and in the row 17, column 19 position of the coefficient matrix.

- **a.** Provide a verbal definition of x_{17} and x_{19}. (Hint: Consider the problem of production versus sales.)
- **b.** What is the effect of adding these variables?
- **c.** Why might the objective coefficients of these variables be zero?
- **d.** Why might the objective coefficients of these variables be positive or negative?

***10.** Show how the addition of x_{17} and x_{19} (as specified above) changes the dual formulation.

***11.** Solve the original model formulation (without x_{17} and x_{19}) using a linear programming computer package.

- **a.** What is the optimal production and marketing strategy and total profit? Using the interpretation of the dual price as measuring the marginal worth of increasing a resource by one unit, explain why it makes sense for the dual price of constraint 9 to be equal to -1.3.
- **b.** Change the upper bound on heavy crude sales demand to 5500 and re-solve the model. How did total profit change? Show another way to predict this change using the dual price of the heavy crude sales demand constraint from part (a).
- **c.** Using the original model, change the upper bound on the fuel oil finished product sales demand to 9000. Why is the change in total profit not equal to the dual price of the fuel oil demand constraint

from part (a) times 1000? In this new solution, why is the dual price of the kerosene demand constraint greater than the selling price of kerosene? (Hint: Note the change in crudes used and outside sale of crudes. Also note the change in gasoline production.)

***12.** Solve model formulation from problem 9 with objective coefficients of zero for x_{17} and x_{19} and with the fuel oil demand still at 9000 using a linear programming computer package.

   a. How has the solution and total profit changed and why has this change taken place?

   b. Why has the value of the dual price associated with constraint 22 dropped to 3.8?

EXERCISES

1.1 A problem can often be modeled or represented in more than one way. For example, the following are two equivalent ways of modeling the same problem.

Way 1:

Minimize $\quad 12x_1 + 8x_2 + 2x_3 + 9x_4 + 6x_5 + 13x_6 + 1x_7 + 5x_8 + 8x_9$

Subject to

$$x_1 + x_2 = 1$$

$$x_1 + x_3 = x_4 + x_5$$

$$x_2 = x_3 + x_6$$

$$x_4 + x_7 = x_8$$

$$x_5 + x_6 = x_7 + x_9$$

$$x_8 + x_9 = 1$$

$$x_1, x_2, \ldots, x_9 \geqslant 0$$

Way 2: Find the shortest path from A to B, following directions permitted by the arrows, where distances are shown in boxes in the figure:

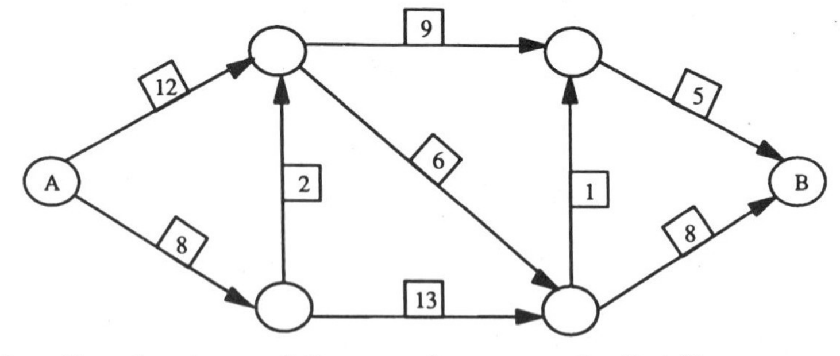

Describe advantages of the second way over the first if you are modeling this problem for a company that wants to send its trucks from A to B, and must review your model with management. What kinds of circumstances might arise in the real world that would accentuate or diminish these advantages?

1.2 Evaluate the following remarks: Computers have not only encouraged the growth of mathematical optimization by providing a tool to carry out computations efficiently, but have also allowed methods to become simpler. In many practical problems, the simplest and easiest approach is to examine all relevant solutions and pick the best (e.g., the one with least cost or greatest profit). The computer can itemize alternatives faster than it takes a human to write a mathematical problem formulation. As a result, a sophisticated mathematical technique is unnecessary.

1.3 Today there exist well-established computer methods, and specially trained groups capable of developing new methods, for solving a variety of problems efficiently. In view of this fact (or in spite of it), which of the following do you think are most important for a management science practitioner to learn? For a production, marketing or financial manager to learn?

a. Mathematical logic and principles underlying the major methods
b. A variety of "cookbook" solution techniques, with minimum emphasis on theory
c. Modeling techniques able to express real-world problems in a way that permits good solution methods to be applied
d. The ability to recognize which problems can be formulated and solved by management science techniques
e. How to use the computer to manage the data that a management science model might use
f. The phone number of someone who is competent in the management science field.

REFERENCES

Adams, F. and J. Griffin. 1972, "Economic-Linear Programming Model of the U.S. Petroleum Refining Industry," *American Statistics Association Journal*, Vol. 67 (September), pp. 542–551.

Ahuja, R., T. Magnanti, and J. Orlin. 1989, "Network Flows," in G. Nemhauser et al. (eds.) *Handbook in OR & MS*, Vol. 1, Elsevier North-Holland, New York, pp. 211–369.

Ahuja, R., T. Magnanti, and J. Orlin. 1992, *Network Flows*, Prentice-Hall, NJ.

Armstrong, R. and E. Frome. 1979, "Least-absolute-value Estimators for One-way and Two-way Tables," *Naval Research Logistics Quarterly*, Vol. 26, No. 1, pp. 79–96.

Aronofsky, T. and A. Lee. 1957, "A Linear Programming Model For Scheduling Crude Oil Production," *Transactions AIME*, Vol. 69, No. 4, pp. 389–403.

Aronofsky, T. and A. Williams. 1962, "The Use of LP and Mathematical Models in Underground Oil Production," *Management Science*, Vol. 8, No. 4, pp. 394–407.

Baker, T. 1981, "A Branch and Bound Network Algorithm for Interactive Process Scheduling," *Mathematical Programming Study*, Vol. 15, pp. 43–57.

Bausch, D., G. Brown, D. Hundley, S. Rapp, and R. Rosenthal. 1991, "Mobilizing Marine Corps Officers," *Interfaces*, Vol. 21, No. 4, pp. 26–38.

Bhaumik, G. 1973, "Optimum Operating Policies of a Water Distribution System with Losses," Ph.D. Dissertation, University of Texas at Austin.

Bodington, C. and T. Baker. 1990, "A History of Mathematical Programming in the Petroleum Industry," *Interfaces*, Vol. 20, No. 4, pp. 117–127.

Causey, B., L. Cox, and L. Ernst. 1985, "Applications of Transportation Theory to Statistical Problems," *Journal of the American Statistical Association*, Vol. 80, No. 392, pp. 903–909.

Charnes, A. 1954, "A Model for Programming and Sensitivity Analysis in an Integrated Oil Company," *Econometrica*, Vol. 22, No. 2 (April), pp. 193–217.

Charnes, A. and W. Cooper. 1961, *Management Models and Industrial Applications of Linear Programming*, Vols. I and II, Wiley, New York.

CONDOR, 1988, "Operations Research: The Next Decade," *Operations Research*, Vol. 36, No. 4, pp. 619–637.

Dantzig, G. 1951, "Application of the Simplex Method to a Transportation Problem," in T. Koopmans (ed.), *Activity Analysis of Production and Allocation*, Wiley, New York, pp. 359–373.

Dantzig, G. 1963, *Linear Programming and Extensions*, Princeton University Press, Princeton, NJ.

Durrer, E. and G. Slater. 1977, "Optimization of Petroleum and Natural Gas Production—A Survey," *Management Science*, Vol. 24, No. 1, pp. 35–43.

Fitzsimmons, J. and L. Allen. 1983, "A Warehouse Location Model Helps Texas Comptroller Select Out-of-State Audit Offices," *Interfaces*, Vol. 13, No. 5, pp. 40–46.

Ford, R. and D. Fulkerson. 1962, *Flows in Networks*, Princeton University Press, Princeton, NJ.

Garvin, W., H. Crandall, J. John, and R. Spellman. 1957, "Applications of Linear Programming in the Oil Industry," *Management Science*, Vol. 3, No. 4, pp. 407–430.

Gilliam, G. and J. Turner. 1974, "A Profile Analysis Network Model to Reduce the Size of Microdata Files," Working Paper, Office of Tax Analysis, Office of the Secretary of the Treasury, Washington, DC.

Glover, F., R. Glover, and F. Martinson. 1984, "A Netform System for Resource Planning in the U.S. Bureau of Land Management," *Journal of the Operational Research Society*, Vol. 35, No. 7, pp. 605–616.

Glover, F., G. Jones, D. Karney, D. Klingman, and J. Mote. 1979, "An Integrated Production, Distribution, and Inventory Planning System," *Interfaces*, Vol. 9, No. 5, pp. 21–35.

Glover, F. and D. Klingman. 1977, "Network Application in Industry and Government," *AIIE Transactions*, Vol. 9, No. 4, pp. 363–376.

Glover, F., D. Klingman, and J. Mote. 1989, "Intelligent Decision System for Logistical Support," CBDA 157, Center for Business Decision Analysis, University of Texas at Austin.

Glover, F., D. Klingman, and N. Phillips. 1989, "A Network-Related Nuclear Power Plant Model With an Intelligent Branch-and-Bound Solution Approach," *Annals of Operations Research*, Vol. 21, pp. 317–332.

Golden, B. and T. Magnanti. 1977, "Deterministic Network Optimization: A Bibliography," *Networks*, Vol. 7, pp. 149–183.

Grigoriadis, M. 1980, "Network Optimization Problems and Algorithms: An Annotated Bibliography," DCS-TR-99, Department of Computer Sciences, Rutgers University, New Brunswick, NJ 08903.

Hitchcock, F. 1941, "The Distribution of a Product from Several Sources to Numerous Facilities," *Journal of Mathematical Physics*, Vol. 20, pp. 224–230.

Jackson, B. 1980, "Using LP for Crude Oil Sales at Elk Hills: A Case Study," *Interfaces*, Vol. 10, No. 3, pp. 65–70.

Jensen, P. and W. Barnes. 1980, *Network Flow Programming*, Wiley, New York.

Kantorovich, L. 1939, "Mathematical Methods in the Organization and Planning of Production," Publication House of the Leningrad University, 68 pp. Translated in *Management Science*, 1960, Vol. 6, pp. 366–422.

Kennington, J. and R. Helgason. 1980, *Algorithms for Network Programming*, Wiley–Interscience, New York.

Klingman, D. and J. Mote. 1982, "Generalized Network Approaches for Solving Least Absolute Value and Tchebycheff Regression Problems," *Studies in the Management Sciences: Optimization in Statistics*, Vol. 19, North-Holland, New York, pp. 53–66.

Klingman, D., J. Mote, and N. Phillips. 1988, "A Logistics Planning System at W. R. Grace," *Operations Research*, Vol. 36, No. 6, pp. 811–822.

Klingman, D. and N. Phillips. 1984, "Topological and Computational Aspects of Preemptive Multi Criteria Military Personnel Assignment Problems," *Management Science*, Vol. 30, No. 11, pp. 1362–1375.

Klingman, D., N. Phillips, D. Steiger, R. Wirth, and W. Young. 1986, "The Challenges and Success Factors in Implementing an Integrated Petroleum Products Planning System for Citgo," *Interfaces*, Vol. 16, No. 3, pp. 1–19.

Klingman, D., N. Phillips, D. Steiger, R. Wirth, R. Padman, and R. Krishnan. 1987, "An Optimization Based Integrated Short-Term Refined Petroleum Product Planning System," *Management Science*, Vol. 33, No. 7, pp. 813–830.

Koopmans, T. 1947, "Optimum Utilization of the Transportation System," *Proceedings of the International Statistical Conference*, Washington, DC (also reprinted as a supplement to *Econometrica*, 1949, Vol. 17).

Manne, A. 1958, "A Linear Programming Model of the U.S. Petroleum Refining Industry," *Econometrica*, Vol. 26, No. 1, pp. 67–106.

Murty, K. 1992, *Network Programming*, Prentice-Hall, NJ.

Orden, A. 1956, "The Transshipment Problem," *Management Science*, Vol. 2, pp. 276–285.

Rockafellar, R. 1984, *Network Flows and Monotropic Optimization*, Wiley, New York.

Rudd, A. and M. Schroeder. 1982, "The Calculation of Minimum Margin," *Management Science*, Vol. 28, No. 12, pp. 1368–1379.

Symonds, G. 1955, *Linear Programming; The Solution of Refinery Problems*, Esso Standard Oil Co. Public Relations Department, New York.

Wagner, H. 1969, *Principles of Operations Research*, Prentice Hall, Englewood Cliffs, NJ.

Zierer, T., W. Mitchell, and T. White. 1976, "Practical Applications of Linear Programming to Shell's Distribution Problems," *Interfaces*, Vol. 6, No. 4, pp. 13–26.

2

FUNDAMENTAL MODELS FOR PURE NETWORKS

This chapter introduces the most basic of all netforms, pure network models, which includes classical model types with names such as assignment, transportation, trans-shipment, minimum cost flow, maximum flow, and shortest path.

2.1 FUNDAMENTAL PRINCIPLES

Before we discuss particular components of pure network models, we need to outline some general good modeling practices.

1. Separate the model structure from the problem data. The basic model structure should not change if an input data parameter, such as amount of available resources, changes.
2. Do not aggregate data. For example, create separate model components (variables) to represent production and shipping rather than adding production and shipping costs and having a single variable represent the amount produced and shipped. The model can be simplified later if necessary, but generally the increased model size resulting from disaggregation is more than compensated for by the increased ease in reporting solutions and changing data parameters.
3. Do not anticipate the optimal solution, leaving out nonoptimal alternatives. If the input data parameters change, previously nonoptimal alternatives may become optimal.
4. Use a general model structure, rather than trying to force the problem to fit a structure for a particular model type.

Finally, implementing a network model is greatly facilitated by developing an entire system which supports functions such as data collection and report generation in addition to solving the network model. There are a number of requirements which we view as absolutely necessary for the successful implementation of such a system. These are detailed in our discussion of decision support systems in Appendix B.

2.1.1 Modeling Inputs and Outputs (Supplies and Demands)

To develop the foundations of pure network models, we start by examining the representations of inputs and outputs. Through their diversity, netform problems encompass a wide variety of inputs and outputs. Common examples of inputs include available resources, raw materials, energy sources, cash, labor, and arriving passengers, while common examples of outputs include finished goods, energy, cash, labor, delivered passengers, and so forth. Inputs and outputs are often referred to as supplies and demands (taking slight liberty with economic terminology).

For modeling purposes, it is convenient to introduce certain "transfer points" that represent the entities or processes by which inputs enter the system and outputs leave the system. (They may also represent intermediate junctions by which inputs make their way from one location, or stage, in the system to another.) In network terminology, these transfer points are called ***nodes***, depicted pictorially by circles. Supplies and demands are depicted by arrows, called ***arcs***, that enter and leave such nodes. These structures are illustrated as follows.

Supply Representation: To denote a fixed supply, a triangle pointing toward a node is used with an arrow leading into the node. Inside the triangle is shown the amount of supply; below, 200 units are required to be input as supply.

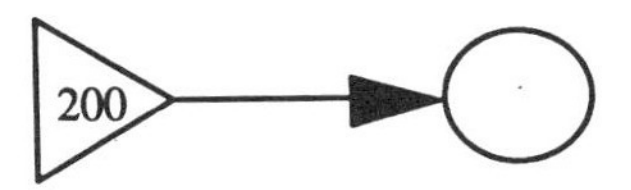

Demand Representation: To indicate a fixed demand, an arc leads from a node to a triangle pointing away from the node. The amount of demand is shown inside the triangle; below, 60 units are required to be output as demand.

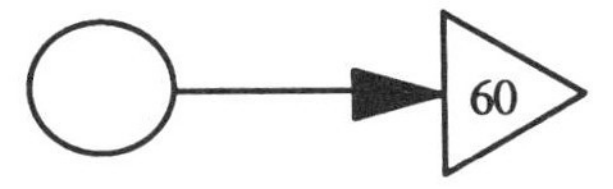

Supplies and demands, of course, are not usually free of restriction, but are generally subject to lower and upper limits. Such limits, called ***bounds***, are treated notationally by enclosing them in parentheses and attaching them to the arcs whose supplies or demands they qualify and omitting the triangles, as shown below. In this case the first number in parentheses on the arc indicates the minimum supply or demand (lower bound) and second number the maximum (upper bound). This is also called a ***bounded arc***.

Constrained Supply: At least 0 and at most 200 units are required to be input as a supply, below.

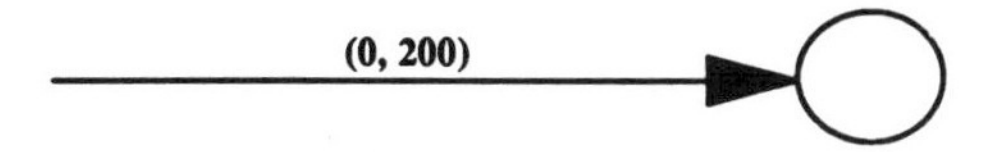

Constrained Demand: At least 60 and at most 830 units are required to be output as a demand, below.

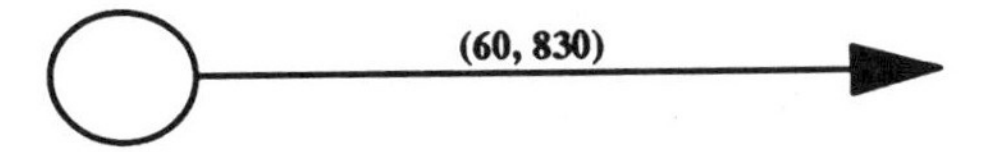

When lower and upper bounds are not explicitly indicated in this fashion, their values are understood to be 0 and infinity, respectively. A supply arc without explicit bounds is called a ***slack arc*** (because it can make up for slack, i.e., a deficit, by providing additional input), and a demand arc without explicit bounds is called a ***surplus arc*** (because it can compensate for surplus by removing it as output). The supply and demand triangles may also be modeled as constrained supply and demand arcs where the upper and lower bounds are set equal to each other as for the supply to node A in Figure 2.1.

2.1.2 Modeling Connecting Activities

Supplies and demands are not isolated from each other, but are characteristically linked by ***activities***. The simplest type of activity constitutes a direct transfer of goods (or people, etc.) from one point to another, as for example,

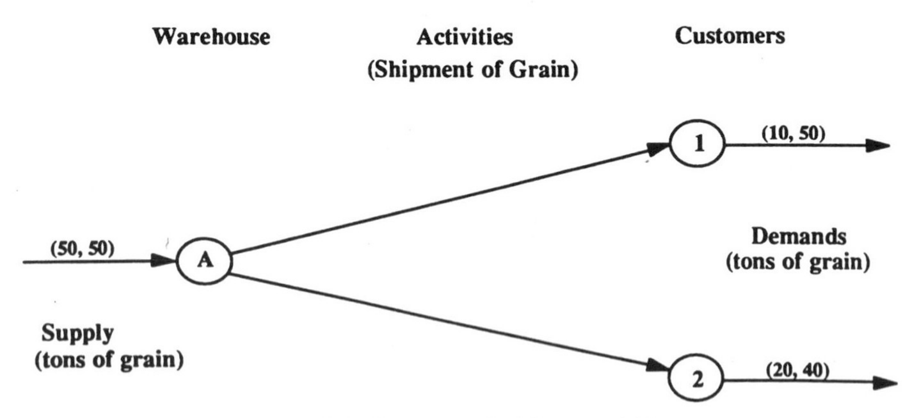

Figure 2.1 Pure network with two activities.

where a product is shipped from a supply center to a distribution outlet. Such a direct transfer activity is represented by an arc, just as arcs are used to represent the "rudimentary transfers" of supplies and demands into and out of the system.

Arcs representing activities, in contrast to those representing supplies and demands, extend between (and hence connect) two different nodes. For example, consider the situation involving two activities, each consisting of shipping grain from a warehouse to one of two customers. This situation is illustrated in Figure 2.1.

This example employs the common practice of inserting a symbol or a number inside a node to help identify the entity it refers to. Thus the diagram indicates that exactly 50 tons of grain must be shipped from Warehouse A to Customers 1 and 2, who must respectively receive between 10 and 50 tons and between 20 and 40 tons. In standard terminology, the two activities, represented by the arcs (A, 1) and (A, 2) (i.e., by the arc from node A to node 1 and the arc from node A to node 2), are said to carry the ***flow*** (tons of grain shipped) from the initial endpoint of each arc to its terminal endpoint. Likewise, the quantities of grain that are carried on the supply and demand arcs are called the flows on these arcs. Flows are normally conceived as variable quantities, until assigned specific values (as by some optimization process), but supplies and demands sometimes constitute an exception, as in the present example where the only acceptable flow on the supply arc is exactly 50 tons. (Common usage often construes the terms "supply" and "demand" to mean exact (i.e., constant) supply and demand values, unless they are qualified as bounded.)

At the same time, it is not uncommon to attach bounds to linking arcs, thereby constraining their flows. A flow does not qualify as ***feasible*** unless it satisfies these bounds. By extension, a ***solution***, which is a specific assignment of flows to all arcs, does not qualify as feasible unless all arc flows satisfy their bounds.

Another type of feasibility condition is also implicit in our discussion. Note that to distribute the 50 tons of grain supplied at node A, the flow on arc (A, 1) and the flow on arc (A, 2) must sum exactly to 50. In general, a solution can be considered feasible only if the total flow into each node is exactly matched by the total flow out. These conditions are important enough to deserve special emphasis.

Feasibility Requirement 1. The flow on every arc must satisfy its lower and upper bounds.

Feasibility Requirement 2. The sum of flows into every node must equal the sum of flows out.

This latter requirement is also commonly called the ***flow conservation requirement*** or the ***node conservation requirement***, because it conveys the idea (from an "engineering" or "systems" point of view) that flow at a node is conserved, i.e., is neither lost nor augmented. (One can easily cause flow to be lost or augmented at a node by introducing an additional surplus or slack arc, but within the context of the system that includes this additional arc, flow is once again conserved!)

Question: Specify three different solutions (assignments of specific flows to the arcs) that satisfy Feasibility Conditions 1 and 2 for Figure 2.1.

2.1.3 The Problem Objective

So far omitted from consideration is the ***problem objective***, which makes it possible to tell when one feasible solution is better than another. Such an objective is typically expressed in terms of optimizing (minimizing or maximizing) a particular quantity of interest. For example, one may be concerned with minimizing costs, customers lost, distance travelled, quantity of scarce resources consumed, hours exposed to dangerous conditions, or, alternatively, one may be concerned with maximizing profits, share of market, crop yields, patients served by medical facilities, and so forth. In cases where strict optimization is inapplicable as where one seeks a "good" but not "best" answer, or seeks a balance between competing goals, there are a variety of techniques to recast the problem in an optimization framework. For instance, instead of maximizing profits, one can select a range of acceptable profit values, and settle for any such value by minimizing the amount by which actual profit falls outside the range. The problem is still one of optimization, but the effect is to pursue a "satisfying" objective.

To capture these and other types of objectives, we make use of a "figure of merit," which we refer to by the generic term ***profit***, or a "figure of demerit," which we refer to by the generic term ***cost***. In this fashion, we may refer to all optimization problems as falling in the category of cost minimization problems or profit maximization problems.

Cost (or profit) figures are associated with each arc of the network, and depicted by enclosing their values in ***boxes*** attached to the arc. In a cost minimization problem, revenues appear as negative costs, since they decrease total cost. Similarly, in a profit maximization problem, costs appear as negative profits, since they decrease total profit. This type of representation is illustrated in Figure 2.2, which modifies and extends the earlier "two activity" network.

By convention, an arc without an attached cost figure is assumed to have a 0 associated cost. It should be stressed that ***all costs are unit costs***, i.e., they apply to each unit of flow on an arc. (Corresponding remarks apply to profits.) For example, the cost of shipping 10 tons of grain from Warehouse A to Customer 2 via the arc (A, 2) incurs a cost of $17 \times 10 = 170$ (in units of 100 dollars).

Note that costs may be attached to supply and demand arcs as well as to linking arcs. For example, each unit (ton) of grain supplied at Warehouse B costs 300 dollars, and each ton delivered to Customer 1 brings in a revenue (negative cost) of 400 dollars. This way of expressing total cost is a ***linear measure***. (Some nonlinear cost measures that arise in practice, and ways for dealing with them, are discussed in later chapters.) The total cost of all flows through the network is just the sum of costs incurred on the individual arcs (i.e., "cost per unit of flow" times "value of flow" summed over all arcs). The pure network problem in Figure 2.2 is also frequently called a ***minimum cost flow*** problem.

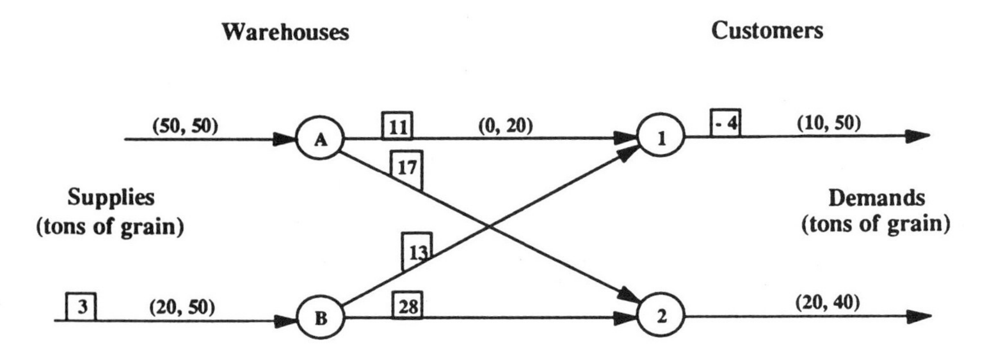

Figure 2.2 Pure network: cost minimization example (costs expressed in hundreds of dollars).

An ***optimal*** solution is a feasible solution which optimizes (minimizes or maximizes) the problem objective, i.e., it is the best feasible solution. To amplify the ideas behind feasible and optimal solutions, we will illustrate an intuitive process for seeking an optimal solution to the problem in Figure 2.2. First, to assure that the solution generated will be feasible, we reiterate the two feasibility requirements:

Requirement 1. The flow of every arc must satisfy its lower and upper bounds.

Requirement 2. The sum of flows into every node must equal the sum of the flows out.

As previously noted, Requirement 1 dictates that the flow on the supply arc into node A must be exactly 50 units (both the lower and upper bound equal 50). It follows then, by Requirement 2, that 50 units must flow out of node A on the arcs (A, 1) and (A, 2). Since the goal is to minimize cost, we are motivated to send the flow the cheapest possible way. Thus, ideally we would send all 50 units on arc (A, 1), which has a unit cost of 11 (in hundreds of dollars). However, the bound on this arc limits the amount that can be sent to 20. Sending these 20 units produces Figure 2.3.

The 20 enclosed in a ***half circle*** attached to each of the arcs identifies this 20 units as a flow value. Because this flow value was sent on a "complete path," i.e., beginning with a supply arc and ending with a demand arc, the flow conservation requirement (flow in = flow out) is satisfied at each node of the path.

However, the bound requirement is still unsatisfied on the supply arc into node A because 30 units remain to be sent in order to meet the lower bound of 50. Again to assure that flow conservation will be satisfied at all nodes, we will

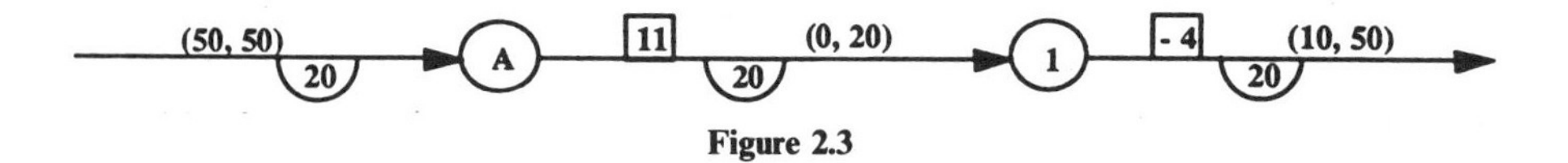

Figure 2.3

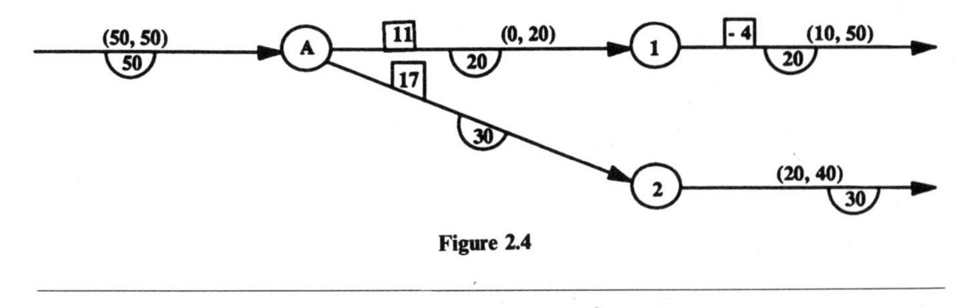

Figure 2.4

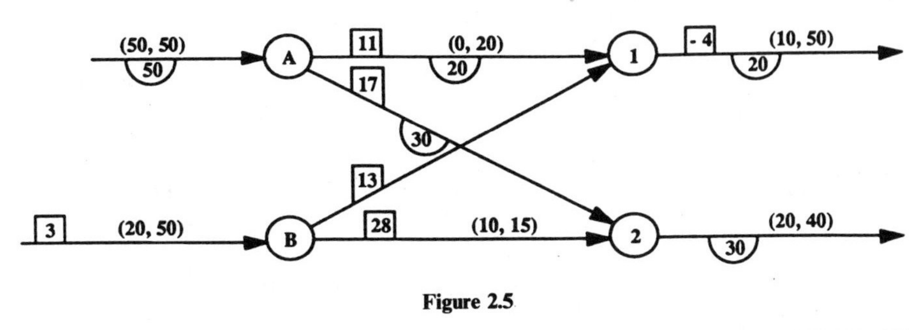

Figure 2.5

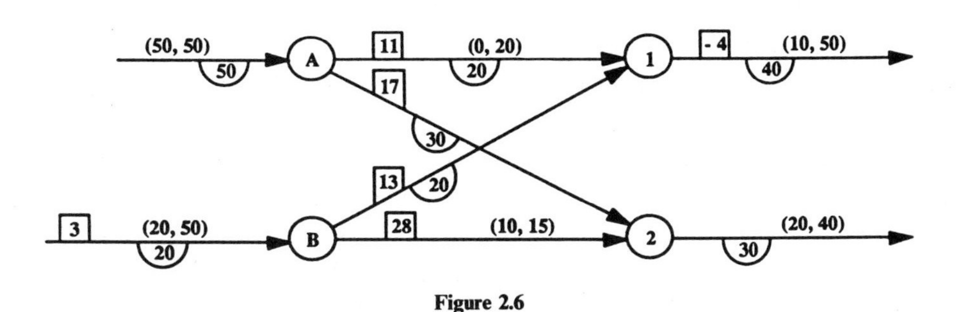

Figure 2.6

send this flow on a complete path, in this case the path that begins with the supply arc into node A and includes arc (A, 2) and the demand arc out of node 2. This increases the flow into node A to 50, producing Figure 2.4. The flow conservation and bound requirements are not satisfied for the portion of the diagram shown. The complete network, with the trial flows determined so far, appears in Figure 2.5. No flow has yet been sent on the supply arc into node B, which is required to carry between 20 and 50 units. There are two possible complete paths that begin with this arc, the path through node 1, with a total path cost of $3+13-4=12$, and the path through node 2, with a total path cost of $3+28+0=31$. Sending the minimum required flow of 20 units for the supply arc into B along the cheaper of the two paths yields Figure 2.6.

We are still not done, however, because arc (B, 2) still has an unsatisfied bound. To meet the lower bound of 10 on this arc and to maintain conservation

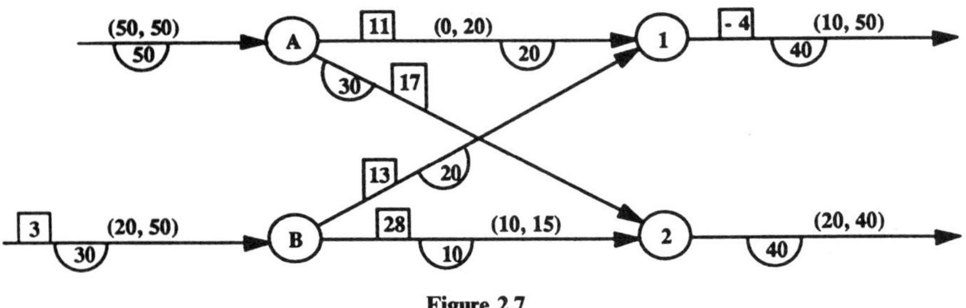

Figure 2.7

of flow at the nodes, an additional flow of 10 units can be sent on the path that begins with the supply arc into B, and ends with the demand arc out of node 2. This yields Figure 2.7. Examining this diagram discloses that all bound and flow conservation requirements are satisfied, thereby providing a feasible solution. The total cost of the feasible solution may be tallied as follows:

Arc	Cost	Flow	Cost × Flow
Into A	0	50	0
Into B	3	30	90
(A, 1)	11	20	220
(A, 2)	17	30	510
(B, 1)	13	20	260
(B, 2)	28	10	280
Out of 1	−4	40	−160
Out of 2	0	40	0
Total Cost of Solution			1200

Although feasible, this solution is not optimal. A better feasible solution, yielding a smaller total cost, may be identified by observing that it is possible to decrease flow by 10 units on the path that begins with the supply arc into node B and ends with the demand arc out of node 1. This reduction continues to satisfy flow conservation (because it is applied to a complete path) and, in addition, leaves all bounds on the path satisfied. Since the cost for each unit of flow on this path is $3+13-4=12$ (adding the costs on each arc of the path), the decrease of 10 units will improve total cost by $10\times 12=120$. The resulting improved solution is indicated in Figure 2.8.

Question: Identify the total cost of the solution shown in Figure 2.8 by tallying all component costs in the manner illustrated earlier.

The flows in Figure 2.8 in fact provide an optimal (minimum cost) solution for the example problem. However, the process of "intelligent eyeballing" used to generate these flows does not always work, even for very small and simple networks. For larger networks, it may be exceedingly difficult or impossible to identify a feasible solution in this fashion, let alone an optimal one.

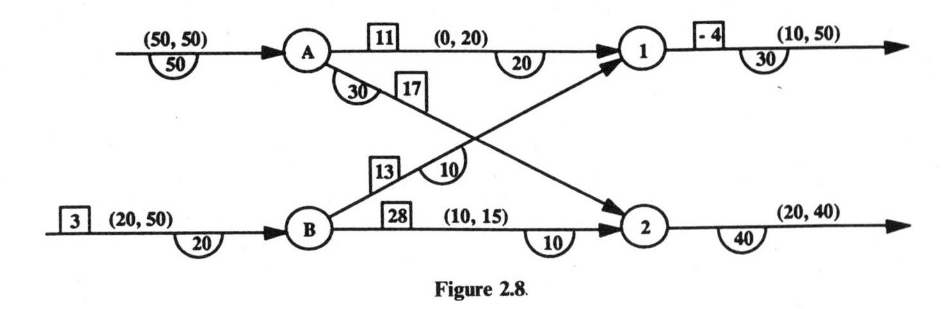

Figure 2.8.

2.2 FORMULATING A NETWORK MODEL FROM A WORD PROBLEM

We will now begin with a verbal description of an example network problem (instead of starting with a diagram) and show how the problem may then be translated into a network formulation.

A shipping company is concerned with maintaining an acceptable number of empty containers at each of four ports, so that these containers may be used to hold cargo of scheduled future shipments. After allocating containers already on hand to serve these needs, Ports 1 and 2 have a net surplus of containers on hand, while Ports 3 and 4 have a net demand for additional containers. Because of warehousing limits, Ports 1 and 2 must each send a certain minimum number of their net containers on hand to Ports 3 and 4 (so that the containers kept on hand throughout the entire period will not exceed storage capacity). Similarly, Ports 3 and 4 each have upper limits on the number of containers they can receive, so that their warehouse capacities will not be exceeded. This information is given in Table 2.1.

Table 2.1

Ports with net containers on hand		
Ports	Net containers on hand	Minimum to be shipped
1	12	10
2	9	5

Ports with net containers needed		
Ports	Minimum Needed	Maximum that can be received
3	6	11
4	9	14

Viewing containers on hand as supplies and containers needed as demands, the information given so far can be modeled as follows:

To complete the model we require information about the costs of shipping, shown in Table 2.2. Adding arcs from the supply ports to the demand ports with those costs attached yields the completed network shown in Figure 2.9.

Table 2.2

Shipping Cost Per Container

From \ To	Port 3	Port 4
Port 1	8	17
Port 2	2	7

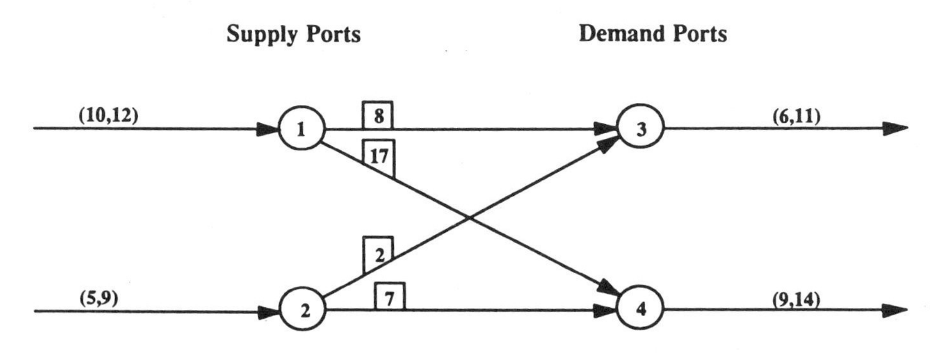

Figure 2.9

2.3 INTUITIVE PROBLEM SOLVING

We have already pointed out that trying to generate optimal solutions even to simple network problems may not work out well. However, the "intelligent eyeballing" used in Figure 2.8 was not really very refined. To gain a fuller appreciation of how an insightful intuitive analysis might proceed, we will examine some approaches based on more thoughtful and systematic foundations. To develop these approaches we will take advantage of our experience in generating a solution to the network example problem in Figure 2.2, which showed that it is not a good idea to ship more than a minimum amount on any path when seeking to minimize cost. (For example, while undertaking to satisfy a lower bound on a supply arc using the cheapest path, more flow was sent on this path than necessary in view of other lower bounds, and the flow on this path later had to be reduced.) Thus, as a basis for developing more effective approaches in the shipping problem shown in Figure 2.9, note that the minimum that must be supplied, $10+5$, just matches the minimum demanded, $6+9$. As a result, it is possible to find solutions (for this particular problem) where none of the flows exceed the supply and demand lower bounds. To keep total cost down by not shipping more than is necessary, we seek ways to send containers that will not generate excess flows on any supply or demand arc.

Some intuitively reasonable "solution methods" for incorporating the preceding observations are as follows.

Method 1. Send a minimum number of containers (that will just satisfy the smaller of a supply lower bound or a demand lower bound) on the least cost path, then on the next least cost path, and so on, until all lower bounds for supplies and demands are satisfied.

Applying this method to the network diagram in Figure 2.9 first sends 5 containers on the path from Port 2 to Port 3 (beginning with the supply arc into node 2 and ending with the demand arc out of node 3). This yields a cost of 10 and satisfies the minimum supply at Port 2. Then the method sends 1 container on the path from Port 1 to Port 3, incurring a cost of 8 and satisfying the minimum demand at Port 3. Finally, it sends 9 containers from Port 1 to Port 4, at a cost of 153, satisfying the minimum supply and demand at these ports. Total solution cost $= 10+8+153=171$.

As it happens, this is not the minimum cost solution. Some thought suggests that perhaps it is wrong to start with the least cost route, but would be better to consider the tradeoffs between costs of alternative routes. For example, since there are two possible routes leading from each supply port, we may consider the relative costs of these two routes, and then select a route that gives the best relative advantage. This principle is embodied in the next method.

Method 2. Send flow as in Method 1, but as long as there remain two candidate paths from a supply port, compute the cost ratio of the minimum cost path to the maximum cost path for this port. Then select the path that gives the best min to max cost ratio, instead of just the minimum cost (until there are no ratios left to calculate).

Unfortunately, this method gives exactly the same solution as Method 1 in the case of the simple example problem in Figure 2.9. Consequently, we seek another way to apply the principle on which it is based.

Method 3. Modify Method 2 by calculating the difference between the minimum and maximum cost path at each port, selecting the path that gives the best min to max difference.

This method yields a different solution than before. First it sends 6 containers from Port 1 to Port 3 (beginning with the supply arc into node 1 and ending with the demand arc out of node 3). This gives a cost of 48, and satisfies the minimum demand at Port 3. Next (taking the cheapest remaining alternative, because the lower bound demand is satisfied for the minimum cost paths from both supply ports), the method sends 5 containers from Port 2 to Port 4, at a cost of 35, and satisfying the minimum supply at Port 2. Finally, 4 containers are sent from Port 1 to port 4 at a cost of 68, completing the solution. The total cost is $48+35+68=151$. This improves substantially on the cost obtained by Methods 1 and 2, but is still not optimal.

Rather than continue to apply intuitively reasonable methods to this problem until finally hitting on one that yields the best cost, we note that the optimum solution (obtained by "inspection" of the various alternatives in the simple example in Figure 2.9) is to send 10 containers from Port 1 to Port 3 and 9 containers from Port 2 to Port 4, for a total cost of $80+63=143$.

Note, incidentally, that the optimal solution sends a total of 19 containers from Ports 1 and 2 to Ports 3 and 4, though the minimum total required to be sent and received is 15 containers. Thus, paradoxically, a smaller cost resulted in this case by shipping more containers than the total required. Small wonder then that the intuitive approaches to this problem failed, although based on a principle which was essential to the solution of the earlier example problem in Figure 2.2 (since sending more on a path than was required produced a nonoptimal solution in that case).

These examples should make clear the risks involved in trying to generate solutions on intuitive grounds, even in extremely simple settings. They should also help the reader to understand the difference between a ***model*** and a solution ***method***; e.g., a model's optimal solutions do not change regardless of how a method goes about finding them. This is an important difference that beginning modelers sometimes have difficulty grasping.

2.4 STRUCTURAL VARIATIONS

The two preceding example problems contain every element found in the most general of pure network problems, except of course for their size and structure. In fact, the structure embodied in the examples is extremely common, and is sometimes called the ***transportation*** structure, because it readily represents the situation of transporting goods from warehouses to customers (or from plants to warehouses). The more technical ***bipartite*** ("two-piece") terminology used to describe this structure comes from the fact that the nodes are divided into the two categories of shipping nodes (***sources***) and receiving nodes (***sinks***). As noted in Chapter 1, this structure has figured prominently in the history of network models and methods. Another one of the classical bipartite problems is the ***assignment*** problem, where every supply and every demand are bounded to be exactly 1, as, for example, in assigning people to jobs, where each person (with a supply of 1) can be put in exactly 1 job, and each job (with a demand of 1) can receive exactly one person. The illustration in Chapter 1 about the astronomical amount of work required to examine all possible assignments of 70 people to 70 jobs was an example of such a problem. (A numerical example is also provided in the exercises at the end of this chapter.)

More complex practical problems typically involve intermediate junctions, or ***trans-shipment*** nodes, through which goods may be shipped (and perhaps temporarily stored) en route to their final destination. A simple yet extensively studied example of a trans-shipment network problem is the classical ***shortest path*** problem, where an individual seeks a minimum distance route from a single origin (source) to a single destination (sink). This problem is expressed as a pure network by introducing a supply of exactly one unit at the origin and a demand of exactly one unit at the destination (since one person enters the system at the origin and leaves at the destination). Because arcs are directed, they may naturally be viewed as "one way" routes or road segments. "Two way" segments can be represented by a pair of oppositely directed arcs. Costs attached to arcs can correspond to the lengths of the routes they represent (to minimize distance traveled) or can more generally represent such things as mileage costs.

An example of a shortest path problem is shown in Figure 2.10.

Question: Identify two feasible solutions to the network problem in Figure 2.10, one of them optimal. Can you think of a situation where the cost of traveling in one direction would not be the same as the cost of traveling in the other?

It is possible to formulate the problem of finding a shortest (minimum cost) path from the source to *every* other node in the network in a similar fashion. To illustrate, in the preceding example there are 5 nodes other than the source. Thus, if the source is given a supply of exactly 5 (giving bounds to the supply arc of (5, 5)), then one unit will be available to be shipped to each of the 5 other nodes. Correspondingly, each of these other nodes is given a demand of exactly 1. The resulting network is shown in Figure 2.11.

To understand the reason why the solution to this problem will implicitly identify the shortest paths from the source to all other nodes, consider the optimal flow values, shown in Figure 2.12. As before, nonzero flows are indicated by numbers inside half circles. The solution may be analyzed a little more clearly by showing only the arcs with nonzero flows, as in Figure 2.13.

From Figure 2.13 it may be verified that all arc bounds are satisfied and flow in = flow out at each node, confirming that the solution is feasible. (Optimality may be confirmed by inspection due to the simplicity of this example.)

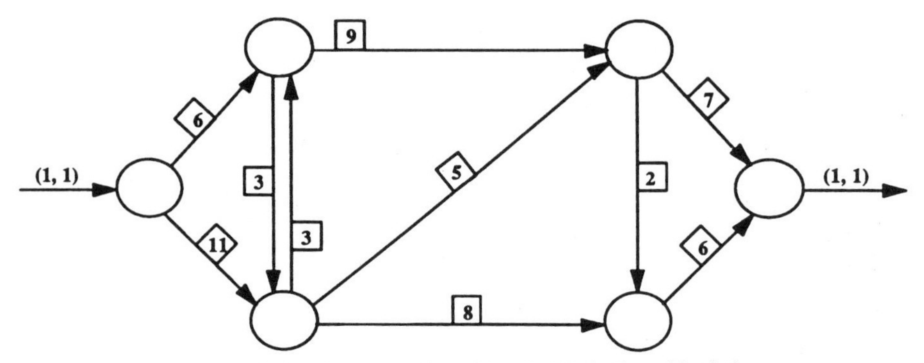

Figure 2.10 Shortest path problem (minimization objective).

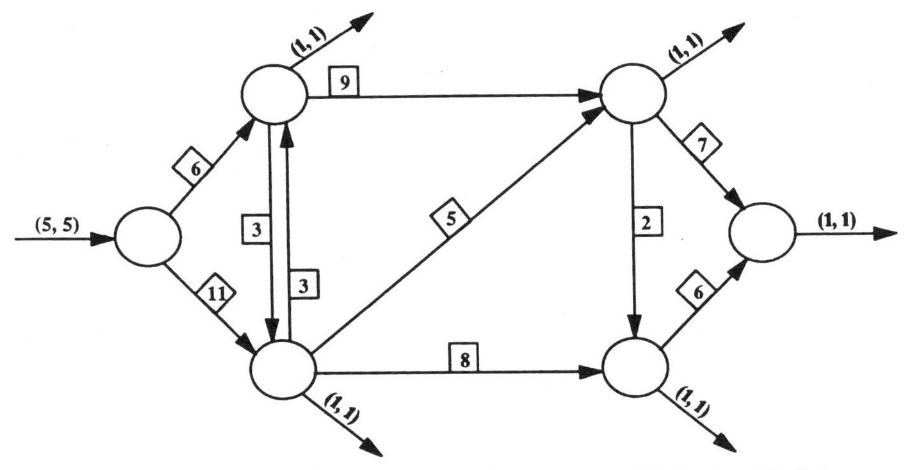

Figure 2.11 Shortest path from source to all other nodes (minimization objective).

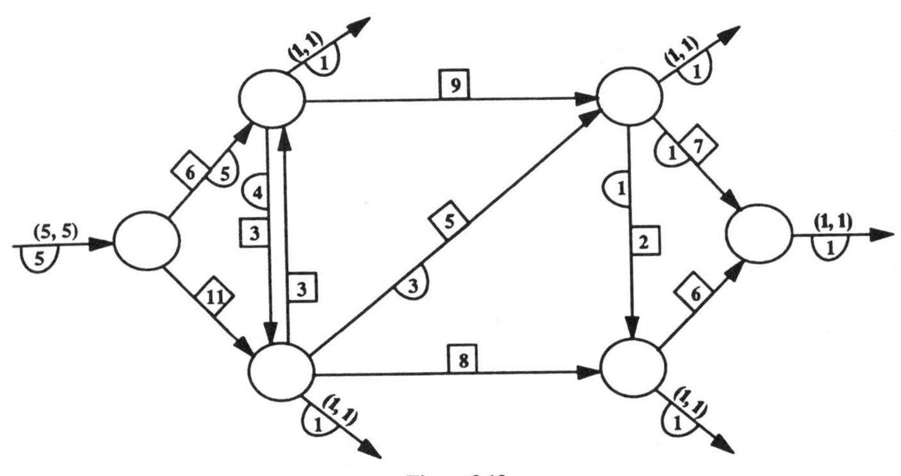

Figure 2.12

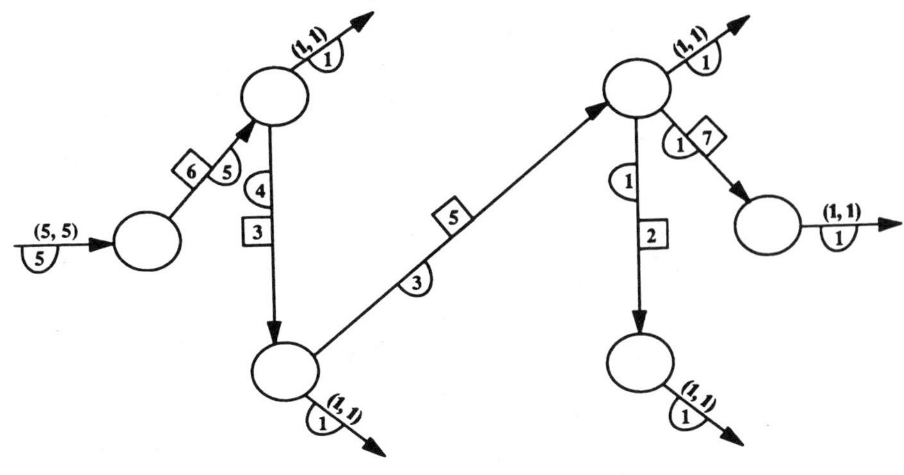

Figure 2.13

It should also be evident from the diagram that a flow path from the source to each other node can easily be identified. But must this path necessarily be a shortest (least cost) path? That is, given that the flows shown minimize total cost for the complete network, could there possibly be a path to any node that is shorter (costs less) than the path by which this node receives flow? Definitely not, because if there were, one could simply re-route one unit of flow from the current path to the shorter path, and the total cost of the solution would be improved—contrary to the fact that the current solution is optimal.

At first, this argument may lead one to suspect that many minimum cost flow problems can be solved just by solving simple shortest path problems. For example, suppose the supply at the source node is changed to 10 and each of the other nodes is given a demand of 2. Do you think it will still be true that the flow received by a given demand node will be sent by the shortest path to that node? And will it be true if the demand nodes have different demands (as long as total supply = total demand)?

The answer is "yes" to both of these questions, for this example. But the example has some very special characteristics: there is a single source, all supplies and demands are exact, and all arcs except supply and demand arcs have lower bounds of 0 and upper bounds of infinity.

Question: Show that by giving an appropriate upper (or lower) bound to one or more appropriate arcs in Figure 2.11, the optimal solution will no longer send flow only on shortest paths.

There are special methods for finding shortest paths from a source to all other nodes, and these are faster than the methods for general minimum cost flow problems. Consequently, whenever a problem can be solved by identifying shortest paths, it may be worthwhile to apply a special rather than a general procedure. (The same is true for other problems with special structures discussed

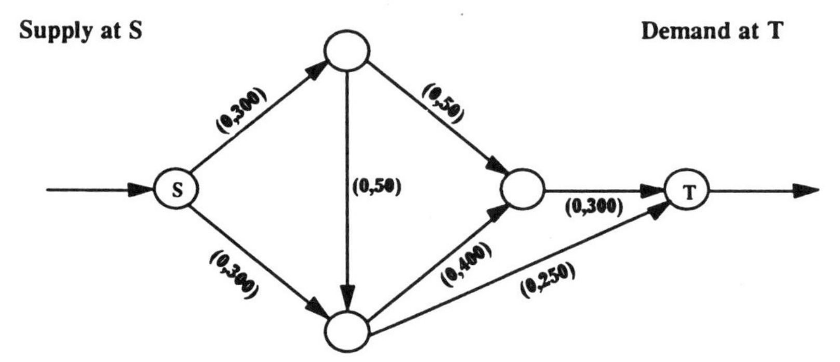

Figure 2.14 Maximum flow example: preliminary.

in this section. However, general pure network methods are so efficient that unless a problem with special structure is very large, or must be solved many times, there may be little motivation to use a special purpose method.)

The classical ***maximum flow*** problem is not concerned with finding the shortest or cheapest route(s), but simply with sending as much flow as possible from the source to the sink. For example, suppose that a natural disaster (such as an earthquake, flood, etc.) has created the need to send emergency supplies to a target area T from a source area S. The available roads from S to T go through three intermediate outposts, as shown in Figure 2.14. The upper bounds on the arcs identify the maximum number of trucks that can safely be handled on a given road per day, and the goal is to maximize the total number of trucks (carrying emergency supplies) sent from S to T.

No bounds are shown on the supply arc for area S or the demand arc for area T in this example, because the only effective limitations are on the number of trucks that can travel each road (segment) per day.

In fact, as a network, this diagram is incomplete, because there are no costs or profits shown. More accurately, no costs or profits are explicitly shown, and therefore it would be assumed that all are 0. This cannot achieve the desired effect, since sending 0 or 1 or 20 trucks from S to T would all be valued equally (because all yield a total cost or profit of 0). Thus, it is essential to be more careful in representing this problem. Since the goal is to maximize the number of trucks that actually reach the demand area T, we place a profit of 1 on the demand arc for node T. (Then 1 truck will be valued as 1, two trucks will be valued as 2, and so forth.) We also explicitly identify the objective to be maximization, so there will be no danger of misinterpreting the profit as a cost. (The name "maximum flow" tends to suggest a maximization objective anyway, but it is not uncommon to see a max flow problem represented with a cost minimization objective, attaching a cost of -1 rather than a profit of 1 to the demand arc.) Therefore, specifying the objective and profit figure explicitly, the *correct* form of the example max flow problem appears in Figure 2.15.

Question: Explain why it would be equivalent to place a profit of 1 on the supply arc into node S instead of on the demand arc out of node T. Also, explain

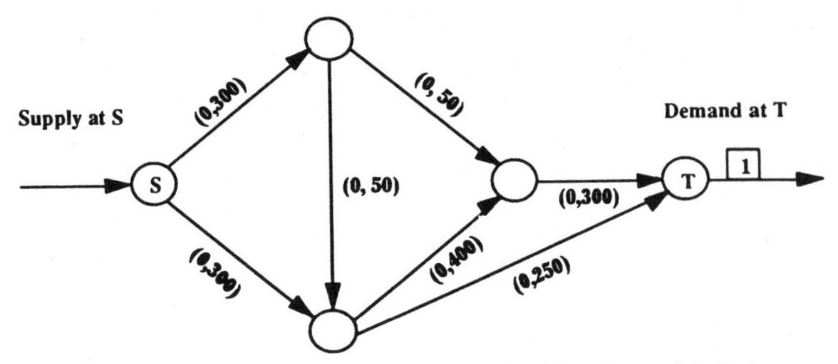

Figure 2.15 Maximum flow example: completed (maximum objective).

why the diagram would be incorrect if the supply arc or demand arc were eliminated. (For example, if the supply arc into S were dropped, what would flow conservation (Feasibility Requirement 2) compel at this node?)

Question: Apply the following intuitively reasonable solution method to the problem in Figure 2.15 and show that it does not yield an optimal solution. At each step, identify a path from S to T (starting with the supply arc and ending with the demand arc) that allows the greatest amount of flow to be sent on that single path. Send as much flow as possible on this path. Then repeat (identifying the path that now allows the greatest flow) until no more paths exist that can carry flow from S to T.

2.5 MORE GENERAL NETWORKS

Pure network problems more general than bipartite, shortest path, and maximum flow problems can have any number of sources, sinks and transshipment nodes. For example, the diagram in Figure 2.16 represents a simplified version of a common type of physical distribution problem, where a product can be produced at any of several plants, and these plants may in turn ship the product on to warehouses and finally to customers. Note the objective of this problem is expressed as one of maximization, hence the revenues (amounts paid by customers) are shown as positive values while shipping costs are shown as negative values.

Real-world problems of this type may involve dozens of plants and warehouses, and hundreds or thousands of customers. (Frequently, in larger problems, customers are aggregated by regions or customer zones.)

Another commonly encountered problem is one in which a company produces several kinds of products, and ships these to various centers for distribution. The trick to representing these multi-product problems is to separate the arcs for different product types at each location where they are shipped and received. To illustrate, suppose a company makes 3 different

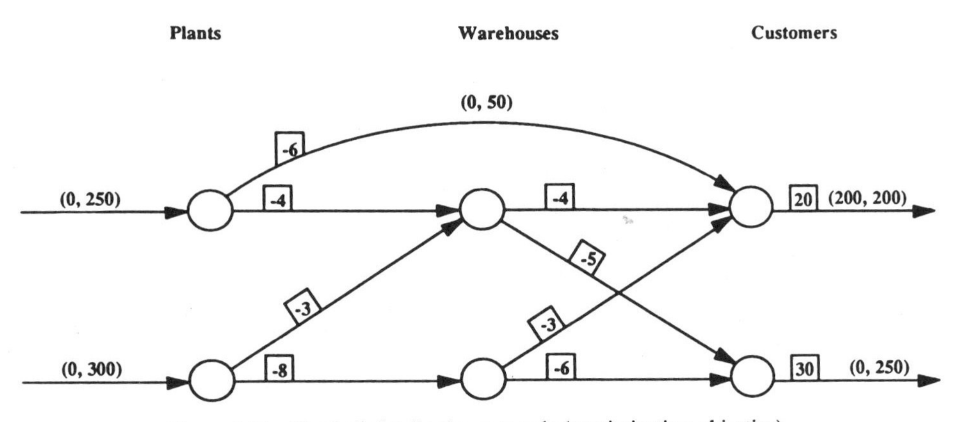

Figure 2.16 Physical distribution example (maximization objective).

products which it ships from 2 plants to 2 distribution centers. The diagram in Figure 2.17 depicts a model of this situation, where exactly 2 of the 3 products are produced at each plant. Since our goal is only to show the structure of this type of model, costs and bounds are not shown.

An interpretation of Figure 2.17 is as follows. The arc into node P1 represents the total combined quantity of both products made at Plant 1. The two arcs from node P1 to nodes 1 and 2 (on the manufacturing side) represent respectively, the quantities of products 1 and 2 made at Plant 1. (Product 3 is not made at this plant.) The arcs out of the "first" node 1, on the path through P1, represent the amount of product 1 that is shipped from Plant 1 to each of the 2 distribution centers. The arc from node 1 to node D1 (on the distribution side) represents the amount of product 1 received at Distribution Center 1, from both Plants 1 and 2 combined. Note that it is not possible in the network structure to place a restriction on quantities such as the total amount of product 1 made at both plants.

The interpretation of the rest of Figure 2.17 proceeds in a similar fashion. Duplicating the product nodes is essential where, for example, different products may have different production limits and costs (on the manufacturing side) or different sales limits and revenues (on the distribution side).

Question: Give an interpretation of a lower and upper bound on the arc:

into P2
from P2 to node 3
from node 2 to node 2 on the path from P1 to D1
from node 3 to node 3 on the path from P2 to D1.

Question: Create a diagram that gives the problem structure when all three products can be made at both plants (and shipped to both centers).

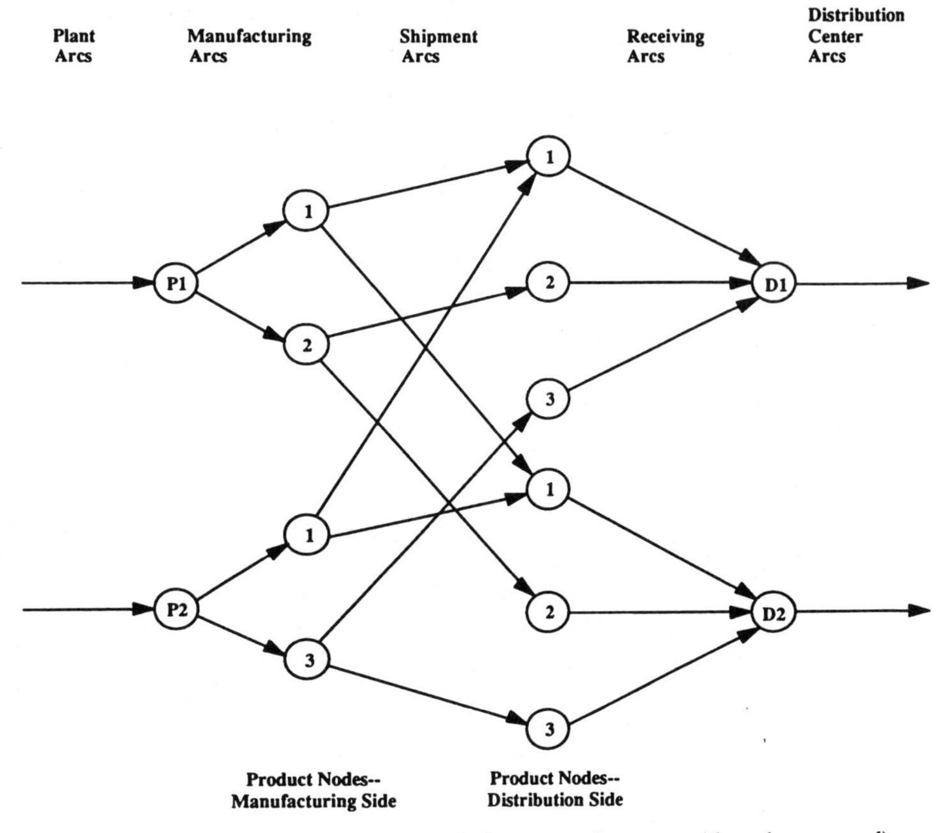

Figure 2.17 Multi-product/distribution example (structure only, costs and bounds suppressed).

It is important to note that Figure 2.17 would not work to model a multi-product problem where each product uses resources at a different rate and where there are joint restrictions on the total flow of all products at plants and distribution centers. For example suppose we have available 400 h of processing time at Plant 1 and 300 h at Plant 2 and that Distribution Center 1 has 200 ft^3 of storage capacity available while Distribution Center 2 has 400 ft^3. The following table gives information about the products.

	Product		
	1	2	3
Production time (h)	2	4	8
Size (ft^3)	2	5	10

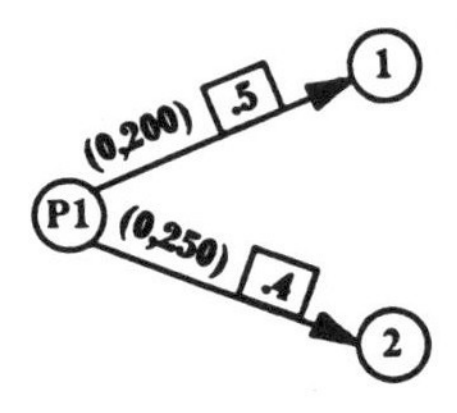

Figure 2.18

This revised problem cannot be modeled as a network. However, if all products required the same amount of production time and the same amount of storage space, then we could convert the constraints on processing time and storage capacity into bounds on the total number of *units* produced at each plant and stored at each distribution center (the arcs into nodes P1 and P2 and out of nodes D1 and D2) and model the problem using the network structure in Figure 2.17. Likewise, if there was only one joint restriction, say the storage capacity constraint at the distribution centers, then we could still model the problem with different production time and storage space requirements per unit of product using Figure 2.17 by converting all bounds to ft^3 and all costs to cost per ft^3. For example, if Plant 1 could produce at most 100 units of product 1 and 50 units of product 2 at a cost of $1 and $2 per unit, respectively, then the bounds and costs on the arcs from node P1 to nodes 1 and 2 in terms of ft^3 would be as shown in Figure 2.18. For instance, 100 units of product 1 equals 200 ft^3 of product 1 and costs $.50 per ft^3 ($1 per unit/2 ft^3 per unit). The storage capacity restrictions would be modeled directly in ft^3 as bounds on the arcs out of nodes D1 and D2.

Note that it is not possible in Figure 2.18 to place a further restriction on the total number of units produced at Plant 1 (such as "at most 125 units total") since it would not be possible to convert that restriction to ft^3 (because different products have different sizes, and we do not know a priori how many of each product we will make in the optimal solution.) To add such a restriction, we would have to add a ***side constraint*** to the network model of the form: 1/2(flow on arc from P1 to 1) + 1/5(flow on arc from P1 to 2) $\leqslant$ 125. Side constraints, while they are part of the problem like all other restrictions and relations, may be conceived as listed "to the side", to differentiate them from constraints that are directly incorporated into the network. Optimization software is available that can easily solve networks with one side constraint, and algorithms exist that can handle networks with several side constraints.

A wide variety of practical network problems can be modeled using the types of structures illustrated in the preceding examples. Additional examples will be provided by the exercises at the end of the chapter.

2.6 ALGEBRAIC STATEMENT OF PURE NETWORK MODEL

Any pure network problem can be characterized by a coefficient matrix which has at most one $+1$ and one -1 entry in each column. For the sake of

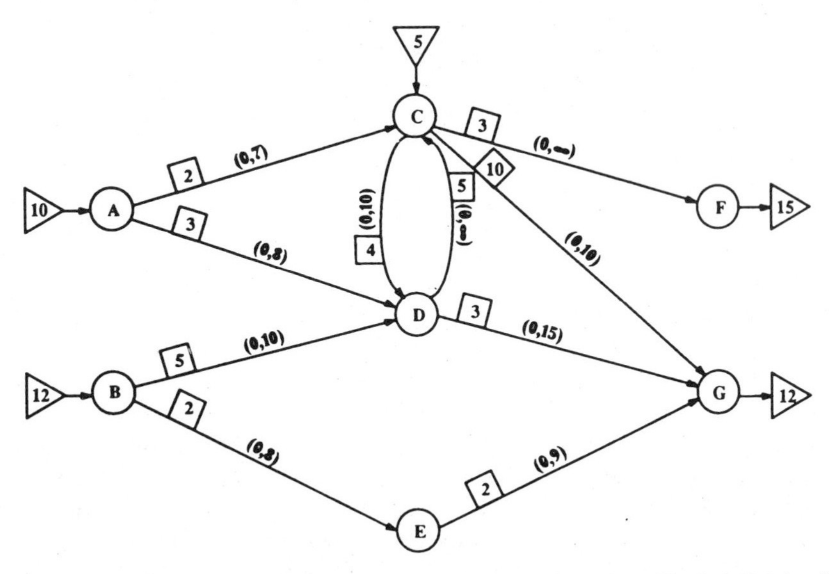

Figure 2.19 Capacitated trans-shipment cash flow model (minimization objective). Reprinted by permission. F. Glover and D. Klingman, *AIIE Transactions*, Vol. 9, No. 4, p. 366. Copyright 1977, Institute of Industrial Engineers.

unification and brevity we will focus attention on the most general of these model types, the ***trans-shipment problem***. An illustration of this model for a cash flow problem is depicted in Figure 2.19.

In this cash flow network, the nodes may be thought of as corresponding to subsidiaries of a central company that operates in different locations. The ***supplies*** and ***demands***, which are shown in the triangles leading into a node for a supply and out of a node for a demand, may be thought of as representing excess or deficit cash positions. Thus, nodes A, B, and C have excess funds, nodes D and E have no funds, and nodes F and G have deficit funds.

The arcs in Figure 2.19 indicate the ways to transfer cash from one subsidiary to another. For instance, the arc from node A to node C indicates that it is possible to transfer funds from subsidiary A to subsidiary C. The absence of an arc indicates that it is not possible to transfer funds directly between the corresponding pair of subsidiaries (though it may be possible to transfer funds indirectly by means of a sequence of arcs through intermediate subsidiaries).

It is quite important to understand how a trans-shipment problem may be stated mathematically in order to be able to recognize when an algebraic statement of a linear programming (LP) problem is actually a network. To state a network problem algebraically, define a variable for each arc. For example, let X_{ij} denote the flow on the arc from node i to node j and c_{ij} denote the unit cost on this arc. Next create the objective function for the problem as an expression involving the costs and variables. For the problem in Figure 2.19, the objective function would be

$$2X_{AC}+3X_{AD}+5X_{BD}+2X_{BE}+4X_{CD}+3X_{CF}+10X_{CG}+5X_{DC}+3X_{DG}+2X_{EG}$$

Upon identifying the objective function, create a constraint for each node which expresses the restriction on flow into and out of the node. Recall from the earlier discussions that we view supply as an inflow and demand as an outflow. Then Feasibility Requirement 2 at each node can be expressed as Total Inflow = Total Outflow, or equivalently, Total Inflow − Total Outflow = 0.

The customary transposition of any constant term of this equation to the right-hand side causes supplies to be denoted as negative quantities and demands to be denoted as positive quantities. To see this, consider constructing the "Total Inflow − Total Outflow = 0" equation for node C in Figure 2.19. The total flow into node C consists of the 5 units of supply and $(X_{AC} + X_{DC})$. The total flow out of node C is $X_{CF} + X_{CG} + X_{CD}$. Thus we have $5 + X_{AC} + X_{DC} - (X_{CF} + X_{CG} + X_{CD}) = 0$, or, after transposing the constant term, $X_{AC} + X_{DC} - X_{CF} - X_{CG} - X_{CD} = -5$. Therefore the supply of 5 becomes expressed as a negative quantity because of its movement to the right-hand side of the equality sign. On the other hand, a demand (which is included in the outflow that is subtracted in the "Inflow − Outflow" equation), appears as a positive quantity when moved to the right-hand side. This also discloses, incidentally, that a supply may be viewed as a negative demand, and vice versa.

Feasibility Requirement 1 is satisfied by adding bound restrictions for the variables. For example, to constrain the flow on the arc from node A to node C to be between 0 and 7 we add the restriction $0 \leqslant X_{AC} \leqslant 7$.

The entire algebraic statement of the capacitated trans-shipment problem shown in Figure 2.19 is as follows:

Minimize: $2X_{AC}+3X_{AD}+5X_{BD}+2X_{BE}+4X_{CD}+3X_{CF}+10X_{CG}+5X_{DC}+3X_{DG}+2X_{EG}$

Subject to:

$$
\begin{array}{llllllllllll}
Ⓐ & -X_{AC} & -X_{AD} & & & & & & & & & = -10 \\
Ⓑ & & & -X_{BD} & -X_{BE} & & & & & & & = -12 \\
Ⓒ & X_{AC} & & & & -X_{CD} & -X_{CF} & -X_{CG} & +X_{DC} & & & = -\ 5 \\
Ⓓ & & X_{AD} & +X_{BD} & & +X_{CD} & & & -X_{DC} & -X_{DG} & & = \ \ 0 \\
Ⓔ & & & & X_{BE} & & & & & & -X_{EG} & = \ \ 0 \\
Ⓕ & & & & & & X_{CF} & & & & & = \ 15 \\
Ⓖ & & & & & & & X_{CG} & & +X_{DG} & +X_{EG} & = \ 12
\end{array}
$$

$0 \leqslant X_{AC} \leqslant 7$, $0 \leqslant X_{AD} \leqslant 8$, $0 \leqslant X_{BD} \leqslant 10$, $0 \leqslant X_{BE} \leqslant 8$, $0 \leqslant X_{CD} \leqslant 10$,
$0 \leqslant X_{DC}$, $0 \leqslant X_{CF}$, $0 \leqslant X_{CG} \leqslant 10$, $0 \leqslant X_{DG} \leqslant 15$, $0 \leqslant X_{EG} \leqslant 9$

It is important to observe that each X_{ij} appears in *exactly two* node equations, i.e., the equation for node i and the equation for node j. Further, since X_{ij} contributes to the outflow of the node i equation and to the inflow of the node j equation, it appears with a coefficient of -1 in the former and with a coefficient of $+1$ in the latter. Thus, each column of the coefficient matrix has one -1 and one $+1$ entry. If the restrictions at the nodes are stated as inequalities rather than equations (i.e., $\leqslant$ or $\geqslant$), then slack or surplus variables may be added or subtracted to convert these constraints to equations, thus satisfying Feasibility Requirement 2, and it is admissible for columns to contain a single nonzero entry. (That is, each slack or surplus variable will only appear in one equation.) In the netform, these columns represent slack arcs (when the coefficient is +1) or surplus arcs (when the coefficient is −1). (See the examples in the next section.)

Note that, by reversing the above steps, it is possible to represent any LP problem whose coefficient matrix has at most one +1 and at most one −1 entry per column as a trans-shipment problem. That is, to construct a graph that corresponds to such a problem, create a ***node*** for each ***constraint*** and an ***arc*** for each ***variable***, affixing supplies, demands and bounds in the manner indicated. The arcs are each directed from the node represented by the constraint with the −1 coefficient to the node represented by the constraint with the +1 coefficient with single-entry columns representing slack and surplus arcs as discussed earlier.

An important observation is that even though these algebraic statements do not restrict the variables to be integer valued, a solution obtained by a standard extreme point solution method (such as a variant of the simplex method) will be an integer as long as the right-hand sides and bounds are integers. This is due to the structure of the coefficient matrix and is a valuable attribute of network models (since integer restrictions often make other models much harder to solve—in some cases impossible to solve within reasonable time limits).

2.7 ALTERNATIVE CONVENTIONS FOR NETWORK DIAGRAMS

Inequality constraints generally arise by stipulating that supplies or demands must be "at least" or "at most" a certain amount. Thus another way to model these restrictions, in addition to using slack or surplus variables (arcs), is by using constrained supply and demand arcs. For example, if the demand at node F in Figure 2.19 had been at most 16, then the constraint for node F would be $X_{CF} \leqslant 16$. Two equivalent algebraic statements and netform models are shown in Figure 2.20. (Only relevant arcs incident to node C are shown. No equation is shown for node C.)

Analogous equivalent formulations for other types of bounded demand and bounded supply are shown in Figure 2.21. The algebraic statements may be constructed using the rules stated earlier.

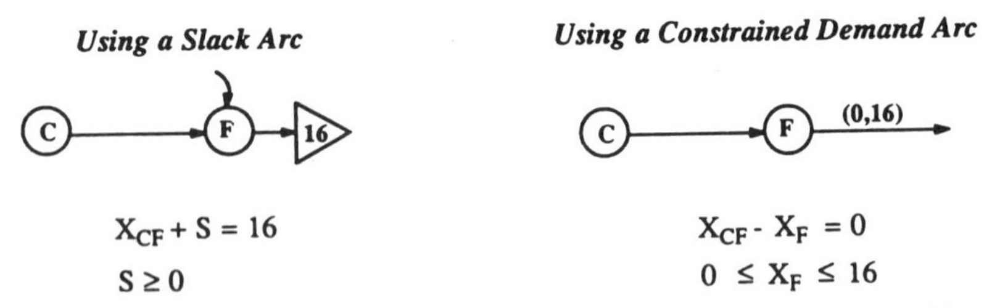

Figure 2.20 Equivalent ways to model $X_{CF} \leqslant 16$ (constrained demand at most 16).

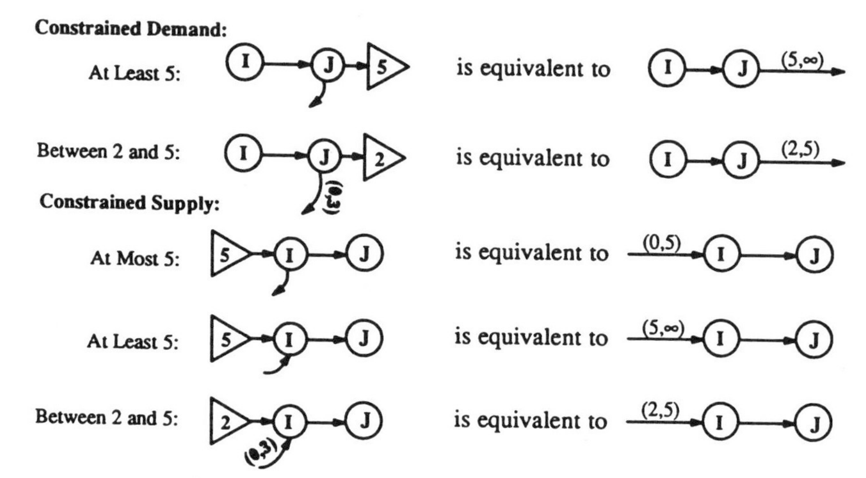

Figure 2.21 Equivalent formulations using slack and surplus arcs.

Question: Write the algebraic statements for the supply and demand nodes in the netform constructions shown in Figure 2.21. The answer is shown below.

Constrained Demand:

At Least 5: $X_{IJ} - S = 5$ or $X_{IJ} - X_J = 0$
$S \geqslant 0$ $\quad 5 \leqslant X_J$

Between 2 and 5: $X_{IJ} - S = 2$ or $X_{IJ} - X_J = 0$
$0 \leqslant S \leqslant 3$ $\quad 2 \leqslant X_J \leqslant 5.$

Constrained Supply:

At Most 5: $-X_{IJ} - S = -5$ or $X_I - X_{IJ} = 0$
$S \geqslant 0$ $\quad 0 \leqslant X_I \leqslant 5$

At Least 5: $-X_{IJ} + S = -5$ or $X_I - X_{IJ} = 0$
$S \geqslant 0$ $\quad 5 \leqslant X_I$

Between 2 and 5: $-X_{IJ} + S = -2$ or $X_I - X_{IJ} = 0$
$0 \leqslant S \leqslant 3$ $\quad 2 \leqslant X_I \leqslant 5$

In a network where the supplies and demands are exact and there are no slack or surplus arcs, then total supply must equal total demand in order for the problem to have a feasible solution. Some network solution codes require input in this ***balanced*** format. (The network in Figure 2.19 is an example of a balanced network.) If, on the other hand, the network has constrained supply or demand or slack or surplus arcs, then it is not necessary for total supply to equal total demand. However, if your network solution code requires balanced input, the conversion process is straightforward. Model any constrained supply and demand using fixed supply and demand and slack and surplus arcs. Connect all slack and surplus arcs to a fictitious or "dummy" node. If total fixed supply exceeds total fixed demand, then the dummy node is a demand node whose demand equals the excess supply. If fixed demand exceeds supply, then the dummy node is a supply node with analogous characteristics. Figure 2.22 illustrates the process.

This conversion process may also be applied to the algebraic statement of the model. After all inequalities are converted to equations by adding slack or subtracting surplus variables, sum all existing equations and multiply both sides by -1. This creates a redundant equation which represents the dummy origin or destination (depending on whether the resulting right-hand side is negative or positive) and results in a balanced network where total supply again equals total demand and every column has one -1 and one $+1$ entry. (See Exercises 2.9 and 2.10 at the end of the chapter. Also, if the coefficient matrix has more than two 1's in any column and the 1's are in consecutive rows, then this matrix may be converted to the balanced network form using elementary row operations. See Exercise 2.14.)

Another specialized network structure that is required by some network solution codes is a ***circularized*** format. For this form, all fixed supply and demand must be represented with constrained supply and demand arcs. Create a ***super supply node*** connected to all supply nodes by the constrained supply arcs and a ***super demand node*** connected to all demand nodes by the constrained demand arcs. An arc with zero cost is added which flows from the super demand node to the super supply node.

Original Network

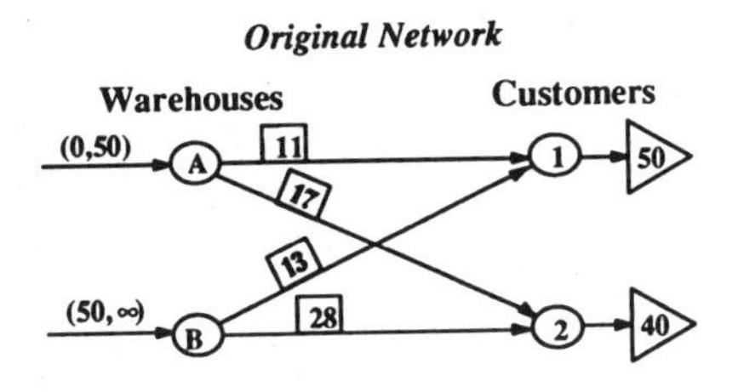

Balanced Network

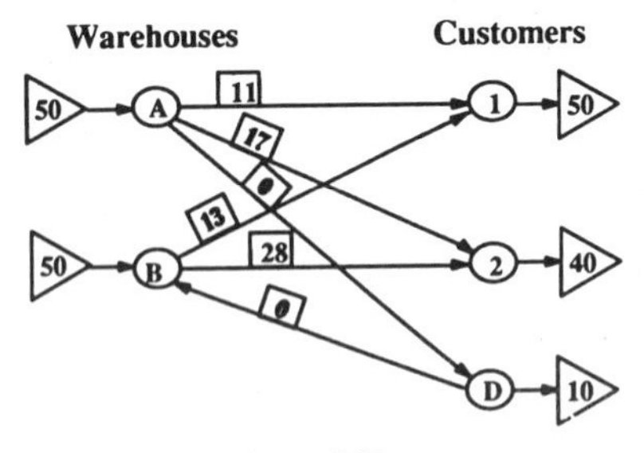

Figure 2.22

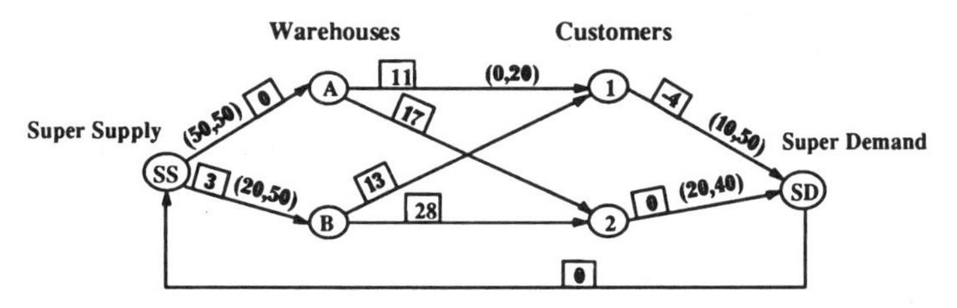

Figure 2.23 Circularized network with constrained supply and demand.

For example, the circularization of the cost minimization network in Figure 2.2 is shown in Figure 2.23.

Frequently in the literature, and in textbooks that consider only the more elementary kinds of network examples, supplies and demands are not shown by using constrained supply and demand arcs or supply and demand triangles, but by simply writing down supply and demand values. An illustration of this appears in Figure 2.24.

This type of representation works acceptably for bipartite problems in which the supply and demand nodes are conveniently separated, provided all supply and demand quantities are exact. In more complex networks, however, greater flexibility is required. One convention sometimes encountered is to differentiate supplies and demands by the letters S and D, as in Figure 2.25.

Here again supplies and demands are exact. Accordingly this notation is sometimes extended to write $20 \geqslant S \geqslant 10$, $15 \geqslant D \geqslant 9$, etc., for supplies and demands that lie between upper and lower bounds.

Such notation is entirely consistent and acceptable. However, using bounded arcs with single endpoints to represent supply and demand quantities has several advantages. (1) Cost and profit figures can be attached directly to supply

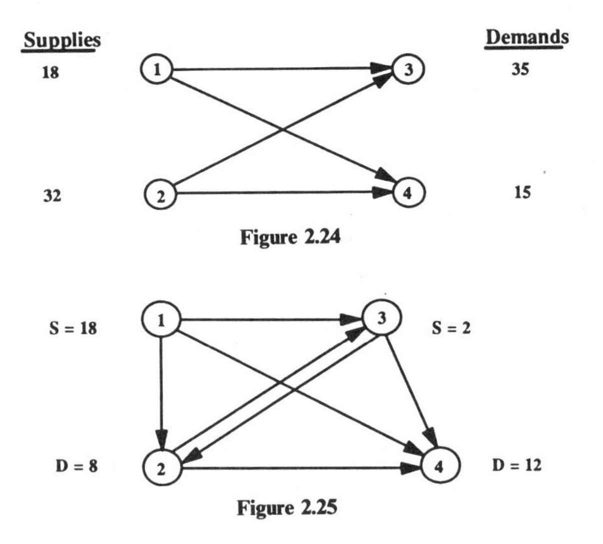

Figure 2.24

Figure 2.25

and demand flows. (This is not possible in the other notations except by adding additional arcs with supply and demand nodes at the start and end respectively.) (2) The basic feasibility requirements can be expressed in a natural and direct way, without making exceptions for supply and demand conditions. (3) All variable quantities are represented by means of arcs. At the same time, it is worth being aware of alternative supply and demand representations, to be able to translate freely from one framework to another and to be able to use network solution codes that require input in a particular format.

APPLICATION: US DEPARTMENT OF THE TREASURY

The US Department of the Treasury has two statistical data base files, *Current Population Survey* and *Statistics of Income*. These files are extensively used to analyze the effect of various policy changes, such as welfare payment levels, social security benefits, income tax rates, etc., on federal revenue. However, to fully analyze the effect of these changes, it is necessary to use some of the information in both of these files. Thus, it is desirable to have a method which limits the amount of information lost as these original files are merged for a particular policy evaluation.

The statistical files contain records. Each record of the files contains descriptions of a family type. For instance, a record in the Current Population Survey file contains items like number of children, head of household (male, female), religion, race, nationality, welfare status, etc. Similarly, a record in the Statistics of Income file contains items describing head of household (male or female), number of family members, amount of work income, amount of interest income, etc. Additionally, each record contains a *weight* indicating the number of families in the US population having these characteristics.

It is important to note that the records in the two files do not contain exactly the same information; however, the records do contain some common information. Thus when policy questions are being evaluated it is important to use the data available in both of the files. To do this it is necessary to merge the files. There are, however, an infinite number of ways to accomplish a file merger. Obviously some ways are much better than others for evaluating certain types of policy questions. Consequently, one must carefully select what types of records to merge. In making this decision one also implicitly determines the total number of records to be in the merged file. Another important issue when merging the records involves how the weights for the merged records will be determined; that is, how many families each merged record will characterize.

Figure 2.26 indicates how this problem can be formulated as a transportation problem. This formulation is derived by thinking of each record in each file as a node. The supply or demand associated with a given node corresponds to the weight of the associated record. Figure 2.26 depicts the current Population

---

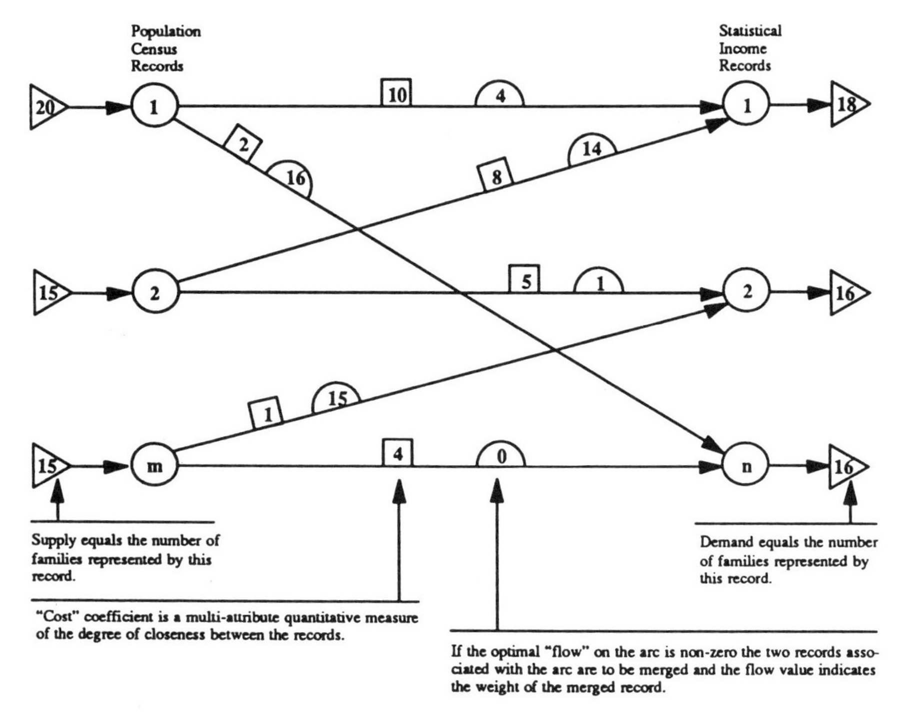

Figure 2.26 Microdata set file merge problem. Reprinted by permission. F. Glover and D. Klingman, *AIIE Transactions*, Vol. 9, No. 4, p. 365. Copyright 1977, Institute of Industrial Engineers.

Survey records as the origin nodes and the Statistics of Income records as the destination nodes. Note that total supply equals total demand since the sum of weights in each file equals the total number of US families.

The nodes in Figure 2.26 are connected by arcs. The presence of an arc indicates that it is reasonable to consider merging the records associated with its pair of nodes. The "cost" coefficient of each arc is a multi-attribute measure of the "distances" between the records in terms of the common information characteristics of the two records associated with the arc. While there is not a unique procedure to use for calculating these distance coefficients, the problem of determining distances between records in a microdata file is similar to the problem of determining distances between coordinates in a multi-dimensional space for multi-variate regression. Thus, a weighted sum of squared deviations can reasonably be used to estimate this distance (or information loss).

The solution to the transportation problem indicates: (a) the number of records to be contained in the merged file; (b) which records are to be merged with each other; and (c) the weight of each merged record. To illustrate, the number in the semicircles in Figure 2.26 indicates the flow on the arcs. If the flow is nonzero, then the two records connected by this arc are to be merged. For example, record 1 of the Population Census file is to be merged with record 1 and record n of the Statistical Income file, creating two new records (one for each arc with nonzero flow emanating from node 1). The weight (i.e., the number of families represented by each merged record) is equal to the flow on the arc. Consequently, solving the transportation problem yields the specifications for a merged file which minimizes the information lost with respect to the distance function used.

Glover and Klingman [1977] implemented an extended transportation system for the US Department of the Treasury which is capable of solving a transportation problem with 50,000 nodes and 62.5 million arcs on a UNIVAC-1108 computer (even though the processing power of this machine is less than a personal computer with a 326 processor today). This system was used to merge the Current Population Survey file and Statistics of Income file on a routine basis given appropriate cost coefficients for the policy to be evaluated. A prototype of this problem with 5000 nodes and 625,000 arcs was solved using a special code developed for the Department of the Treasury in less than 4 min of central processing time and less than 9 min of total job time (including all input and output processing) on a UNIVAC-1108. Because a full size problem of 50,000 nodes and 62.5 million arcs was not generated for solution, the amount of time required to solve it is unknown. In problems of this magnitude, the intimate coordination of modeling and computer solution effort is indispensable. In fact, the value of such coordination for problems at all levels can spell major differences in overall solution efficiency. This type of coordination in the file merger application was instrumental in achieving the unprecedented efficiency of 4 min of CPU time to solve a 625,000 variable problem, and further led to an iterative procedure for bounding the optimal objective function value. This procedure can be extremely valuable for identifying near optimal solutions (within a stated percent) for larger applications where exact optimality may not be required.

CASE: ANGORA

Angora is a tiny Western European country located high in the mountains and known for its production of mohair. The country's government is a monarchy ruled by a semibenevolent Empress (semibenevolent means she keeps half of the proceeds of the mohair business).

The citizens of Angora are employed in two basic functions, agriculture (raising crops and goats) and commerce (arranging for the export of mohair). Traditionally, the agricultural function has carried much more prestige and financial reward than the commercial function which has been viewed as menial.

The Empress has come to realize that commercial dealings with mohair customers are extremely important for the future growth and success of the business. Therefore, in order to make the commercial jobs more attractive to highly qualified citizens, she is considering a new tax reduction and retirement package for commercial employees. The tax benefits depend upon family income characteristics whereas retirement benefits will be controlled by family and head-of-household demographic characteristics as well as income.

The Empress has instructed the Royal Statistician to give her an estimate of the cost of several alternative plans before she announces the new policy. (She does not want this endeavor to cut too much out of her half of the proceeds). The statistician realizes that this is no small task primarily due to his record keeping system.

He has kept updated citizen data faithfully over the years, but it is in two separate books. He calls these books "Income Data File" and "Population Data File." (No one knows exactly why he uses the term "File" for a book, but he is an eccentric old man so they do not worry about it.) The Income Data File has 40 pages, which the old man calls "records" (very strange man). Each record describes certain income characteristics and gives the number of families in Angora who possess these characteristics. Similarly, the Population Data File has 40 pages, each of which describes demographic characteristics and gives the number of Angorian families who possess them. The contents of five of these records from each file are summarized below:

Income Data File

Record	No. of Families	Gross Family Income	No. of Family Members	Source of Income	Interest Income
1	20	10,000	3	Commerce	0
2	30	15,000	2	Commerce	1,000
3	25	20,000	5	Agriculture	1,000
4	18	25,000	4	Agriculture	3,000
5	32	15,000	3	Commerce	500
⋮	⋮	⋮	⋮	⋮	⋮

Population Data File

Record	No. of Families	Gross Family Income	No. of Family Members	Head of Household			No. of Family Members Under 18
				Age	Education	Sex	
1	25	15,000	4	40	12	M	2
2	30	15,000	2	25	16	M	0
3	18	20,000	1	30	18	F	0
4	27	25,000	2	35	16	F	1
5	25	20,000	4	25	12	M	1
⋮	⋮	⋮	⋮	⋮	⋮	⋮	⋮

The old man realizes that he must merge the information in the two files in order to get an estimate of how many families will qualify for the various levels of benefits proposed by the Empress. He also realizes that in performing the merger some information may be lost and that this loss should be kept to a minimum. In other words, in matching families from the two files together, he should strive to keep their common characteristics as "close" as possible. His files have two common characteristics, "Gross Family Income" and "Number of Family Members." Being mathematically oriented, the statistician contrives the following function as a measure of the "distance" (i.e., lack of closeness) between the family types described by the two files:

$$d_{ij} = \sqrt{\frac{(\mathrm{GFI}_i - \mathrm{GFI}_j)^2}{S^2_{\mathrm{GFI}}} + \frac{(\mathrm{NFM}_i - \mathrm{NFM}_j)^2}{S^2_{\mathrm{NFM}}}}$$

where

GFI_i = gross family income on record i of income data file (in 1000s)
GFI_j = gross family income on record j of population data file (in 1000s)
S^2_{GFI} = variance of gross family income data items
NFM_i = number of family members on record i of income data file
NFM_j = number of family members on record j of population data file
S^2_{NFM} = variance of number of family members data items

(This distance measure is the square root of the sum of squared deviations of common data items between files scaled by their variances. A variety of other distance measures could be appropriate.)

The old man begins to calculate the distance measure for all possible record combinations. The results of his calculations are shown in the following table:

$$S^2_{\mathrm{GFI}} = 21, \quad S^2_{\mathrm{NFM}} = 1.1$$

	Distance d_{ij}					
	Population Records					
Income Records	1	2	3	4	5	⋯
1	1.449	1.449	2.899	4.681	2.381	⋯
2	1.907	0.000	1.449	2.182	2.197	⋯
3	1.449	3.061	1.907	3.061	0.953	⋯
4	2.182	2.899	3.061	1.907		⋯
5	0.953	0.953	2.197			⋯
⋮	⋮	⋮	⋮	⋮	⋮	⋮

Unfortunately, the strain of this effort caused the old man to suffer a mental breakdown before he could finish the last three calculations or tell anyone what he planned to do with them once completed.

Case Questions

1. Calculate the remaining three d_{ij}'s.
2. Explain what is meant by "Information Loss" as a result of merging the two files.
3. Assume that the old man meant d_{ij} to be the distance incurred by *each* family of Income Record i that gets matched with a family of Population Record j. Create a model formulation (netform) which will estimate how many families of each combined type should exist based upon the criterion of minimizing total distance incurred in the matching process.
4. Suppose in the optimal solution, $X_{45} = 10$. What is the interpretation of this result? How does it help to find the costs desired by the Empress?
5. What is the maximum number of combined family types, or "records," which will exist in the merged file (i.e., in the solution)?

EXERCISES

2.1 ***Assignment Problem.*** Three astronauts must carry out three tasks on a space shuttle, so that exactly 1 person is assigned to each task. Efficiency ratings of each astronaut in each task are shown in the following table.

Efficiency Ratings

	Task		
Astronaut	1	2	3
1	30	28	20
2	35	36	30
3	40	36	20

The goal is to maximize the total value of the assignment, expressed as the sum of the efficiency ratings over the three specific astronaut-to-task assignments made. Formulate an assignment model for this problem.

2.2 ***Transportation Problem.*** The Flint N' Chips Gravel Company has received a contract to supply gravel for two new construction projects, located in the towns of Pantego and Hurst. Construction engineers have estimated the required amounts of gravel needed.

Project	Location	Weekly Requirement (Truckloads) (Acceptable Range)
A	Pantego	60–70
B	Hurst	75–85

The Flint N' Chips Company has two gravel pits located in the towns of Grapevine and Irving. The chief dispatcher has calculated the amounts of gravel which can be supplied, according to limits set by union contract and the operating facilities at the two towns.

Pit	Location	Amount Available (Truckloads) (Acceptable Range)
1	Grapevine	Exactly 70
2	Irving	80–90

The delivery costs are as follows:

	Cost ($) per Truckload	
From	To Project A	To Project B
Pit 1	8	7
Pit 2	10	11

Also, because of union contracts, no more than 40 truckloads may go from Grapevine to Hurst and at least 40 truckloads but no more than 50 truckloads must be shipped from Irving to Pantego. Formulate a network to minimize cost.

2.3 ***Maximum Flow Problem.*** A gem collector in Venezuela is trying to send the greatest possible value in stones to his banker in Zurich, but wishes to make the transfer in secret and knows that he can only use certain channels to avoid detection. Further, he is aware of limits that the dollar value of stones sent through each channel must satisfy if he is to escape notice by the wrong parties. The following table identifies each channel by two letters coding its origin and destination (with V = Venezuela, Z = Zurich, and A, B, C = trans-shipment points), and contains entries identifying upper bounds on the dollar value of stones that can be carried on these channels. (An × indicates that the channel does not exist.)

Upper Bounds (in Ten Thousand Dollar Units)

	To			
From	A	B	C	Z
V	60	90	×	×
A		×	20	30
B	×		50	20
C	×	×		60

Identify a network formulation to maximize the dollar value of gems sent to Zurich.

2.4 Take the figures for Exercises 2.1, 2.2, and 2.3 and create "reversed" figures by reversing the direction of all arcs. (Make all previous demand arcs into supply arcs, and vice versa, causing all flows to go in the opposite direction.) Do not change any bounds, costs or profits. Can you give an interpretation to the "problem" expressed by each of these figures? Is this interpretation as meaningful as that of the original figures? Are the problems actually equivalent to the original problems, in the sense of having the same solutions, or are they changed? Does this suggest a flexibility in what is viewed as "supply" and what is viewed as "demand"?

2.5 ***Trans-shipment Problem.*** The Macho Nacho Company decides to expand its product line and manufacture a guacamole-lined ski boot (advertised as the ultimate in comfort and great for apres ski snacks). The company has 2 plants, 3 warehouses and 3 customers. The following tables provide relevant data for the goal of minimizing total cost. (An × means no shipment is possible. All shipments are in carloads.)

Shipping Costs

	Warehouses		
Plants	1	2	3
1	80	110	×
2	140	×	130

Shipping Costs

	Customers		
Warehouses	1	2	3
1	×	15	25
2	20	30	×
3	30	×	20

Upper Limits on Amounts that can be Shipped (Imposed by Transportation Connections)

	Customers		
Warehouses	1	2	3
1	×	100	50
2	80	70	×
3	80	×	70

Both plants must produce at least 60 carloads to stay in business, but are limited by production capacity to a maximum of 120 carloads each. On the other hand, Warehouse 2 already has 60 carloads of inventory on hand available for shipping and Warehouse 3 has 30 carloads on hand. Customer 1 demands exactly 80 carloads, Customer 2 demands exactly 90 carloads, and Customer 3 demands exactly 100 carloads. Formulate the problem as a trans-shipment problem. (Suggestion: consider inventory on hand as an upper bounded supply, i.e., as a quantity available to enter a warehouse node via a supply arc, where the amount used can vary between 0 and the total on hand).

2.6 ***Multi-product Problem.*** An auto manufacturer makes 3 models of cars, known unofficially within the industry as Slug, Joy Ride, and Kamikazee. The car models are made at two production plants and then sent to two Customer Zones. Monthly production capacities and costs at the plants are shown in the following tables (Kamikazee is not made at Plant 1).

Upper Limits on Cars Produced

	Models			
Plants	Slug	Joy Ride	Kamikazee	All Combined
1	500	900	–	1100
2	400	700	500	1400

Production Costs ($) per Car

	Models		
Plants	Slug	Joy Ride	Kamikazee
1	1450	1800	–
2	1620	1750	2500

Shipping costs are independent of the car model, as shown below:

Shipping Cost ($) per Car

	To Zone	
From Plant	A	B
1	130	220
2	190	120

Finally, monthly lower limits on car models required at the customer zones and selling prices for the cars in these zones are as follows:

Lower Limits on Cars Required

	Model			
Zone	Slug	Joy Ride	Kamikazee	All Combined
A	400	500	200	1200
B	300	600	200	1200

Selling Price ($) Per Car

	Model		
Zone	Slug	Joy Ride	Kamikazee
A	5200	5700	7100
B	5500	5800	7000

a. Prepare a network model to maximize the monthly profit of the combined production and distribution operation.

b. Replace the upper limits on the total number of cars produced (1100 cars at Plant 1 and 1400 cars at Plant 2) with upper limits of 20,000 production hours at Plant 1 and 30,000 production hours at Plant 2. The production hours required for the Slug, Joy Ride, and Kamikazee are 10, 15, and 20, respectively, independent of the plant. Prepare a network model to maximize the monthly profit of the combined production and distribution operation, ignoring the "All Combined" lower limit on cars required.

c. Ignoring the modification in part (b), replace the lower limits on the total number of cars required in Zones A and B in the original problem (1200 cars for each Customer Zone) with upper limits of 60,000 ft^2 of parking space at the car lots in each zone. The square feet of parking space for the Slug, Joy Ride, and Kamikazee are 50, 70, and 60, respectively. Prepare a network model to maximize the monthly profit of the combined production and distribution operation, ignoring the "All Combined" upper limit on cars produced.

d. Why did we have to ignore the indicated "All Combined" limits in parts (b) and (c)? Can you represent both the production hour restriction and the square feet storage restriction in a single pure network model?

2.7 Draw the network corresponding to the model below.

$$\begin{aligned}
\text{Minimize} \quad & 3x_{12} + 5x_{14} + 2x_{23} + x_{24} + 5x_{52} + 3x_{53} + 4x_{54} \\
\text{Subject to} \quad & -x_{12} - x_{14} = -10 \\
& x_{12} - x_{23} - x_{24} + x_{52} = 0 \\
& x_{23} + x_{53} = 6 \\
& x_{14} + x_{24} + x_{54} = 7 \\
& -x_{52} - x_{53} - x_{54} = -3 \\
& x_{12}, x_{14}, x_{23}, x_{24}, x_{52}, x_{53}, x_{54} \geqslant 0 \\
& x_{12} \leqslant 3, \quad x_{54} \leqslant 5
\end{aligned}$$

2.8 The Magnum Force Distillery needs to figure out the best way to ship its famous spirits to meet customer demands. Magnum has two innovative distribution centers in Fort Worth and Phoenix from which it pipes its product through giant conduits (under a special joint use agreement with Exxon) to its primary marketing centers in Los Angeles and Denver. Magnum's product is so popular (some of its customers describe its effect as "high octane high") that its demand is expressed in terms of lower bounds, rather than upper bounds. Similarly, to maintain operational efficiency, Magnum supplies its product at levels that are only constrained on the underside. The entire distribution operation is organized with the goal of minimizing cost. The following tables give the relevant statistics:

Distribution Cost for Each 100,000 gallons (in $10,000s)

From	To: Los Angeles	Denver
Fort Worth	20	11
Phoenix	7	6

Lower Bounds (in 100,000 gallons)

Fort Worth supply	10
Phoenix supply	5
Los Angeles demand	9
Denver demand	5

Prepare a network model for this problem, labeling nodes associated with the cities by their first initials (i.e., F, P, L, D). Use the units of the tables (e.g., $20 instead of $200,000).

2.9 J. Foster Smith was aggrevated. The production manager of CDC, to which his farm belonged, has just informed him that beginning next week, the delivery of the 20,000 gallons per day of milk from his farm should be made to Weirton Processing Facility, instead of to Beaver Valley Facility close to his farm. Each facility has the capacity to process up to 50,000 gallons per day of milk. It appeared that since the retirement of the previous production manager there had been nothing but change. In Mr. Smith's opinion, there was no real reason why this change was needed. Beaver Valley Processing Facility was only 2 miles from his farm while Weirton was 5. When Smith questioned the new production manager about the change, he had been told that the change was an attempt to save money since Jones Dairy, which supplies 80,000 gallons of milk to the cooperative, would be using all of Beaver Valley's capacity. Beaver Valley was 4 miles from Jones Dairy, whereas Weirton was 8 miles distant. Since the cooperative divided overall profits among the farmers in proportion to ownership, and costs were reported and reimbursed centrally, Smith reasoned that the change was probably reasonable, but he wanted to think about it and satisfy himself completely that the change would not lose money overall.

a. Formulate a netform to minimize total cost (miles) for the dairy cooperative and write the algebraic statement of the netform. Notice that total supply equals total demand, so that supplies and demands can be represented as exact.

b. Suppose that the capacity of the Beaver Valley Facility is increased to 60,000 gallons per day. Edit the algebraic statement of your model to incorporate this change. Convert the algebraic statement to balanced form by converting the inequalities to equalities and adding a redundant equation. Draw the netform that this balanced algebraic statement represents. What does the redundant equation correspond to in the netform?

2.10 Consider the model shown below.

$$\text{Minimize} \quad 3x_{12} + x_{13} + 2x_{14} - x_{23} - x_{24} + x_{32}$$

$$\text{Subject to} \quad -x_{12} - x_{13} - x_{14} \geqslant -8$$

$$x_{12} - x_{23} - x_{24} + x_{32} = 0$$

$$x_{13} + x_{23} - x_{32} = 0$$

$$x_{14} + x_{24} \geqslant 6$$

$$x_{12}, x_{13}, x_{14}, x_{23}, x_{24}, x_{32} \geqslant 0$$

a. Draw a network using constrained supply and demand arcs.

b. Draw a network using slack and/or surplus arcs.

c. Draw a balanced network. (Hint: First convert the algebraic statement to balanced form by converting the inequalities to equalities and adding a redundant equation.)

d. Draw a balanced and circularized network.

2.11 Write the algebraic statement for the network model shown in Figure 2.2.

***2.12** Draw a balanced network corresponding to the constraints below. (Hint: Convert the inequalities to equalities, multiply the appropriate constraints by -1 to get the proper $+1/-1$ form, and then add a redundant restriction corresponding to a fictitious node.)

$$x_1 + x_2 = 6$$

$$x_1 - x_3 + x_4 \geqslant 5$$

$$-x_2 - x_3 \leqslant 6$$

$$x_1, x_2, x_3, x_4 \geqslant 0$$

2.13 A company has two manufacturing plants and three distribution warehouses located in different areas. This month's supply at Plants 1 and 2 are 100 and 200, respectively. The sales potentials at Warehouses 1, 2, and 3 are 150, 200, and 250, respectively. As you can see, potential demand vastly exceeds available supply, and consequently some demand will go unsatisfied. Assume the cost of shipping one unit from Plant i to Warehouse j is t_{ij}, and that the sales revenue per unit at Warehouse j is p_j. (The company is able to charge different prices for its product in different parts of the country).

a. Formulate a netform model to obtain a profit-maximizing solution. Assume that the supply at each plant represents an upper limit on what can be sold to maximize profit.

b. Assuming supplies represent a fixed amount which must be sold regardless of profit, draw your network from part (a) in balanced format. (Hint: Add a super demand node with a fixed demand of 300).

c. Draw your network from part (a) in circularized format.

***2.14** Suppose an LP formulation has the following coefficient matrix:

$$\begin{bmatrix} 1 & 0 & 0 & 0 & 0 & 0 & 0 \\ 1 & 1 & 0 & 1 & 0 & 1 & 0 \\ 1 & 1 & 1 & 1 & 0 & 0 & 1 \\ 0 & 1 & 1 & 0 & 0 & 0 & 1 \\ 0 & 1 & 1 & 0 & 1 & 0 & 1 \\ 0 & 0 & 1 & 0 & 1 & 0 & 0 \end{bmatrix} = \begin{bmatrix} R_1 \\ R_2 \\ R_3 \\ R_4 \\ R_5 \\ R_6 \end{bmatrix}$$

a. Identify a special structure exhibited by the columns of the coefficient matrix. (Hint: Note the configuration of zeros and ones.)

b. Using elementary row operations, attempt to convert this formulation to a network structure. (Hint: Begin by setting row 6 equal to row 5 − row 6; row 5 = row 4 − row 5, etc. Then create a redundant constraint.)

c. Suggest a general, direct procedure (not involving row operations) for converting any LP formulation with the special structure you identified in part (a) to a network formulation. (Hint: Observe the similarities and differences in the original coefficient matrix and the network coefficient matrix).

2.15 Colonel Cutlass, having just taken command of the brigade, has decided to assign people to his staff based on previous experience. His list of major staff positions is adjutant (personnel officer), intelligence officer, operations officer, supply officer, and training officer. He has 5 people he feels could occupy these 5 positions. The table gives their years of experience in several fields.

	Adjutant	Intelligence	Operations	Supply	Training
Muddle	3	5	6	2	2
Whiteside	2	3	5	3	2
Kid	3	0	4	2	2
Klutch	3	0	3	2	2
Whiz	0	3	0	1	0

Formulate the problem (algebraically) as an assignment problem to fill all positions and to maximize the total number of people years experience and draw its netform.

2.16 Suppose that after solving a transportation problem with positive shipping costs c_{ij} along all arcs, we increase the supply at one source and the demand at one destination in a manner that will maintain equality of total supply and total demand.

a. Would you expect the total shipping cost in the modified problem with a larger total shipment of goods to be higher than the optimal shipping plan from the original problem?

b. Solve the following transportation problem. (You may use a network optimization or LP software package.)

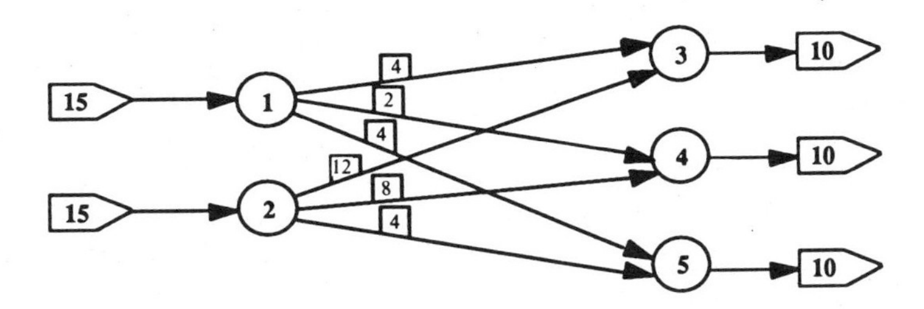

c. Increase the supply at source ① by 1 unit and the demand at destination ⑤ by 1 unit and resolve the problem. Has the cost of the optimal shipping plan increased? Explain this behavior.

***2.17** The example below represents a multicommodity network with three commodities, A, B, and C, where a -1 indicates a supply of one unit and a $+1$ indicates a demand of one unit. For example, at node 1 there is a supply of one unit of A and a demand of one unit of B. Write an algebraic description of this model including the restriction that the *total* flow of *all* commodities on an arc cannot exceed one unit. (Hint: First draw a separate network for each commodity with the appropriate supplies and demands. Then write the algebraic statement for each together with the restrictions on total flow.) How does the tableau structure differ significantly from that of a pure network? Can you identify a special structure?

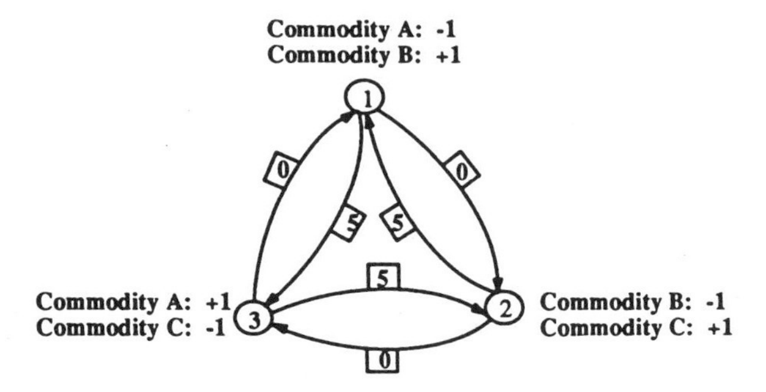

2.18** ***Maze Problem. Draw a network corresponding to the following maze:

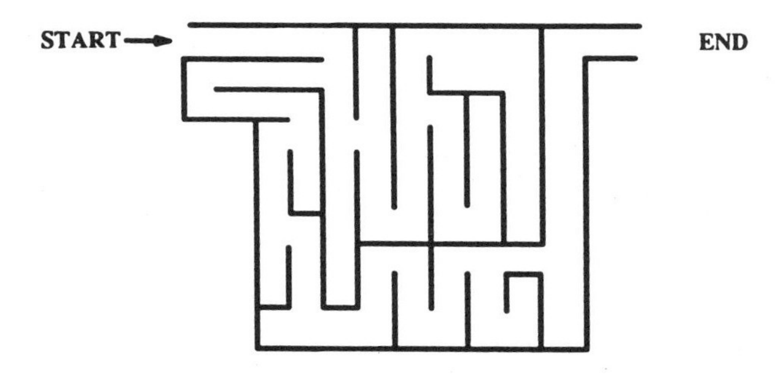

To create a network, a node should be introduced at each intersection, corner, and dead end. The maze entrance can be represented by a "supply arc" and the exit by a "demand arc." Make sure the possibilities for movement are the same in the network and maze (but do not bother with arcs that retrace footsteps), using the general format shown in the following example.

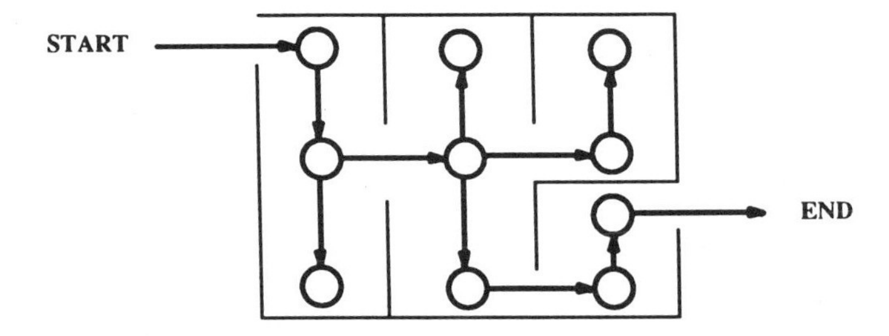

a. Identify an appropriate supply and/or demand so that a feasible solution to the network will give a path through the maze. (Does it matter if both supply and demand values are specified or not?)

b. Do you imagine you could similarly convert any standard maze problem into a network problem? What if the maze were "3-

dimensional", where corridors could lead to various levels above and below the "ground floor" level?

c. Consider a "colored" maze problem that has more than one path from start to end, where you must pay a toll of $2.00 for each red corridor you take, $1.00 for each blue corridor you take, and nothing for each yellow corridor you take. Your goal is to get from the start to the end at the cheapest cost. Identify the class of network problems such a maze problem belongs to.

d. Suppose in the colored maze problem you have a choice of beginning from any of several "start locations" and of finishing at any of several "end locations". How would you represent this with a network model? (Since you are not required to choose a specific entrance or exit, how can you set the model up so that a single unit of supply can get to any of the alternative entrances, or a single unit of demand can be met from any of the alternative exits?) Give an example with 2 entrances and 3 exits.

***2.19** Explain how a standard (unbounded) transportation problem can be converted into a (large) assignment problem. Illustrate your technique using the following example:

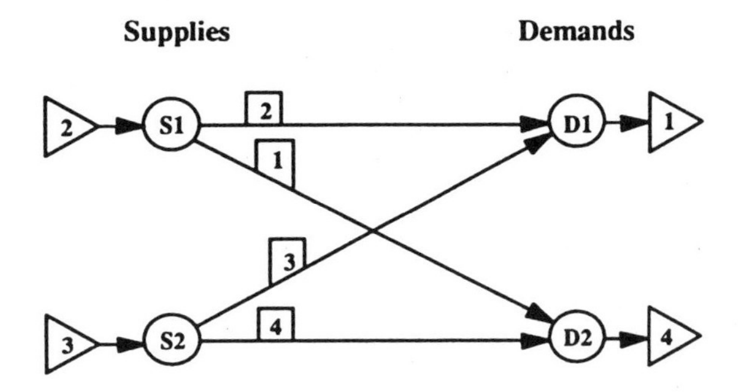

2.20 a. Randomly place six nodes on a page. Construct a connected network using only five arcs. What happens if you try to include a loop when constructing the network?

b. Repeat part (a) with seven nodes and six arcs.

***2.21** Explain how to convert a trans-shipment network with capacity restrictions on flow through nodes into an equivalent (enlarged) trans-shipment problem with capacity restrictions on the flow through arcs.

REFERENCE

Glover, F. and D. Klingman. 1977, "Network Application in Industry and Government," *AIIE Transactions*, Vol. 9, No. 4, pp. 363–376.

3

ADDITIONAL PURE NETWORK FORMULATION TECHNIQUES

In this chapter we expand on the techniques for formulating pure networks. Our focus will be on ways to model a variety of useful conditions that arise in practical settings.

3.1 A CORE EXAMPLE

We illustrate the additional model techniques by reference to a single "core" example, which we modify in stages. This problem is an analog of one that occurs in practical applications, though we have taken the liberty of simplifying it to allow additional structures to be introduced in a convenient fashion. The problem may be expressed as follows.

A clothing manufacturer makes three garments, jeans, blouses, and shirts, which must be allocated to one or another of four sew-lines for stitching. The cost of allocating a particular garment to a particular sew-line is shown in Table 3.1. (Blank spaces indicate that a garment cannot be handled on a particular sew-line.)

Based on the quantities of garments that the company's outlets typically stock and sell, the company has determined it should make from 300 to 400

Table 3.1 Cost per Garment Stitched on a Given Sew-Line

	Sew-Lines			
Garment	1	2	3	4
Jeans	3	4		
Blouses	2		3	
Shirts		5	4	4

jeans, 150 to 200 blouses, and 400 to 550 shirts each week. Any amounts in these ranges are acceptable, as long as the number of garments handled weekly by sew-lines 1–4 do not exceed 250, 300, 300, and 200, respectively. In addition, to meet union requirements, and to avoid certain expenses of under-utilization, each sew-line must turn out at least 100 garments each week. Within these restrictions, the company seeks to minimize its total weekly operating cost. A network diagram of the problem appears in Figure 3.1.

As pointed out in Chapter 2 (see especially Exercise 2.4), it is entirely possible to reverse the "supply-demand orientation" of a network and obtain a meaningful alternative representation. Thus, in the present case, the network could just as well be shown proceeding from sew-lines to garments, with the interpretation of "supplying production capacity" on the sew-lines, allocating the capacity to garments, and finally meeting garment demands. It is useful to keep in mind in the following that "supplies" and "demands" are easily reversible designations. The language of certain conditions may suggest a "demand role" for a quantity portrayed as a supply, and a "supply role" for a quantity portrayed as a demand. There is nothing wrong or inconsistent with this, but it underscores the value of the modeling skill that enables one to shift comfortably from a supply orientation to a demand orientation and back.

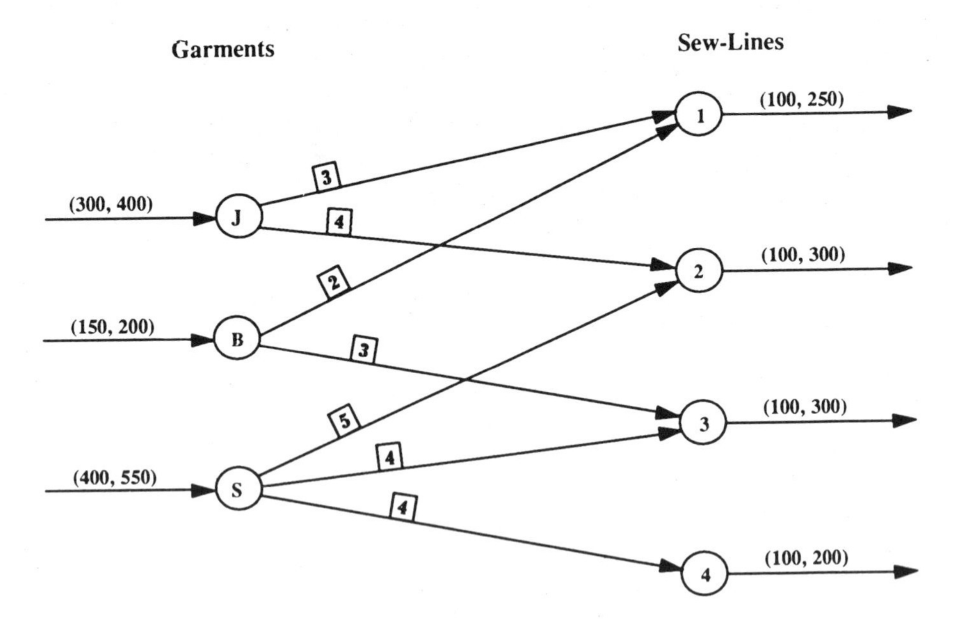

Figure 3.1 Clothing manufacturer (cost minimization objective).

3.2 GOAL PROGRAMMING MODEL CONDITIONS

We first extend the preceding example by a series of conditions often called "goal programming" conditions [see, e.g., Charnes and Cooper 1961, Cohon 1978, Zeleny 1973, 1974, Zionts 1974]. Each involves establishing a target, or goal, and some sort of penalty for failing to meet it. The penalty may be expressed as an increased cost or a decreased profit.

Condition (1): The operator that works sew-line 1 can only turn out the minimum 100 garments per week. To make more than this requires calling in outside help, adding $2 to the cost of each garment made on sew-line 1 beyond the first 100 (up to the 250 maximum).

To handle this condition we split the quantity of garments handled on sew-line 1 into two parts, the minimum 100 "target" quantity and the remaining garments which incur the added $2 cost and which are limited to at most 150 (making a total limit of 250 with the first 100 included). A construction that handles this is shown in Figure 3.2. (The remaining portion of the network diagram is unchanged.)

Condition (2): On sew-line 2 there is no difficulty in producing the maximum of 300 garments with the operators available. Instead, turning out fewer than 300 garments incurs an additional expense due to running the machines on this line below normal capacity, resulting in a disruptive stop-and-start kind of operation. After some consideration, the company figures the cost of this disruption roughly equates to $.80 for each garment short of the 300 quantity targeted. That is, if only 290 garments are turned out on the sew-line, the disruption cost is $8.00; if 280 garments are turned out, the cost is $16.00, and so forth. Figure 3.3 handles this condition.

A clearer understanding of the construction in Figure 3.3 results by comparing it with the construction in Figure 3.2. In both cases, note that the

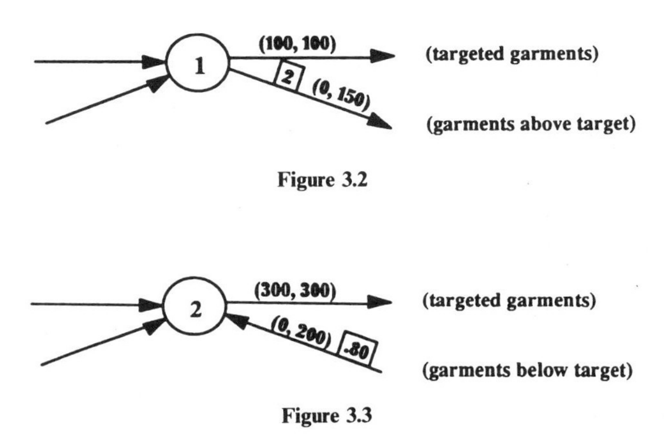

Figure 3.2

Figure 3.3

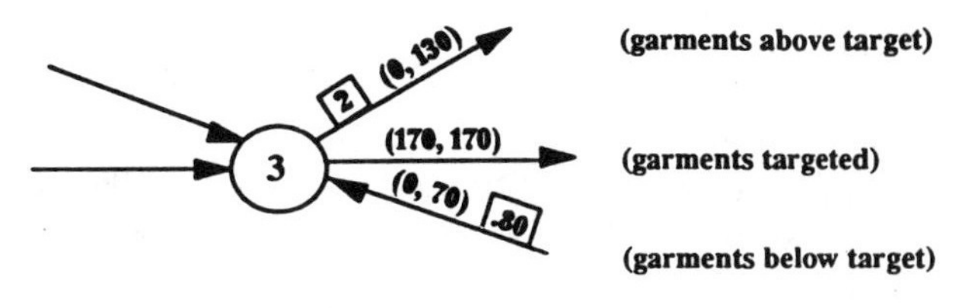

Figure 3.4

quantity targeted was handled by a constant-flow arc (lower and upper bound equal to the constant representing the targeted quantity). Flow above the target, as in Condition (1), is carried on an arc with the same direction as the target arc. Flow below the target, as in Condition (2), is carried on an arc with the opposite direction.

To see that the "below target" construction is correct, note that the targeted 300 garments may be thought of as divided into two parts: garments actually handled on the sew-line, and garments not handled (hence below target). These two parts (the flow into node 2 from the garment nodes plus the flow into node 2 on the "below target" arc) must add up to the 300 total, which accounts for the orientation of the arcs. Also note that the upper bound of 200 on the "below target" arc automatically assures that the minimum of 100 garments will be produced on the sew-line.

Condition (3): Sew-line 3 shares features of both sew-lines 1 and 2. The actual targeted output for this sew-line is 170 garments, with below-target production incurring an added cost of distruption of $.80 for each garment short of the target. If more than 170 garments are handled on this line, outside help is required that increases the cost of turning out these additional garments by $2.00 each. A diagram for this situation appears in Figure 3.4.

Figure 3.4 seems to allow an "absurd" possibility, where both the "above target" and "below target" arcs carry positive flows at the same time. For example, consider the arc flows shown in the diagram in Figure 3.5.

All bounds are satisfied on the arcs shown, and the "flow in = flow out" requirement is met at node 3. Thus, the arc flows are entirely feasible (relative to

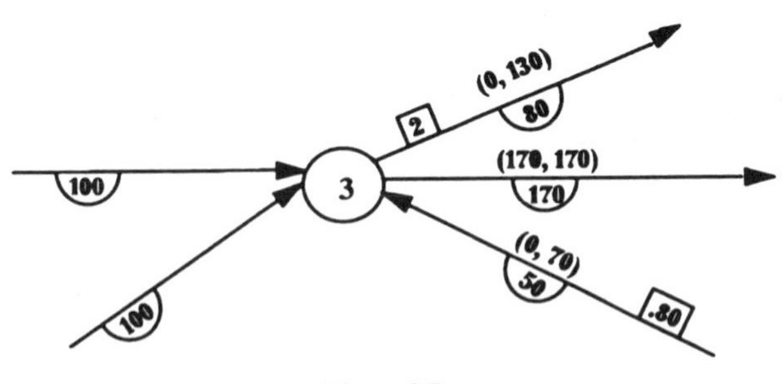

Figure 3.5

this portion of the network). Obviously, though, they do not make sense, because there is no way to have flows above target and below target at the same time. The 200 units entering node 3 from the garment allocation arcs on the left should translate into a 0 flow on the "below target" arc and a flow of 30 on the "above target" arc. How can we be sure this will happen?

Fortunately, the cost minimization objective comes to the rescue. It is unnecessarily costly to have positive flows on both the "above target" and "below target" arcs. Consequently, to minimize total cost, at least one of these arcs will have a 0 flow. (Viewed another way, whenever both flows are positive, flow conservation at the node can always be maintained by decreasing the flow on both arcs by the same amount. This improves total cost and can be continued until one of the two arcs has a 0 flow.)

For example, if the two garment allocation arcs entering node 3 carry a total flow of 140 units, the least costly way to meet this (after extracting the required 170 units on the target arc) is to carry 30 units on the "below target" arc and 0 units on the "above target" arc.

Question: Create an appropriate diagram to model each of the following additional conditions. In each case, identify which arcs refer to target, above target, and below target requirements. Current bound limits of the original problem continue to apply.

Condition (4): Sew-line 4 can turn out anywhere in the range from 130 to 150 garments without disruption or the need for outside labor help. The cost of disruption is $0.80 for each garment short of 130 (to the lower limit of 100), and the cost of outside help is $2.00 for each garment above 150 (to the upper limit of 200).

Condition (5): The company's outlets have agreed to take exactly 300 jeans during the week. Any amount turned out above 300 will incur costs of storing and handling of $0.50 each.

Condition (6): Company outlets require exactly 200 blouses for the week. Any amount turned out below that amount will have to be offset by purchasing blouses outside, at a net cost of $2.30 per blouse.

Condition (7): Company outlets will accept anywhere from 440 to 500 shirts during the week. Shirts above 500 will have to be stored at a cost of $0.40 each, while shirts below 440 will have to be purchased outside at a cost of $7.00 each.

Figure 3.6 incorporates all seven of the preceding conditions. In Figure 3.6, arcs referring to quantities above a target are located above the target arcs, while arcs referring to quantities below a target are located below the target arcs. (The word "target" is used loosely to represent either a constant value or a range of values.) Note that, on the supply side just as on the demand side, an "above target" arc always has the same direction as the target arc, while a "below target" arc always has the opposite direction.

There is one additional feature of Figure 3.6 that deserves mention. Consider, for example, the construction at node 4, repeated in Figure 3.7.

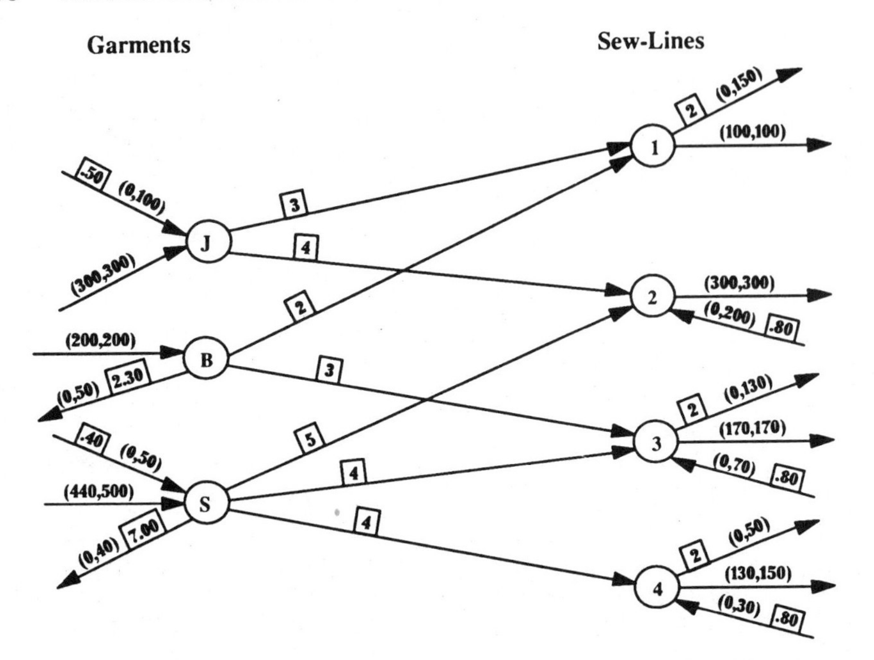

Figure 3.6 Clothing manufacturer: conditions (1)–(7) (cost minimization objective).

In Figure 3.7, a feasible alternative is to meet the 130 lower bound of the target, and then carry flow on the "above target" arc, although this latter arc is intended to carry flow only if the target arc carries a flow of 150. Once again, the cost minimization objective assures that the flows will be as intended. The "above target" arc will never carry flow until the target arc is filled up, because the target arc is cheaper. (The target arc is also cheaper than the "below target" arc, but this does not mean that it will fill up before the "below target" arc is used. Why?)

These same kinds of results can be achieved just as well with a profit maximization objective as with a cost minimization objective. For example, if the preceding problem is expressed as one of maximizing profit, the costs would appear as negative revenues, and Figure 3.7 would become Figure 3.8.

The target arc will fill up before the "above target" arc carries flow because

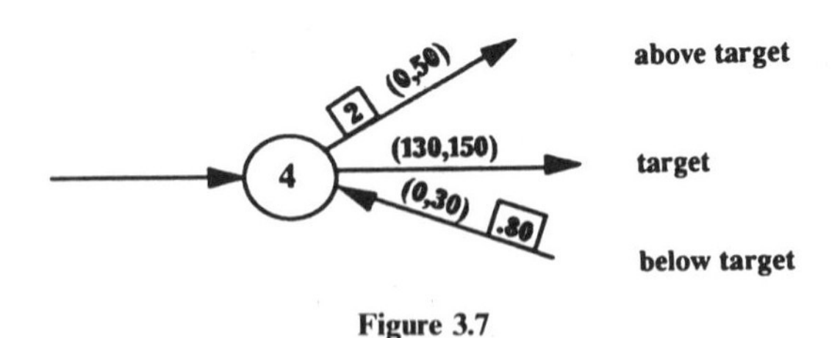

Figure 3.7

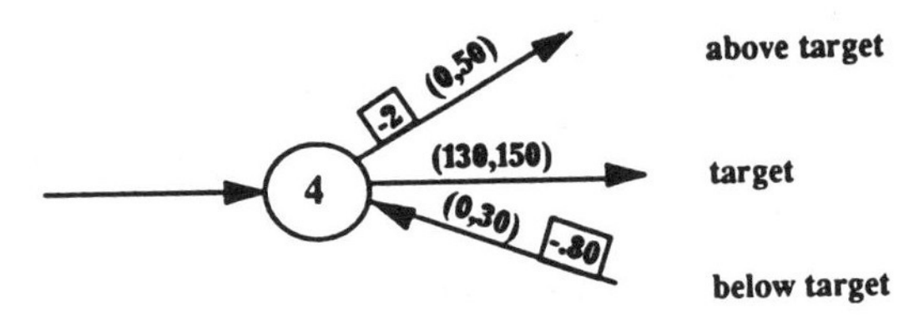

Figure 3.8

the target arc is more profitable. (Similarly, the "above target" and "below target" arcs will not simultaneously carry positive flows because it would be more profitable to decrease both their flows until one reaches 0.) In short, because "more profitable" and "less costly" are equivalent concepts, the same results are achieved whether maximizing profit or minimizing cost.

3.3 THE GOAL PROGRAMMING CLASSIFICATION OF TARGET CONDITIONS AND PRE-EMPTIVE GOAL PROGRAMMING

The kinds of goal programming conditions modeled in Figure 3.6 were historically first considered in the setting of linear programming, and for a number of years it was not widely known that they could be incorporated into network models. The introduction of the term "goal programming", as a way of classifying model types, was motivated by the fact that managers tend at first to view goals as constraints, thereby unconsciously ruling out alternatives that may be worth entertaining. The goal programming philosophy encouraged the view that quantities above and below goals (target values) should not be excluded from consideration, but should be considered admissible subject to appropriate costs. For example, in Figure 3.6, the information in Condition (7) could easily at first be conceived as a constraint to be imposed on the problem, compelling between 440 and 500 shirts to be produced. Further reflection, prompted by the goal programming orientation, would then lead to the conclusion that more or fewer shirts could be turned out after all (by incurring the cost of storing and handling the extras, or of making up the deficit by purchasing shirts outside).

Being on the lookout for constraints that can be relaxed, at a price, is often valuable in practical settings. It is not always easy, of course, to make a good estimate of the cost of this relaxation (for example, attaching appropriate values to such things as "handling and storage costs" and "disruption costs"). Yet most costs, as most supplies and demands, share some degree of fuzziness. Generally, a highly productive way to deal with the fuzzier ones, in goal programming as in other settings, is to conduct "what if" analyses. The efficiency of methods for network and network-related problems makes it easy to resolve these problems using different parameter estimates. Many companies have found the rewards of such an approach to be considerable.

Goal programming has also had another historical motivation. In some situations, one may have a number of goals which are imperfectly compatible. For example, it may not be possible to simultaneously minimize cost and maximize the utilization of human resources, or to maximize profit and minimize waste byproducts. The goal programming approach is to set targets for each of these objectives, and to attach costs (or "penalties") to the amounts by which these targets fail to be met. This approach can also be highly useful in establishing various regulatory policies. Coupled with "sensitivity" and "what if" analyses, for example, it can be used to determine appropriate surcharges and taxes to levy on particular activities so that the most profitable alternatives are to operate in ways that the society at large deems responsible.

Caution should be exercised, however, in interpreting penalities as reflecting the relative importance of several incompatible goals. For example, making a penalty for missing a particular objective's target twice as big as the penalty for another objective's target does not necessarily mean that the deviations from these targets will occur in a 2:1 ratio. To do that, one must add extra constraints to the model, specifying the required ratios for the deviations.

When the multiple goals or objectives are pre-emptive in nature, then a simultaneous solution is possible. In pre-emptive goal programming problems, the objectives are strictly ordered. Each objective is optimized (maximized or minimized) subject to keeping the prior (more important) objectives at their optimal values. This is equivalent to sequentially restricting each problem to the set of alternate optimal solutions to the prior problem.

Alternatively, one could form a single composite objective function by multiplying the objectives by sufficiently large weights and adding them together. The structure of network problems provides an advantage for determining these weights if all problem parameters are integer because this insures integer objective function values. Consequently, differences between objective function values for two solutions will always be integer. In particular, the difference between an optimal solution and some other solution will always be at least one unit. Thus it suffices to insure that the weights are large enough so that a one unit increase in the first (most important) objective is worth more than changing each of the other objectives from its worst to its best value. These weights may be determined iteratively, beginning with the least important objective (whose weight is equal to 1). However, for many network problems with pre-emptive multiple objectives, the weights need not be this large, as illustrated in the example below.

Suppose that the clothing manufacturer's primary objective is to utilize the sew-lines as much as possible. As a second objective, priority is given according to the sew-line number. That is, it is most important to utilize sew-line 1, then sew-line 2, etc. The third most important objective is to minimize total weekly operating costs. These three objectives may be incorporated in a single netform without using extremely large weights as shown in Figure 3.9.

Note that the garment costs are written as negative numbers since the composite problem objective is maximization. The priority objective function coefficients are one digit (a factor of ten) larger than the cost coefficients, so that

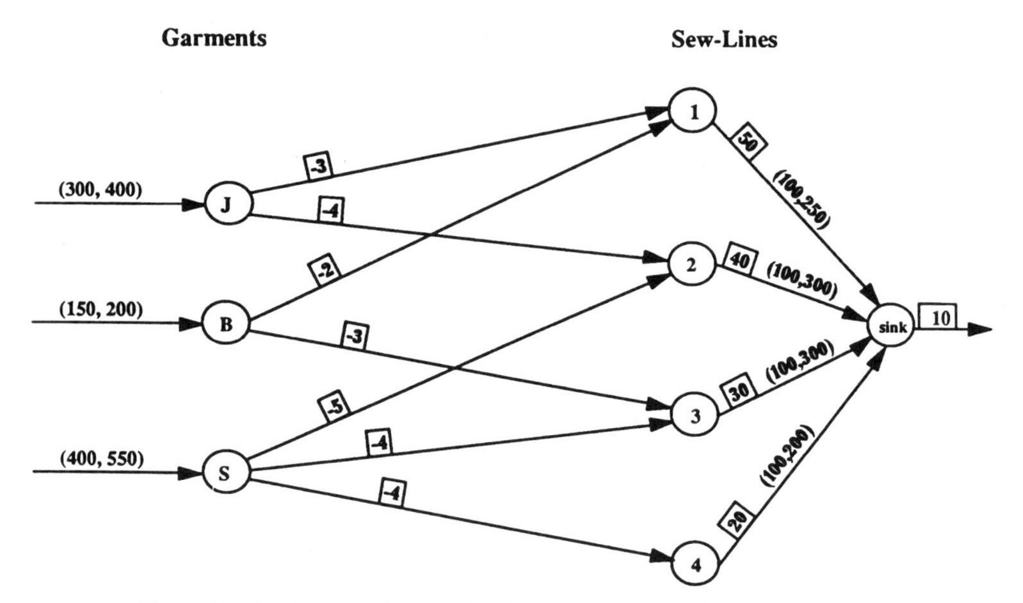

Figure 3.9 Clothing manufacturer (multiple pre-emptive maximization objectives).

they will pre-empt any assignment made based on cost alone. The utilization objective is accomplished by adding a "sink" node (super demand node) and arc on the right. All flow in the network must travel across this arc which has an objective function coefficient larger than the largest cost coefficient. There is no need to make this coefficient one digit larger than the priority coefficients because as long as it is larger than the cost coefficients, this arc will pull as much flow as possible through the network since there will be a path through the network with a net positive profit. This flow will be routed first through node 1, then when the arc out of node 1 has received as much flow as possible, flow will be routed through node 2, etc. To see why this is true, notice that the difference between routing flow through node 1 and node 2 is \$50–40 = \$10. This difference is larger than any cost savings we could incur by making a garment on sew-line 2 as opposed to sew-line 1. In particular, the largest garment cost savings is only \$3 (between shirts on line 2 and blouses on line 1). Thus by making the priority coefficients one digit larger than the garment cost coefficients, we have insured that the priority order will be preserved.

3.4 TARGET FLOWS ON ARCS WITH TWO ENDPOINTS

The goal conditions treated so far have applied exclusively to supply and demand arcs (which have a single node as an endpoint). Now we consider the situation where such conditions are imposed on arcs with two endpoints. We continue with the preceding Clothing Manufacturer example.

Condition (8): The company decides that it must allocate between 60 and 180 jeans to sew-line 2 to avoid upsetting operations. No adjustment of operations is required as long as 80–150 jeans are allocated to this sew-line, and hence any number of jeans in this range can be handled without incurring additional cost. However, an allocation of more than 150 jeans to the sew-line incurs an extra cost of $0.70 to handle each additional pair. An allocation of fewer than 80 jeans incurs a cost of $0.50 for each pair below 80.

To make a diagram of this condition, we will consider two versions, first where the "base cost" of allocating jeans to sew-line 2 is 0, and second where this cost is 4 (as in the original example). In both cases the limiting lower and upper bounds of 60 and 180 apply.

In Figure 3.10, the "above target" arc is shown above the target arc, and the "below target" arc is shown below. Note that, exactly as in the case of the supply and demand arcs, the "above target" arc has the same direction as the target arc, while the "below target" arc has the opposite direction.

Because the "above target" arc costs more than the target arc, it will not be used until the target arc is filled up, but then may carry up to 30 additional units (to the upper limit of 180) at a cost of $0.70 each. The "below target" arc requires a bit more attention. The upper bound of 20 on this arc accords with the fact that the flow from node J to node 2 should not drop below 60 (80–20). However, it may seem strange at first that this arc should "carry flow back" from node 2 to node J. The construction can be understood by viewing it as simultaneously sending flows on both the target and "below target" arcs, thus sending a net amount of flow from node J to node 2. This net amount is the "true" amount sent from node J to node 2. The flow on the "below target" arc tells exactly how much the net amount lies below the target range (just as the flow on the "above target" arc tells how much the net amount lies above the target range). The "below target" and "above target" arcs will not both have positive flows, and the "below target" arc will not carry flow unless the target arc (flow) is at its lower bound, by the same kinds of reasoning given earlier. (Otherwise, the same net flow could be achieved at a cheaper cost.) The three diagrams in Figure 3.11 provide a concrete illustration of how a particular flow from node J to node 2 (arbitrarily chosen to be 75) can be carried progressively more cheaply and most cheaply by the "correct" flows on the component arcs.

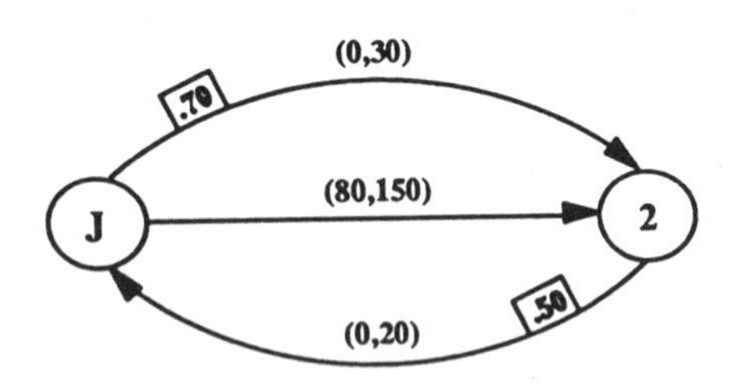

Figure 3.10 Version 1: "base cost" is 0.

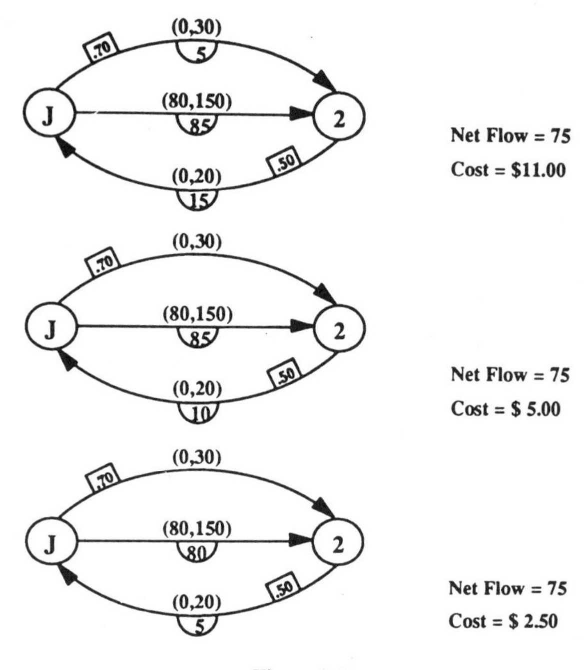

Figure 3.11

We now turn to the second version of Condition (8), where the "base" cost of allocating jeans to sew-line 2 is not 0 but \$4.00. In this case, the added cost of \$0.70 per unit above 150 must translate into a total unit cost of \$4.70 (since these extra jeans are also being allocated). On the other hand, the cost of \$0.50 per unit below 80 must translate into a total unit cost of −\$3.50. (The lower bound of 80 and cost of \$4.00 on the target arc means \$4.00 per unit will be paid for 80 units in any case. The −\$3.50 cost for each unit short will yield a net cost for these units of \$4.00–3.50 = \$0.50 as desired.) This is demonstrated in Figure 3.12.

To understand this better, suppose that the net number of jeans sent from node J to node 2 is 70. This would produce flows as shown in Figure 3.13.

The \$285 cost is proper because it corresponds to paying \$4/unit for the net of 70 units sent, and paying \$0.50/unit for the 10 units short of 80. If we break the flow on the target arc into two pieces, we could show the net of 70 being carried

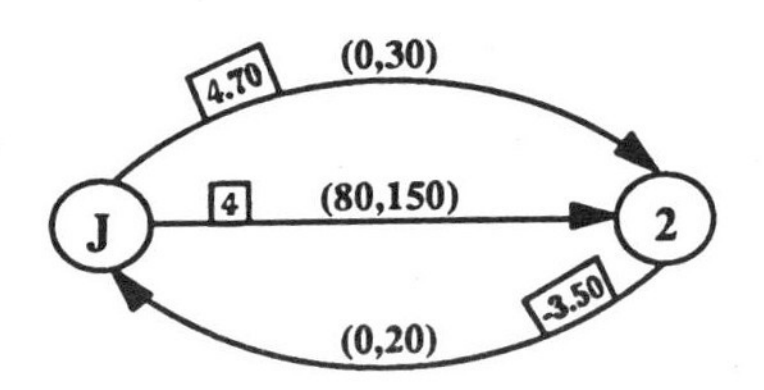

Figure 3.12 Version 2: "base cost" is \$4.00.

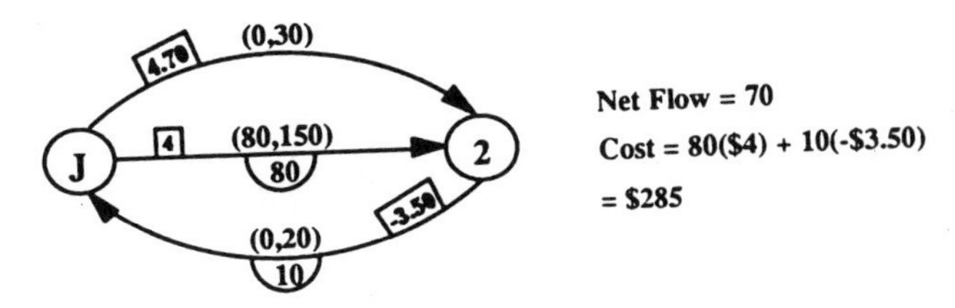

Figure 3.13

at $4 each, and then the remaining 10 (which is the same 10 carried on the "below target" arc) picking up a cost of $4 in one direction and $-$3.50 in the other, as appropriate.

It is possible to use a diagram for Version 2 that includes Figure 3.10 as a component. This alternative diagram for Version 2 has no negative costs, but includes an extra node and arc, as shown in Figure 3.14.

Note that the left portion of this construction is exactly that of Figure 3.10, the case for a 0 base cost. The $4 cost applicable to each unit of net flow from node J to node 2 is attached to the arc that has been added on the right. The node 2′ at the head end of this arc may be imagined as a "copy" of node 2, and has all the arcs meeting it that normally meet node 2 (except for those of the construction itself).

Question: Show two constructions (corresponding to the two alternatives for Version 2 above) for allocating blouses to sew-line 3, where the unit allocation cost is $3/blouse (as originally), the lower and upper allocation limits are 90 and 140, flow below 100 units incurs a cost of $.40 for each unit short, and flow above 120 units incurs an extra cost of $.70 for each unit over. For completeness, show all arcs of the original network that touch any of the nodes of these constructions.

3.5 MODELING DECREASING RETURNS TO SCALE

Situations involving decreasing returns to scale (sometimes called decreasing economies of scale) can be modeled in a manner similar to that of the goal programming constructions. Decreasing returns to scale occur, for example,

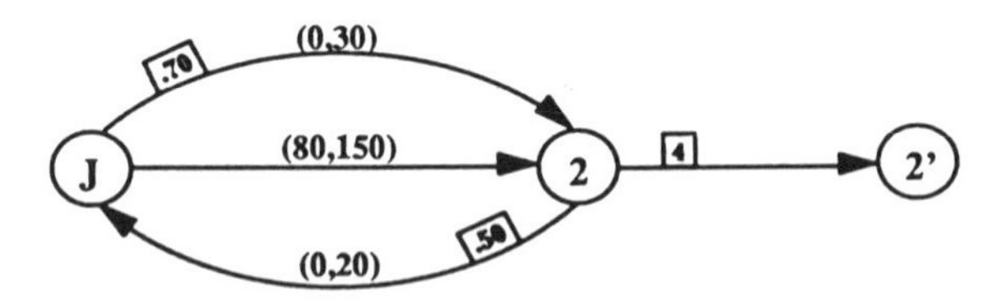

Figure 3.14 Version 2: alternative without negative costs.

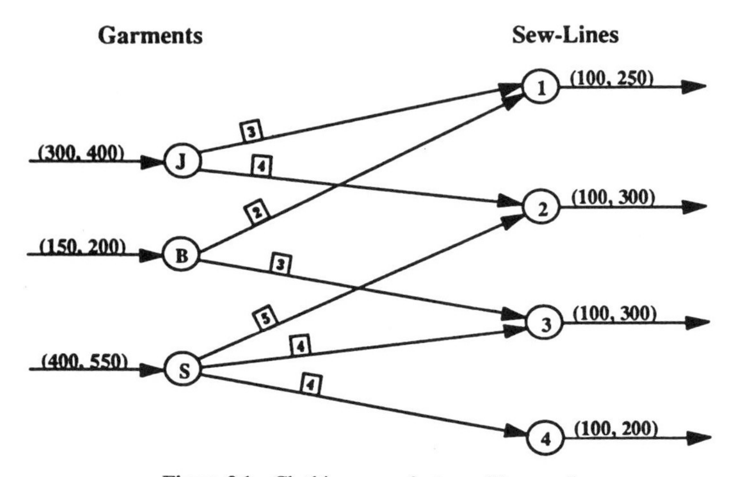

Figure 3.1 Clothing manufacturer (*Repeated*).

where scarcity or limited capacity causes the unit cost (and not just the total cost) to increase with the purchase or production of increasing quantities. Concrete examples will be provided by reference to the Clothing Manufacturer problem discussed previously. For convenience, Figure 3.1 is repeated here.

Conditions representing decreasing returns to scale will be presented as occurring in place of the earlier goal programming types of conditions.

Condition (1′): One sew-line 1, the first 120 garments incur no added cost. (The lower bound of 100 continues to apply.) However, as increasing numbers of garments are handled the efficiency of operations on this line drops. Handling more than 120 up to 150 garments incurs an added cost of \$1.00 for each unit above 120. Handling more than 150 up to 210 garments incurs an added cost of \$1.80 for additional units. Finally, handling more than 210 to the limit of 250 incurs an added cost of \$2.00 for each unit above 210. Figure 3.15 models this condition.

Because the unit costs are increasing, the lowest arc (on the right) of the diagram will fill up before any of the others can carry flow, then the second

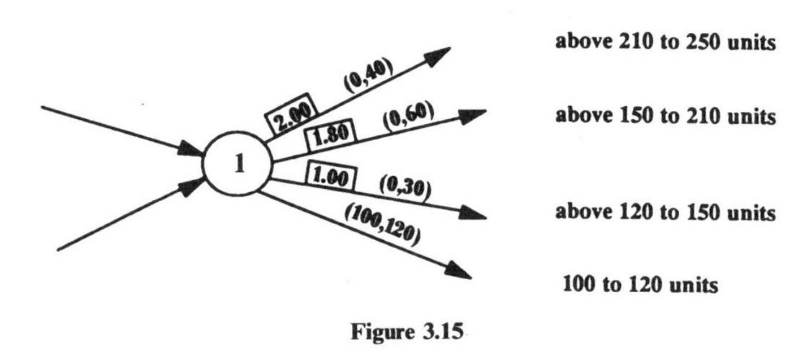

Figure 3.15

lowest will fill up, and so forth. In this fashion, the successive arcs accurately represent the stipulations of Condition (1′).

It is tempting to try to model Condition (1′) by a diagram such as that shown in Figure 3.16. The construction in Figure 3.16 does not work because it requires at least 100 units of flow on the lowest arc and at least 121 units of flow on the next lowest, and so forth, thus pulling a minimum of 583 units of flow out of node 1.

Question: Give an appropriate diagram for each of the following three conditions.

Condition (2′): The company discovers that there is a temporary shortage of materials available for making jeans. If it undertakes to make more than the stated minimum of 300, then each of the jeans above 300 up to 350 will cost an extra $1.20, as a result of having to buy materials from more expensive sources. If the company undertakes to make more than 350 jeans, then it must go to still more expensive sources for material, incurring an extra cost for these jeans (up to the 400 limit) of $1.30 each.

Condition (3′): Sew-line 3 can handle an allocation of up to 120 shirts without any complication. If more than 120 shirts are allocated to sew-line 3, then the cost of each unit above 120 up to 200 is not $4.00 (as for the first 120) but $6.00. In addition, the cost of each unit above 200 is $8.00 (rather than $6.00).

Condition (4′): The current lower bound of 150 blouses that must be turned out represents an amount that the company has committed itself to supply to its usual outlets. There are also two special outlets, each of which will take up to 20 blouses each week. Selling to the first brings in an extra revenue of $1.50 per blouse, while selling to the second brings in an extra $1.00 per blouse. Finally, any blouses left over, up to the maximum of 200 to be produced, can be placed with the usual outlets, though without additional revenue. (In preparing a diagram, treat the $1.50 and $1.00 revenues as negative costs.)

Figure 3.17 incorporates all of Conditions (1′)–(4′). Condition (2′), with increasing costs for producing jeans above the 300 minimum level, is handled much like Condition (1′). Condition (3′), involving increasingly costly allocations of shirts to sew-line 3, is also handled similarly. No upper bound was attached to

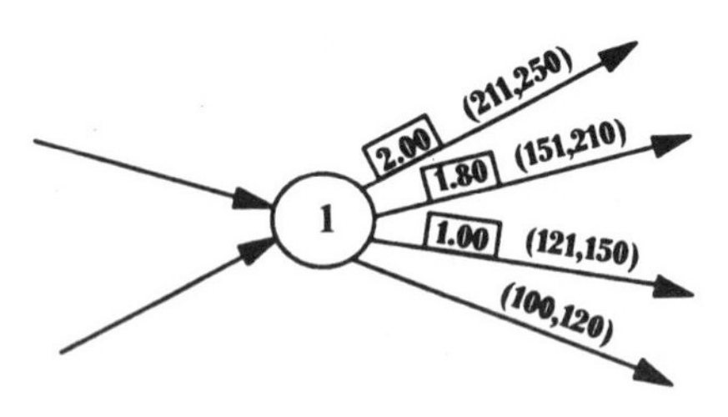

Figure 3.16 Erroneous representation.

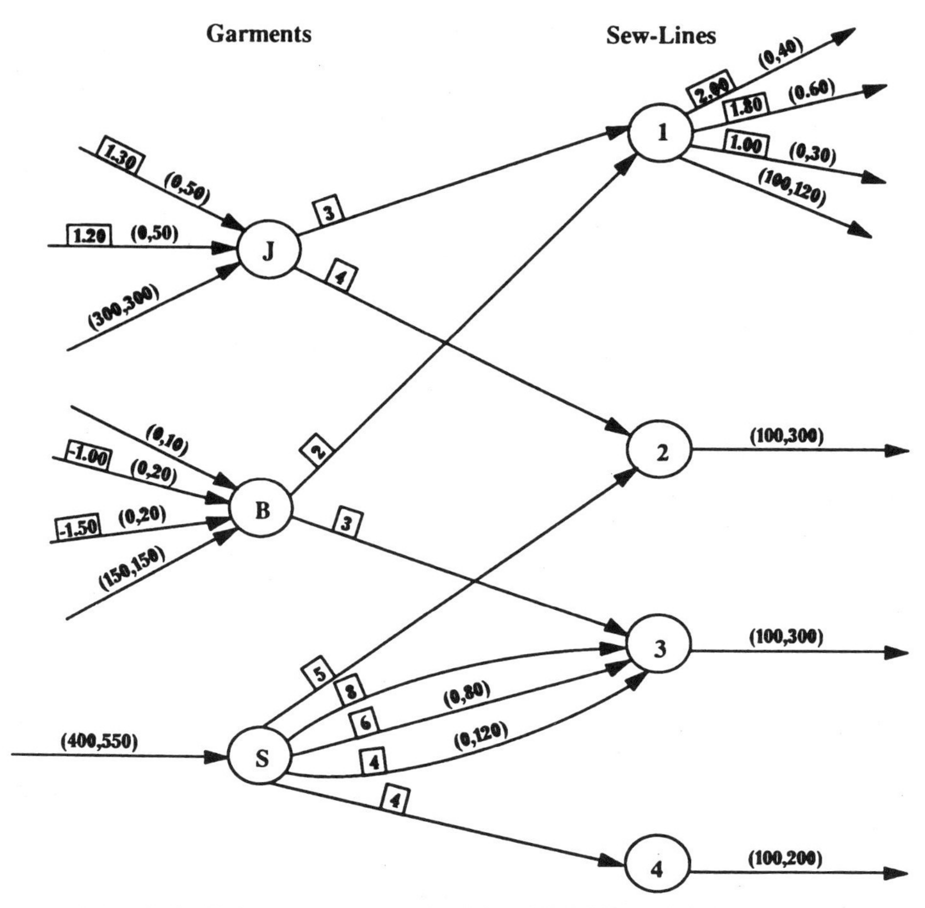

Figure 3.17 Clothing manufacturer: Conditions (1′)–(4′) (cost minimization objective).

the high cost ($8.00) arc in this construction, because its cost is applicable to all flow above 200 (120 + 80). On the other hand, an upper bound of 100 could have been attached (although redundant) since node 3 cannot accept a flow above 300 in any case.

Condition (4′), which represents decreasing revenue options (instead of increasing cost options) for producing blouses, merits extra examination. With revenues shown as negative costs, successively decreased revenues in fact appear as successively increased costs (i.e., −1.50 is smaller than −1.00, which in turn is smaller than 0). Although the 1.50 revenue (−1.50 cost) arc is the most attractive for sending flow, and will fill up before sending flow on either of the arcs above it, there is no danger that it will carry flow "before" the more costly "constant supply arc" below it. The (150,150) bounds will compel 150 units of flow on this lower arc regardless of whether the −1.50 cost arc carries flow.

Since the "bottom" and "top" arcs of this construction both have the same

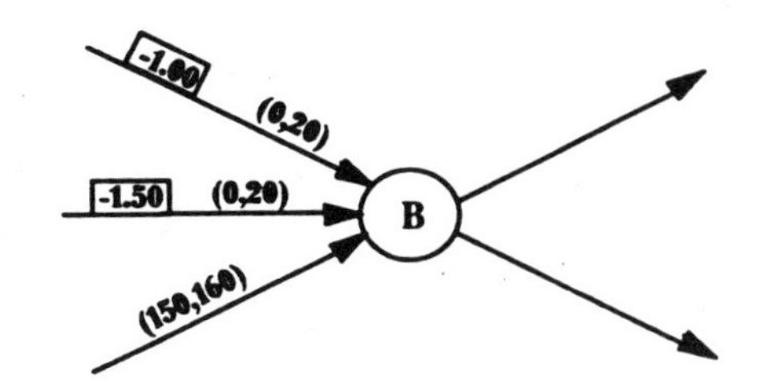

Figure 3.18 Condition (4′): alternative representation.

cost (and in fact represent blouses that will go to the same set of outlets), the diagram of the construction could also be as shown in Figure 3.18.

The form of this diagram does not so clearly portray the "sequential attractiveness" of flow possibilities, since the bottom arc will not carry flow above 150 units until the two arcs above it are filled to capacity. Of course, the positions of the arcs in any of these constructions could be shuffled without changing the fact that cheaper arcs will carry flow before more expensive arcs.

It should also be remembered, as illustrated by examples of Chapter 2, that in general, cheaper arcs may not necessarily carry flows before others. The principle of "cheaper arcs first" applies only under special conditions such as those satisfied in the constructions for goal programming and decreasing returns to scale.

3.6 AN EXTENSION OF GOAL PROGRAMMING CONDITIONS

The decreasing returns to scale constructions can also be used to extend the types of condition previously considered in the goal programming setting. For example, progressively larger deviations both above and below a goal may be increasingly costly (or increasingly less profitable). Too much water or too little water are both bad for crops, but as conditions progress toward flood or drought, the effect is increasingly harmful. Similarly, either too many or too few people assigned to a particular task may result in decreased productivity and reduced profit. (And as critical numbers are passed on either side, the excess may create a hopeless snarl or the deficit make it virtually impossible to perform the task at all, causing a more significant reduction in profitability.) This extended sort of goal programming situation is illustrated in the following condition for the Clothing Manufacturer example.

Condition (5′): Sew-line 2 functions best when turning out between 160 and 240 garments. Operating costs are incurred for handling more or fewer garments as follows: $1 for the first 20 units short of 160 or the first 20 above 240 and $2 for the next 40 units short of 160 or the next 40 above 240 (see Figure 3.19).

Corresponding types of diagrams can be created to handle such conditions

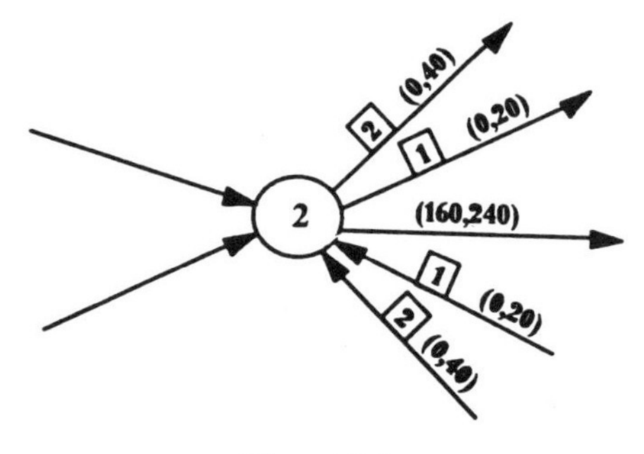

Figure 3.19

for supply arcs and two-endpoint arcs. It may be noted that "extended goal programming" conditions are nothing more than decreasing returns to scale conditions applied in "two directions."

3.7 GRAPHICAL INTERPRETATION OF DECREASING RETURNS TO SCALE

Decreasing returns to scale actually constitute a nonlinear form of cost or profit. Therefore, the preceding diagrams may be viewed as a means of representing this special type of nonlinearity in a linear framework.

The structure of this kind of nonlinearity can be illustrated by the graphs in Figure 3.20. In both of the graphs in Figure 3.20 the unit cost or profit in a particular range is represented by the slope of the corresponding line segment. Thus the four line segments of the first graph, whose slopes are progressively steeper, represent four successive unit costs, each greater than the preceding. Similarly, the four line segments of the second graph, whose slopes are progressively less steep, represent four successive unit profits, each smaller than the preceding. Each segment, possessing a constant slope (and hence a constant unit cost), is a linear segment (linear = "straight line"), and so these graphs are said to depict piecewise-linear relationships. The cost graph is a graph of a convex function, and the profit graph is a graph of a concave function, where the

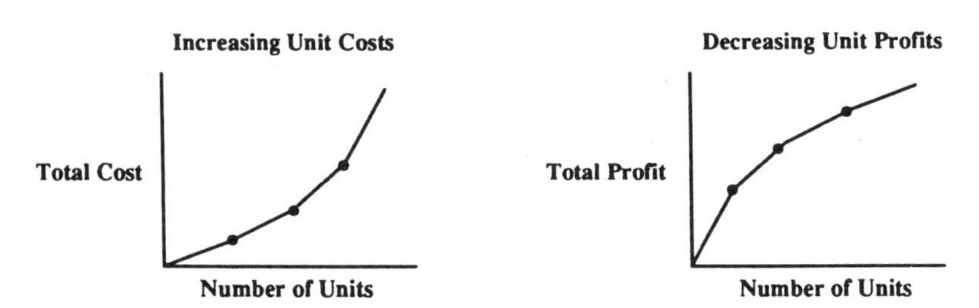

Figure 3.20

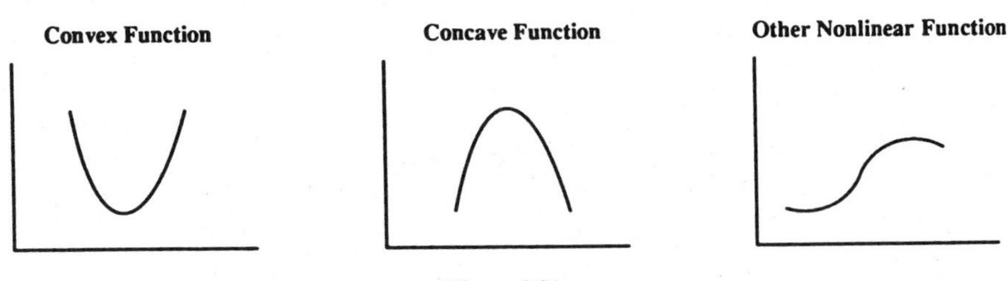

Figure 3.21

"general forms" of convex and concave functions are shown in Figure 3.21. (A concave function looks like a cave when you lie on your back and look up at it!) Convex functions, especially piecewise-linear ones, can be conveniently minimized as illustrated by the arcs into node J in Figure 3.17 (Condition 2′), and concave functions can be conveniently maximized as indicated by the arcs into node B in Figure 3.17 (Condition 4′). Hence conditions where costs are convex and profits are concave may be considered fortunate. (Other kinds of nonlinearities, such as one where the function switches from convex to concave as in the third graph in Figure 3.21, are generally not so easily handled.)

Question: What will happen in a cost minimizing solution of the Clothing Manufacturer example if you undertake to represent the following condition: sew-line 3 handles 100–150 garments at a cost of \$3 each, above 150–175 garments at a cost of \$2 each, and above 175–300 garments at a cost of \$1 each?

Figure 3.22 shows a graph of the cost function in the above question and an erroneous network representation. The slope of each linear segment in the cost function is indicated on the graph. When we solve the erroneous network model, minimizing total costs, then after the lower bound of 100 on the top arc is met, the arc with a cost of \$1 will receive flow first. When it reaches its upper bound, the arc with a cost of \$2 will receive flow, followed by the arc with a cost of \$3.

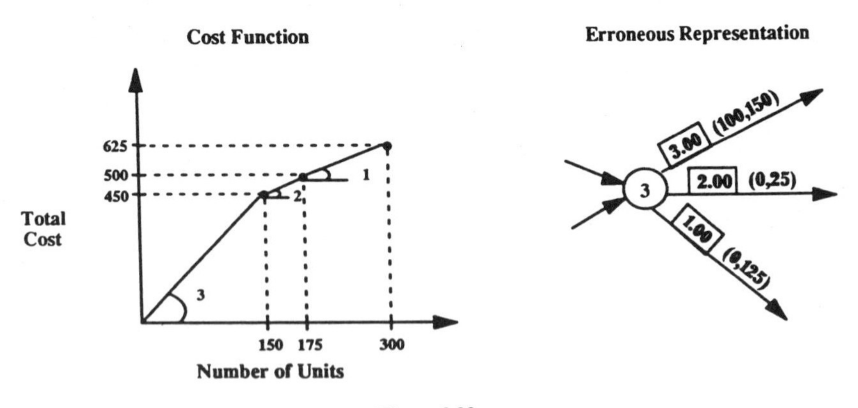

Figure 3.22.

Since you cannot produce the $2 garments before the $3 ones or the $1 garments before the $2 ones, then this model will not work. That is, concave cost functions like the one in Figure 3.22 cannot be handled in a pure network format. We will show in a later chapter how to model such conditions by means of special ***discrete flow*** requirements, but models involving such requirements are often a good deal harder to solve. At the same time, discrete flow models often afford potential gains that more than compensate for their increased difficulty of solution.

3.8 COMBINED FLOW RESTRICTIONS

The last special modeling technique we will consider in this chapter concerns the handling of restrictions on certain combined flows. These restrictions will be illustrated again by a series of conditions applied to the Clothing Manufacturer example. The original (condition free) Figure 3.1 is repeated here for convenience.

Condition (1″): Blouses and shirts are supplied to the same outlet. In addition to the individual bounds on blouses and shirts, the outlet will not accept fewer than 600 nor more than 690 blouses and shirts combined.

There is a correct way and an incorrect way to draw the diagram for this condition. The correct way is shown first. In Figure 3.23 the creation of a new arc and a new node, to which the arcs for both blouses and shirts connect,

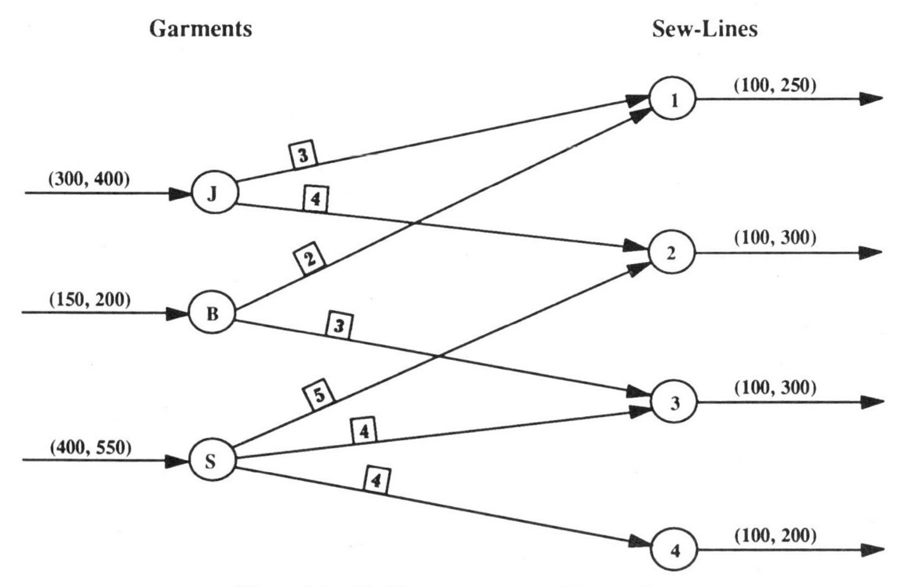

Figure 3.1 Clothing manufacturer (*Repeated*).

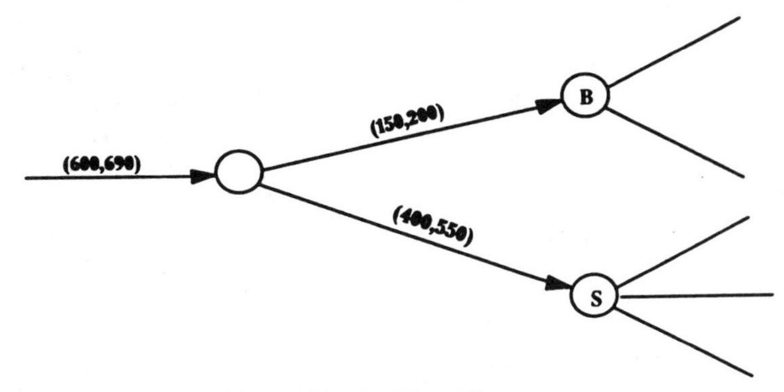

Figure 3.23 Condition (1″): correct.

controls the combined flows on these latter arcs as desired. Specifically, the bounds on the new arc assure that the combined flows of blouses and shirts will lie between 600 and 690.

This same principle seems to be applied in Figure 3.24, but there is a flaw. The arc with bounds (600, 690) properly controls the flow of blouses and shirts to lie within this range, but the difficulty here is that the identity of the flows is lost when they merge in the (600, 690) arc. To see this more clearly, Figure 3.24 is expanded in Figure 3.25 to show its connections in the complete network. We also show flows which provide an optimal solution for this expanded diagram (as determined, for example, by an appropriate computer optimization method).

Although proper arcs with proper costs go to each of the sew-line nodes, the "loss of identity" on the (600, 690) arc makes it impossible to tell whether the flows on succeeding arcs are coming from shirts or blouses. In fact, Figure 3.1 shows that the arcs with flow to nodes 1 and 3 should be carrying blouses. But these two arcs both carry a flow of 250 in the optimal solution of Figure 3.25, while the "blouse supply" arc into node B carries a flow of 200, which is also the upper bound for blouses. In short, a mis-application of the combined flow

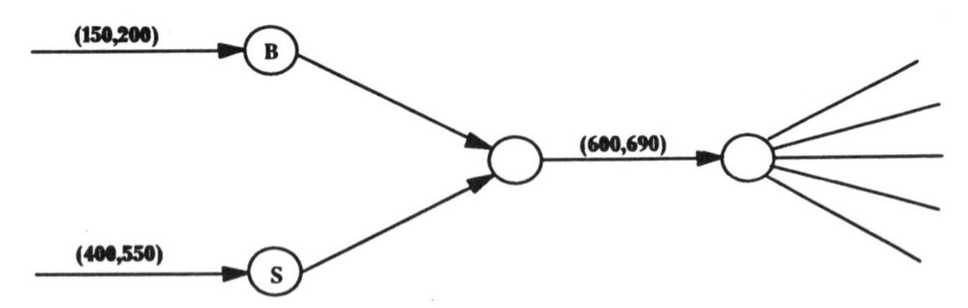

Figure 3.24 Condition (1″): incorrect.

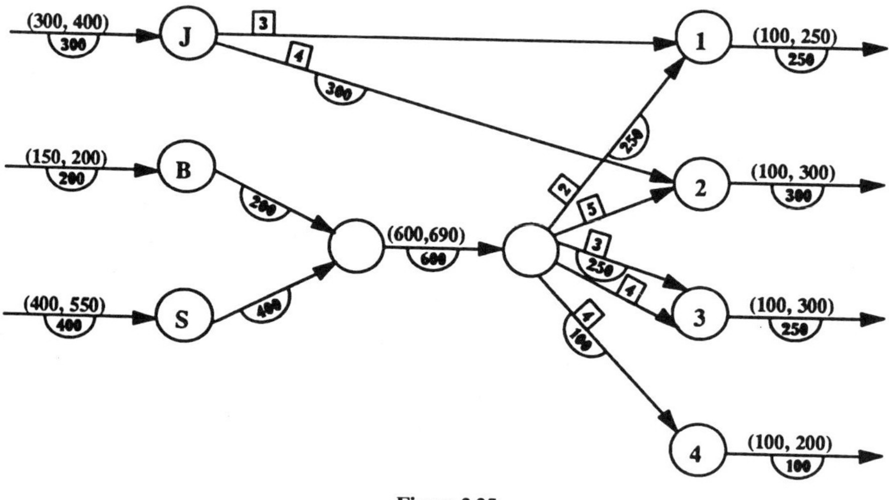

Figure 3.25

technique (losing the identity of component flows) can play havoc with the problem formulation.

Condition (2″): The outlet for blouses and shirts merges with the outlet for jeans, and now requires, instead of the combined limits on blouses and shirts, a minimum and maximum for all three garments combined of 950 and 1050, respectively.

Since this condition replaces Condition (1″), there is no carry-over from the preceding condition, and the appropriate diagram appears in Figure 3.26.

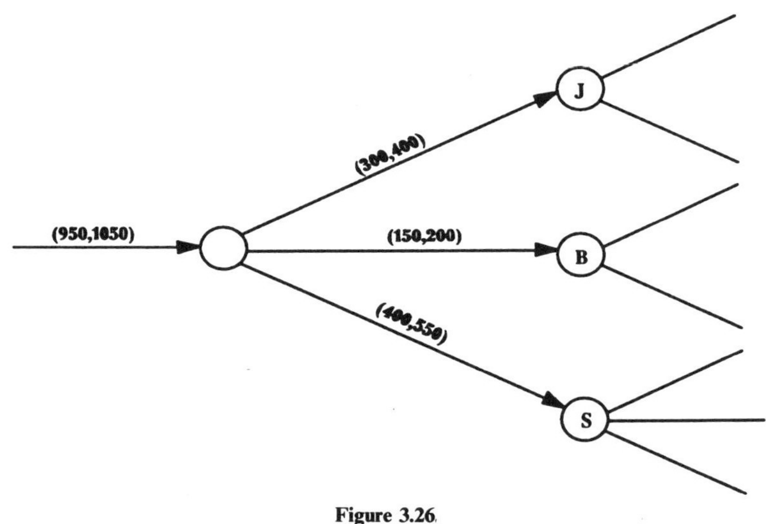

Figure 3.26

Condition (3″): The outlet now decides it made a mistake in discarding Condition (1″). Consequently, it works out an agreement with the clothing manufacturer to make and supply garments in quantities that satisfy the restrictions of Conditions (1″) and (2″) together.

Figure 3.27 unites the restrictions of the two previous conditions.

Question: Provide a diagram to model the combined flow restrictions of each of the following conditions.

Condition (4″): Due to shared facilities with restricted capacity, the maximum number of garments that can be turned out by sew-lines 1 and 2 combined is 530.

Condition (5″): Sew-lines 3 and 4 must turn out a minimum of 220 garments combined, to meet demand for garments that have the special kinds of stitching produced by these sew-lines.

Condition (6″): In addition to both Conditions (4″) and (5″), all sew-lines together must produce between 950 and 1050 garments.

Figure 3.28 models Condition (6″) and hence also includes Conditions (4″) and (5″). The bounds on the (0,530) arc and the (220, ∞) arc could be tightened to (200, 530) and (220, 500), as implied by the bounds that "lead into" these arcs, but this tightening is unnecessary. The advantage of not tightening bounds in this way is that subsequent revisions can be made more readily. For example, suppose the (0,530) bounds are tightened to (200, 530), and it is later decided that the lower bound on the arc out of node 1 should be 50 instead of 100. This means the (200, 530) bound should then also be changed to (150, 530). This second implied change is easy to overlook, and would be wholly unnecessary in any case if the original (0,530) bounds had not been tightened.

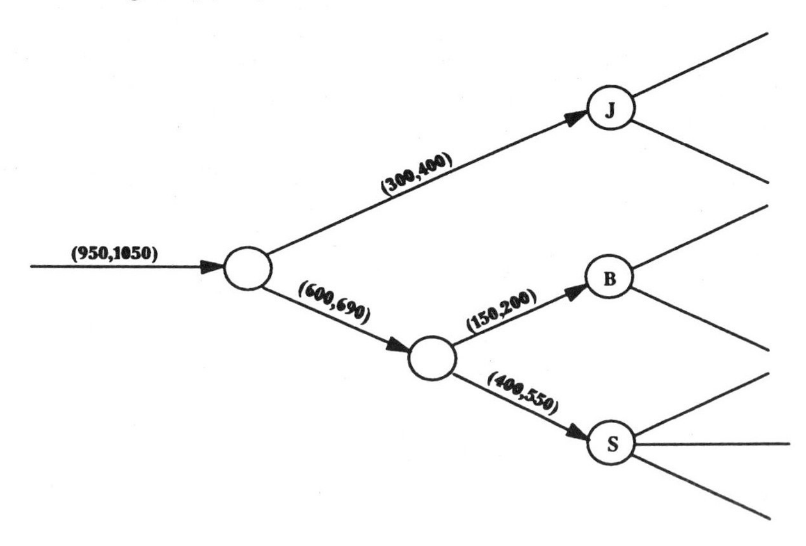

Figure 3.27

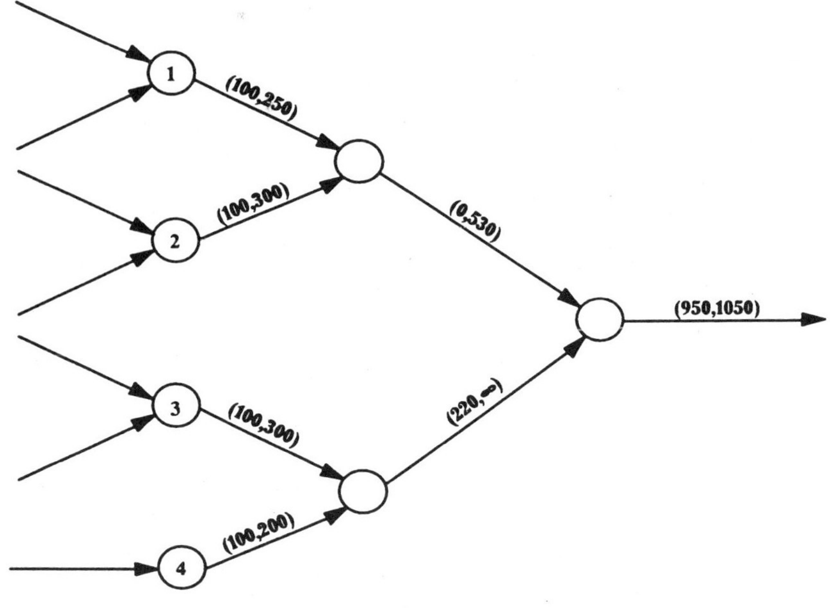

Figure 3.28

There are some contexts in which the use of tightened bounds, even though redundant, can speed up the solution of a problem. However, except in the case of discrete flow (integer programming) types of problems discussed in a later chapter, this effect is usually of negligible consequence, and it is preferable to avoid tightening that may too easily be forgotten when subsequent changes are introduced.

One other useful observation emerges from the preceding example. Condition (6″) actually limits the total garments produced in exactly the same way as Condition (3″). This shows that bounds on total flows can be handled either on the "supply side" or on the "demand side," according to convenience.

The preceding restrictions have all dealt with combined flows of single-endpoint arcs (though some of these arcs acquire a second endpoint as a means of modeling the restrictions). The principles can also be used to model certain combined flow restrictions on two-endpoint arcs, as illustrated next.

Question: Provide a diagram for the following condition.

***Condition** (7″):* The number of shirts allocated to sew-lines 3 and 4 combined must lie between 250 and 300.

Figure 3.29 models for Condition (7″). (Portions of the larger network are also shown to clarify the connections.)

The techniques illustrated are not able to model all types of combined flow restrictions. For example, they do not apply where one might wish to limit the combination of blouses allocated to sew-line 1 and shirts allocated to sew-line 2.

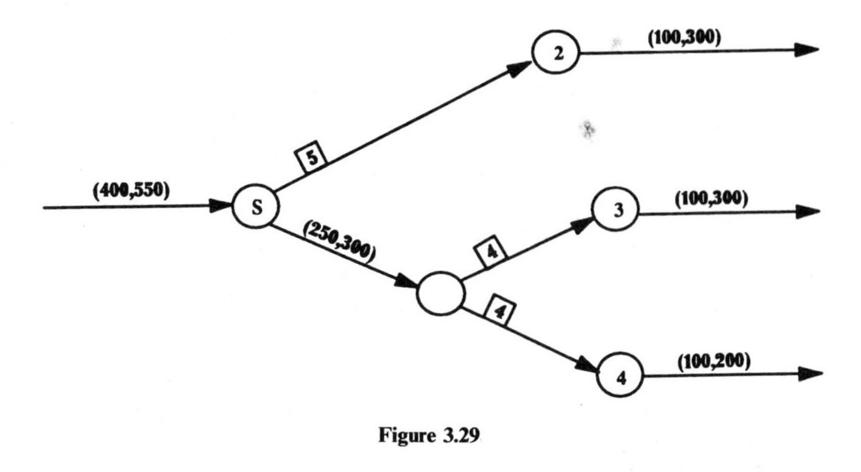

Figure 3.29

To model such constraints that cannot be incorporated directly into the network itself, but which must nevertheless be treated, we use ***side constraints***. As discussed in Chapter 2, side constraints are also used to model joint restrictions on resources where different products use resources at different rates, such as storage space limitations, man power constraints, and machine time constraints.

APPLICATION: ENLISTED PERSONNEL ASSIGNMENT MODEL

The application of optimization models to military manpower planning dates back to the late 1940s when assignment models were first formulated as linear programming models. However, operational applications of the assignment model to military manpower planning substantially increased in the mid 1970s when Assistant Secretary of Defense, William K. Brehem, on December 2, 1975, directed:

The Services will utilize simultaneous rather than sequential assignment selections in programming personnel reassignments.

As a result, several large pure network optimization models have been developed to assign military personnel to jobs simultaneously.

Network optimization models for personnel assignment problems are important for several reasons. First, these problems are extremely large. A US Army personnel assignment problem often involves up to 10,000 persons and positions, over a million eligible person/position matches, and from 3 to 15 preemptive assignment criteria. Thus, a mathematical programming model for personnel assignment may exceed 15,000 constraints and over a million variables. Second, personnel assignment systems are operational systems; most military services generate and solve such problems each month. Therefore, the solutions to these models need to be integer, and the computer resources required to ascertain such solutions must be reasonable. Network optimizers nicely accommodate these requirements.

Each year the US Army seeks to select, classify, and assign thousands of personnel in such a manner as to provide the best US defense at the lowest cost. The armed services attempt to use their limited personnel resources effectively and efficiently by focusing on the decisions which match people and jobs. Assignments are made based on pre-emptive criteria which include the service's need for personnel in different skill classifications, the individual's job preference, and the service's cost of assignment. Large-scale pure network models have been developed which can be used to simultaneously make these assignments and to evaluate the effect of changes in policies and eligibility requirements on the entire personnel pool.

Enlisted personnel are assigned to positions primarily according to three pre-emptive goals or criteria called *maximum fill*, *maximum priority distribution fill*, and *maximum fit*. The maximum fill criterion is to maximize the total number of positions filled. The maximum priority distribution fill criterion seeks to fill positions in a specified priority order such that the total number of positions filled is not reduced, i.e., the priority distribution fill criterion is pre-empted by the fill criterion. In the US Army, the priority order is specified by assigning each position a unique distribution fill priority. The maximum fit criterion seeks to maximize the fit (as determined by a numerical score) between personnel and positions based on such factors as the person's qualification or preference for the position, the service's preference for the person's assignment, and the associated relocation cost. The US Army forms a weighted combination of the fit attribute scores. The resulting fit criterion is pre-empted by the fill criteria.

Figure 3.30 depicts a single criterion network flow model which handles the maximum fill, priority distribution fill, and fit criteria for the US Army enlisted personnel assignment model.

In order to assure that the problem has a feasible solution and to make total supply equal total demand (a requirement of the particular code used), a dummy source node with supply equal to the total number of jobs and a dummy sink node with demand equal to the total number of persons available are added to the netform. Flow on arcs into and out of these two nodes represent men not assigned to jobs and jobs not filled respectively. The maximum fill objective is modeled by connecting these two nodes with an arc which has a "big M" (very large) objective function coefficient. In this way, the maximum number of dummy persons will be assigned to dummy jobs, thus filling as many real jobs with real persons as possible. (This is an alternative to the sink node concept in Figure 3.9. That is, we could omit the dummy nodes by changing the fixed supplies and demands to bounded supplies and demands, with (0, 1) bounds, and add a sink node with an emanating arc which has a profit of 1. However, this would require a different type of solution code.)

The other objective function coefficients model both fit and priority distribution fill objectives. In Figure 3.30 the fit scores appear in the "ones" and "tens" digits. The priority coefficient appears in the "hundreds" digits. (In general the

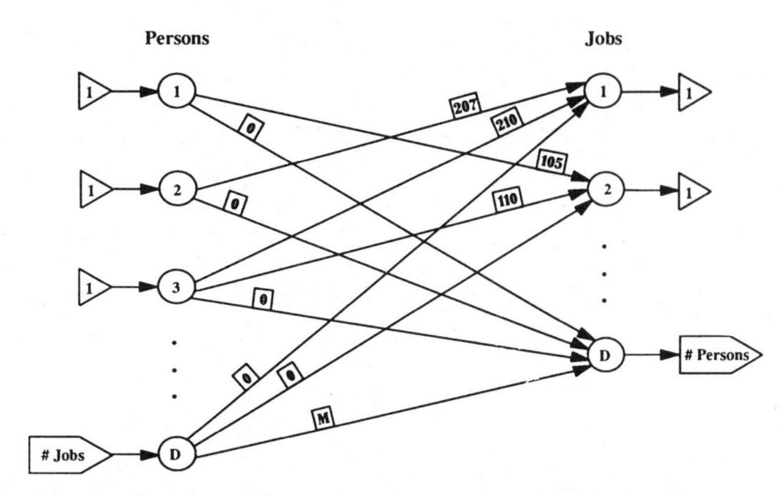

Figure 3.30 Enlisted personnel assignment model (pre-emptive maximization objective).

priority digit is one higher than the highest fit digit.) Thus in Figure 3.30, Job 1 has a priority of 2 while Job 2 has a priority of 1 (i.e., it is more important to fill Job 1 than Job 2). Person 3 fits Job 2 better than Person 1 (since 10 is larger than 5). The priority objective could also be handled by putting decreasing large profits on the arcs from the dummy person node to the job nodes, decreasing in job priority order. (That is, the lowest priority job has the highest profit associated with being unfilled.) These costs should be a factor of 10 less than the profit on the dummy-to-dummy arc and larger than the fit scores.

Solution times illustrate the superiority of the netform approach. An example problem with 8138 persons, 702 jobs, and 1,062,695 eligible matches was solved on a UNIVAC 1108 computer with 125K of memory (comparable to a personal computer today). A network optimization code designed for this application solved this problem in 13.8 min, compared to 7.9 h for a personnel assignment code which the Air Force loaned to the Army. Another example problem with 7239 persons, 557 jobs, and 631,915 eligible matches was solved in 2.3 min by the network code versus 3.7 h by the existing method.

CASE: PERSONNEL ASSIGNMENT BENCHMARK PROBLEM

What follows is an actual problem description which was sent out for bids by the United States Marine Corps [Klingman and Phillips 1984].

The USMC does not desire to fund development of the advanced personnel assignment technology required by this procurement. Therefore, to demonstrate

---

Reprinted by permission. Topological and Computational Aspects of Pre-emptive Multicriteria Military Personnel Assignment Problems, D. Klingman and N. Phillips, *The Institute of Management Sciences*, Vol. 30, No. 11, 1984, pp. 1362–1375.

that those people offering a solution already possess highly sophisticated mathematical methodologies and algorithms, offerers will be required to solve the following benchmark problem within the short time frame permitted by solicitation.

Benchmark Problem Requirement

A hypothetical Marine Corps' assignment problem has been structured such that the major "problem areas" of concern to the USMC in solving personnel assignment problems will have to be addressed by the offerer. These problem areas are:

1. Ensuring that the number of assignments is a maximum (optimal). This is a first priority objective function and may not be "traded off" with any other objective function in the problem.
2. Achieving a distribution of personnel under shortage conditions such that competing demands (quotas) may be accorded: (a) absolute priority and/or (b) proportionate percentage fills as dictated by distribution policies.
3. Ensuring that optimization of "desirable" assignment policies will not alter, in any way, the percentage fill of quotas achieved in accommodating distribution policies.
4. Ensuring that "desirable" assignment policies are optimized in a prioritized order without "trade offs" between these "desirable" assignment policy objectives.

Structure

The matrix characterizing the assignment problem contains 10,000 enlisted personnel (rows) and 240 quotas (columns). The 10,000 row by 240 column matrix is further aggregated into twenty locations. The available personnel, 500 stationed at each of the 20 locations, must be reassigned to one of the 20 locations. Each location, therefore, is also characterized by a requirement expressed via 12 quotas (columns). The structure is provided in Table 3.2.

Each quota specifies a demand for one or more people. Quotas are also characterized by both mandatory and desirable prerequisites for assignment. As a result, a person may or may not be eligible for assignment to a particular quota. And, if eligible, the assignment may or may not be desirable.

Accommodation of Distribution Policies (Fill)

The most important objectives are *fill* or *distribution* related. Primary among these is the requirement to achieve the maximum number of assignments possible, regardless of the impact on sharing policies or the "desirable assignment policies" specified later. Table 3.2 provides a detailed description of the distribution policies to be adhered to in solving the benchmark problem. (Distribution policies are "fill" policies and should not be confused with desirable assignment ("fit") policies.)

Note that the distribution policies, including proportionate sharing policies,

Table 3.2 Matrix Denoting Problem Structure

	Location A: Quota or Column IDs* 1-12	Location B: Quota or Column IDs 13-24	Location C: Quota or Column IDs 25-36	.....	Location T: Quota or Column IDs 229-240
Location A: Row IDs** 1-500					
Location B: Row IDs 501-1,000					
Location C: Row IDs 1,001-1,500					
Location C: Row IDs 1,001-1,500					
. . . .					
Location T: Row IDs 9,501 - 10,000					

* Column ID: Identifies a demand for a class of Marines for which a quota consisting of one or more vacancies exists. Table 3.2 provides quota values (vacancies).

**Row ID: Identifies an individual Marine.

must not be allowed to reduce the overall total number of assignments, a number which must be a maximum. Importantly, desirable assignment policies to be specified below must not be allowed to influence the percent fill of any given quota, or to influence the total number of assignments.

Faced with an overall shortage of personnel as well as the pattern of eligibilities among availables, it will not be possible to meet all of the quotas, and may not be possible to assign all of the personnel. Therefore, the relative importance of satisfying individual demands must be considered and shortages distributed accordingly (see Table 3.2).

Table 3.3 Distribution Policy Objectives

Distribution Policy*	Quota I.D.s (columns)	Quota Values	Distribution Policy Definitions and Objectives
A	1-48	Quota values equal 5 x column number **	Fill each quota in quota group A to the maximum extent possible, regardless of consequences to the fill of quotas in quota groups B and C. ***Proportionate Fill Policy***: If all quotas in this quota group cannot be fully satisfied, shortages must be shared as fairly as possible among the group A competing quotas *without* reducing the number of assignments to the group as a whole.
B	49-192	All quotas equal 40	Fill each quota in quota group B to the maximum extent possible, regardless of consequences to the fill of quotas in quota group C and without influencing the *percent* fill of any of the quotas in quota group A. ***Proportionate Fill Policy***: If all quotas in quota group B cannot be fully satisfied, shortages must be shared as fairly as possible among the group B competing quotas *without* reducing the number of assignments to the group as a whole.
C	193-240	All quotas equal 50	Fill each quota in quota group C to the maximum extent possible without influencing the *percent* fill of any of the quotas in quota groups A and B. ***Disproportionate Fill Policy***: If all quotas in quota group C cannot be fully satisfied, shortages must be shared among the group C competing quotas *without* reducing the number of assignments to the group as a whole. Sharing targets are as follows: 100% fill of columns 193-208, 50% fill of columns 209-224, and 20% fill of columns 225-240.

* Maximize the total number of assignments. Policies A, B, and C provide objectives for distributing personnel under conditions in which one or more quotas cannot be fully satisfied, and in which more than one configuration of fills can be achieved with the same overall total number of assignments.

** Demands are calculated by multiplying five times the column number; i.e., the 1st quota value=5, and the 2nd quota value=10, etc., ranging to the 48th quota value=240.

Accommodation of Desirable Assignment Policies (Fit)

Desirable assignment objectives, as a group, are secondary in priority and importance to distribution objectives. To solve this benchmark problem, the offerer will need to optimize four desirable assignment objectives following solution of the distribution objectives. The following four policies must be optimized in prioritized order (as listed):

1. *First priority desirable assignment objective:* If eligible, a Marine may or may not present a desirable grade/MOS substitution opportunity for a given quota. Maximize assignments of personnel to quotas resulting in desirable grade/MOS substitutions.
2. *Second priority desirable assignment objective:* If eligible, a Marine may or may not be currently located so as to provide a desirable "tour sequence"

to any particular assignment opportunity. Maximize assignments resulting in desirable tour sequence combinations.

3. *Third priority desirable assignment objective:* A Marine, if eligible, may or may not "prefer" the assignment presented by a particular assignment opportunity. Maximize assignments of personnel to preferred assignment opportunities (quotas).
4. *Fourth priority desirable assignment objective:* Finally, each Marine is currently located a given distance from each of the 19 other locations, and the assignment outcome should result in the movement of Marines over a minimum of total miles.

Case Questions

You should approach this case from the point of view of an independent consultant who is going to submit a proposal for the project. The proposal will consist of a model description and a solution approach. As in the actual bid solicitation, proposals will be evaluated using two equally weighted criteria: (1) model validity (the capability of your model to find the correct solution) and (2) practicality of approach. (A valid model which requires 6 months of computer time to solve may not be viewed as highly practical. You are not limited to the employment of network models. However, their proven solution efficiency could greatly enhance the practicality of the approach. Beyond that, the actual structure of different network models will also affect their practicality.)

Your answers to questions 1 and 2 below should include at least the following elements:

1. A graphical representation of your model (even if it is not strictly a network model) with clearly defined notation. By a graphical representation, we mean a diagram which presents the problem conceptually but does not contain every variable and constraint.
2. A short verbal description of the model which clarifies the graphical representation (2 pages maximum).
3. A statement of how all cost parameters are determined and if your model is minimizing or maximizing.

Remember, this is a proposal which you hope will win a contract award. Neatness and readability are prerequisite requirements. (Hint: Ignore Table 1 and all references to it.)

1. First, focus on the highest priority goals, the "fill" goals. The order of priority of these goals is:
   a. Maximize total number assigned.
   b. Maximize total number assigned to distribution policy Group A ignoring proportionate fill policy (see Table 2).

c. Maximize total number assigned to distribution policy Group B ignoring proportionate fill policy (see Table 2).

d. Maximize total number assigned to distribution policy Group C including disproportionate fill policy (see Table 2).

Create a "single criterion" mathematical programming model for the Benchmark Problem which will achieve these objectives. Caution: Although the "distribution" and "fit" criteria need not be explicitly handled at this stage, you should keep in mind that they must be handled eventually. Therefore, a model which is readily enhanceable to accommodate the next level of complexity is better than one which is not (even though the exact configuration of the enhancements has not yet been determined). (Hint: Consider a super sink (super demand node).)

2. The next set of pre-emptive priority goals concerns the proportionate fill policy of assignments within the distribution policy categories A and B as specified in Table 2. Using your model for question 1 as a basis, augment this model to include the proportionate fill policies for all categories A and B. That is, create a "single criterion" model such that an optimal solution to it is optimal to the criteria stated in Table 2.
3. Outline a solution approach for the model.

EXERCISES

3.1 The Fleur-de-Lease Car Rental Agency has had 10 cars dropped off in the town of Aix and 6 dropped off in Beaune, but needs at least 4 in Chamonix and at least 5 in Dijon. No cars can be left in Aix and Beaune, so all must be sent to Chamonix or Dijon in any case. The costs per car of sending them are shown in the table.

	To	
From	Chamonix	Dijon
Aix	\$5	\$3
Beaune	\$4	\$10

a. Set this up as a minimum cost transportation problem (label the nodes A, B, C, and D).

b. Now, suppose whatever number of cars end up in Chamonix and Dijon beyond the minimum required can be painted at once. Car painting costs \$34 in Chamonix and \$36 in Dijon. At least 5 cars must be painted in Chamonix and Dijon combined. Show the network.

c. The company gets a cut rate deal on painting in Chamonix of $20 each for the first 3 cars, but it must paint at least one car there. Model this with the fewest number of added nodes and arcs.

3.2 Five jobs are available for processing on five machines, subject to the restriction that each machine can handle at most one job and each job can be assigned to at most one machine. Based on the quality of the processing for various job machine combinations, the profit that is expected by assigning particular jobs to particular machines is shown in the following table. (A blank space indicates that the assignment is not possible.) It may not be possible to process all jobs.

	Machine				
Job	1	2	3	4	5
1	20	15	30		
2		15	10		15
3	30		25	40	
4		30	35	20	
5				15	20

Provide a profit maximizing network diagram for each of the following conditions (for simplicity in (b)–(f), show only the parts of the network that differ from (a)).

a. No additional restrictions.

b. At most three (but any three) of Machines 1, 2, 3 and 4 are available to process jobs.

c. At least one and at most two of Jobs 2, 3 and 4 can be assigned to the machines for processing.

d. In addition to (b), at least one of Machines 2 and 3 must be used to process jobs.

e. In addition to (d), at least one of Machines 1 and 4 must be used to process jobs.

f. At most four of the five jobs can be assigned to machines, and in addition, at least two of Jobs 3, 4 and 5 must be assigned. Can this condition be coupled with (e)?

g. See how "close" you can come, using the techniques of (b)–(f), to giving a diagram for the following restriction: at most one of Jobs 1 and 2 and at most one of Jobs 2 and 3 can be assigned to machines. Explain why your "close" diagram does not work. (In general, no procedure exists within the pure network framework for constraining the sums of flows into (or out of) two overlapping sets of source (or sink) nodes unless one of the sets is contained in the other.)

3.3 A company must ship between 12 and 25 truckloads from Warehouse A to Customer 1. They can ship from 15 to 20 truckloads at a cost of $60 per load. Shipping more than this, up to 25 truckloads, costs an additional $10 (total $70) for each of the extra loads. If the company ships less than 15 truckloads, it pays $20 less for each load. Identify which of the following netforms model this situation correctly, and explain why each does or does not work. Which of these netforms do you prefer? (Assume a cost minimization objective.)

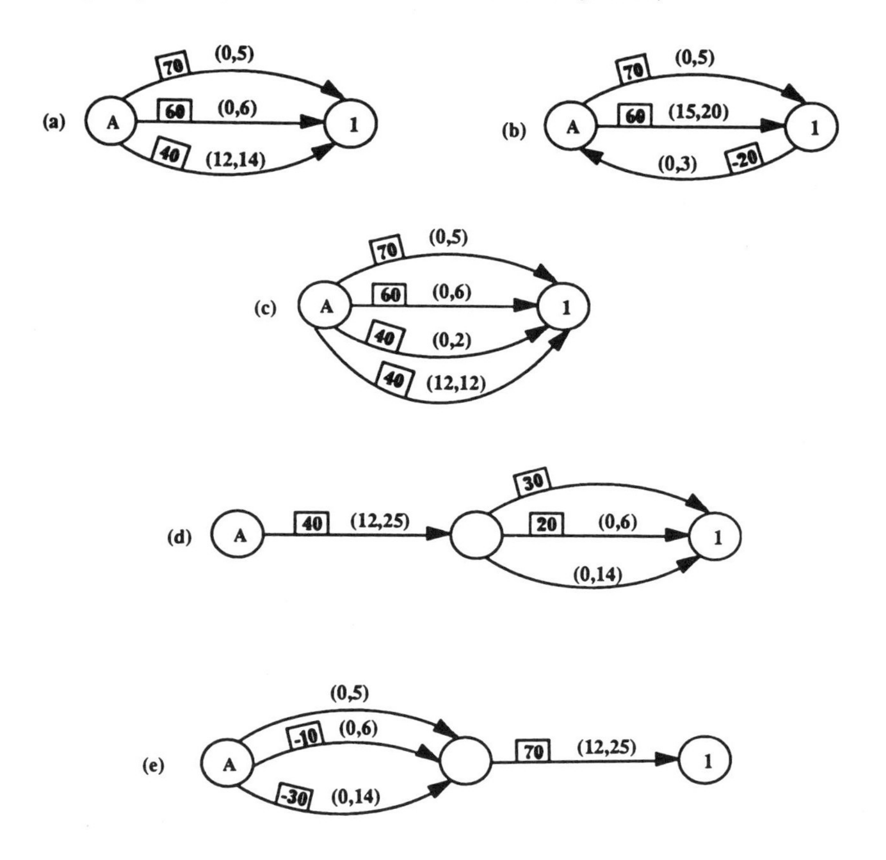

3.4 Assume that the company of Exercise 3.3 has the same costs as previously specified for sending 15–25 truckloads. However, the contract with the trucking firm is based on the expectation that at least 15 truckloads will be shipped. If fewer truckloads are shipped, the trucking firm will agree to rebate $20 for each load not shipped (since the trucking firm can make partial use of the truck elsewhere), provided the shipper agrees to send 12 truckloads at the absolute minimum. Is this problem equivalent to the one of Exercise 3.3? Which of the netforms of Exercise 3.3 correctly models this exercise? Will any of the following netforms also work? (Indicate why or why not.)

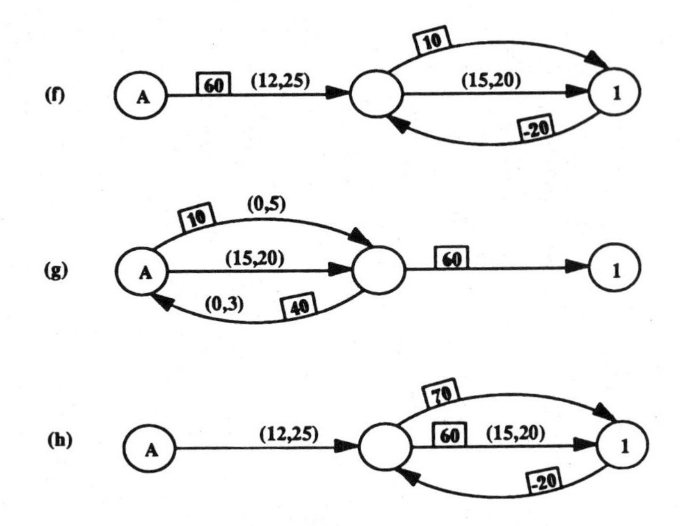

3.5 As a general construct for modeling goals or targets, netform (b) of Exercise 3.3 appears to have a defect when viewed in a larger context. This is illustrated by the following.

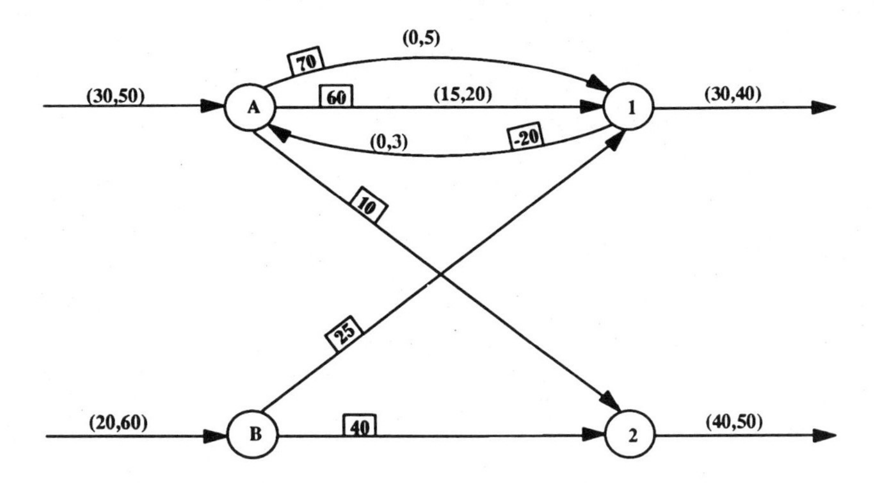

Examination of the preceding shows it would be possible to send flow from node B to node 2 by an "illegitimate" route from B to 1, then back to A, and finally to 2. The cost of this route is $25-20+10=15$, as opposed to the cost of 40 that is supposed to apply to shipping from B to 2.

a. Show two ways to add a single arc and node that will eliminate this route.

b. In spite of the preceding observation, the netform is the correct model for Exercise 3.4. Explain this by showing that if the arc from node 1 to node A has any flow, it can always be interpreted as identifying the number of loads not shipped from A to 1, for which rebates are received. In this view, reinterpret a flow on the illegitimate route, making use of the fact that 15 units are sent from A to 1 on the lower bounded arc in any event.

3.6 Assume a cost minimization objective. Which of the following two netforms, (a) or (b), is the same as (c)?

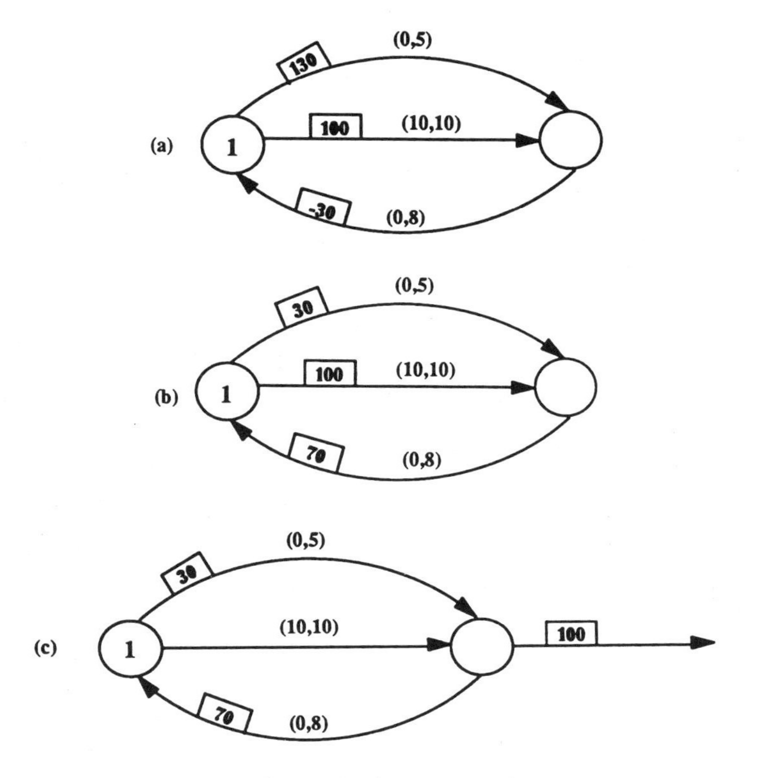

Explain why. Will netform (d) also accomplish the same thing? Why or why not?

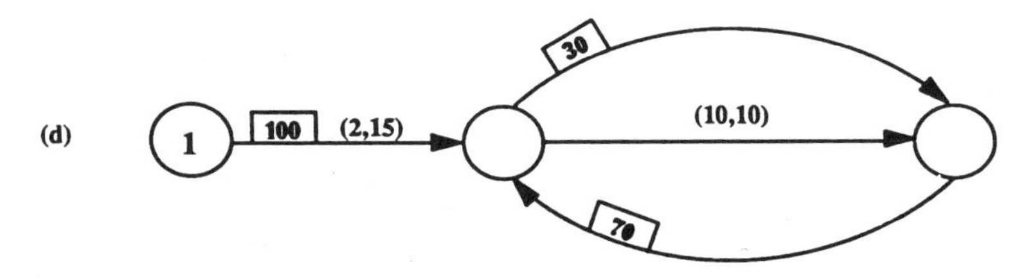

3.7 The Go-For-The-Snow travel agency has been given a special assignment by ski clubs of Kansas City and St. Louis to get their members lodging at the ski areas of Winter Park, Vail, and Beaver Creek. Because of a late start in their planning and the difficulty of finding vacancies at the time of the year selected, the ski clubs have agreed to let the travel agency make whatever arrangements it can for accommodations. The agency in turn has agreed to accept a fixed fee for its services, arranging and paying for all transportation and lodging out of this fee. The agency determines that the best transportation rates (air fare plus ground transportation) are as follows.

Transportation Costs Per Person

	To		
From	Winter Park	Vail	Beaver Creek
Kansas City	300	330	335
St. Louis	360	390	395

Best lodging rates at the areas during the period booked can be obtained only if the number of people actually lodged at each area meet an ideal number, or goal, set by the area. There is an extra charge for each person above the number specified by the goal. If not enough people show up to meet the goal, then there is also a charge for "no shows", i.e., each person that does not show up must still pay, although less than those who do show up. The goals, rates, and extra charges are as follows.

	Goal (ideal) No. of People	Rate per Person	Extra Charge per Person Above Goal	Charge for "No Shows"
Winter Park	60	$120	$20	$80
Vail	55	$130	$65	$40
Beaver Creek	30	$143	$55	$85

For example, if 70 people lodge at Winter Park the total cost is $70 \times \$120 + 10 \times \$20 = \$8600$, while if 50 people lodge there the total cost is $50 \times \$120 + 10 \times \$80 = \$6800$. Each area must lodge no fewer than 10 people below its goal, and no more than 20 people above its goal. Finally, Kansas City has 90 people and St. Louis has 70 people who are going on the ski trip. The travel agency seeks to determine the transportation and lodging arrangements that will minimize total cost.

a. Construct a network diagram under the stated assumptions. Diagram the goal conditions two ways, first treating "no show" charges in the form of rebates from incurring the full charges of meeting the goal, and second so that "no show" charges appear on arcs with the same values they have in the table. In each case, use as few nodes and arcs as possible. Use the rebate form for the following additional conditions.

b. How does the diagram change if the goal at Winter Park is to have anywhere from 55 to 65 people (and the stated extra charges and limits are based on going below 55 and above 65)?

c. Beaver Creek and Vail notify the agency that they are offering a package deal, where, instead of requiring a minimum number to lodge at each area separately (10 below the goal), they require that at least 65 lodge at their two areas combined. The goals of 55 and 30 people at the two areas and the original charges for "no shows" continue to apply. How does the diagram change?

d. Suppose the context of the problem is changed to that of two power plants serving three regions, with specified costs for sending power from each plant to each region, and with specified power goals at each region. Also the amount of power supplied at each plant is a goal, with extra costs attached to delivering more and less than this goal. How would the network diagram represent a supply goal of exactly 1000 units at the first plant, and of 800–1200 units at the second plant, with extra costs of $500/unit for deviations above these goals and of $400/unit for deviations below these goals? How would the diagram change if, in addition, due to shared resources for generating power, the maximum number of units both plants could provide together is 2600 units?

3.8 Consider the following extension of the Macho Nacho Company, Exercise 2.5. As previously indicated, the company makes a special ski boot which it produces and ships in carload quantities from its plants to its warehouses and then to its customers. The following tables, taken from Exercise 2.5, provide relevant data for the goal of minimizing total cost. (An × means no shipment is possible. All shipments are in carloads.)

Shipping Costs

	Warehouses		
Plants	1	2	3
1	80	110	×
2	140	×	130

Shipping Costs

Warehouses	Customers 1	2	3
1	×	15	25
2	20	30	×
3	30	×	20

Upper Limits on Amounts That Can Be Shipped (Imposed by Transportation Connections)

Warehouses	Customers 1	2	3
1	×	100	50
2	80	70	×
3	80	×	70

Both plants must produce at least 60 carloads to stay in business, but are limited by production capacity to a maximum of 120 carloads each. On the other hand, Warehouse 2 already has 60 carloads of inventory on hand available for shipping and Warehouse 3 has 30 carloads on hand. Customer 1 demands exactly 80 carloads, Customer 2 demands exactly 90 carloads, and Customer 3 demands exactly 100 carloads. Building on the minimum cost flow network for the preceding information (as requested in Exercise 2.5), show how to incorporate the following additional stipulations.

a. Warehouse 1 had limited facilities at its receiving dock that compel it to accept no more than a total of 40 carloads from both plants combined.

b. In addition to (a), limited shipping facilities at Warehouse 3 compel the amount shipped from the warehouse (to all customers combined) to be at most 115 carloads.

c. In addition to (b) (and hence (a)), it is possible to trans-ship up to 30 carloads (total) from Warehouse 2 to Warehouses 1 and 3, at no cost. (These trans-shipments are handled the same as direct shipments from the plants as far as combining with (a) is concerned.)

d. In addition to (b) (instead of (c)), it is possible to trans-ship from Warehouses 1 and 3 to Warehouse 2 at no cost, so long as Warehouse 2 receives no more than 40 carloads from the other two warehouses (combined). Warehouse 3 ships from the same limited facilities that it uses to ship to customers.

e. In addition to (d), a volume from 0 to 20 carloads received at Warehouse 2 (from plants and warehouses) incurs a unit handling cost of $200. A volume from above 20 up to 30 carloads incurs a unit handling cost of $230, from above 30 carloads to 60 carloads incurs a unit handling cost of $290.

3.9 The Widget Company maintains manufacturing and distribution facilities for its Four Star Widget at three plants in the United States: Boston, Chicago, and Denver. Each plant can make and sell the widget but, because of capacity imbalances, may have to resort to internal redistribution (i.e., shipments) in order to satisfy anticipated demand. Specifically, the following information is available for the coming month's situation:

Marketing Report

Plant	Expected Demand	Unit Selling Price ($)
B (Boston)	1000	80
C (Chicago)	800	60
D (Denver)	1200	70

The demand and price data are exact since orders are compiled in advance.

Manufacturing Report

	Regular Time		Overtime	
Plant	Unit Cost ($)	Capacity	Unit Cost ($)	Capacity
B	50	900	75	400
C	40	1000	55	500
D	45	800	70	300

The overtime capacity is in addition to the regular figures.

Shipping Cost Per Unit Report

	To Plant		
From Plant	B	C	D
B	0	12	18
C	12	0	10
D	18	10	0

Given these data, construct a netform model which will find a manufacturing and shipping schedule that maximizes the monthly profit.

***3.10** Consider a standard transportation problem shown below. Let x_{ij} represent the flow on the arc from supply node i to demand node j. In each part below, further constraints are added. Suggest a way to reformulate the model so that it will include these nonnetwork constraints within the standard transportation model (unbounded arcs). Draw the netform.

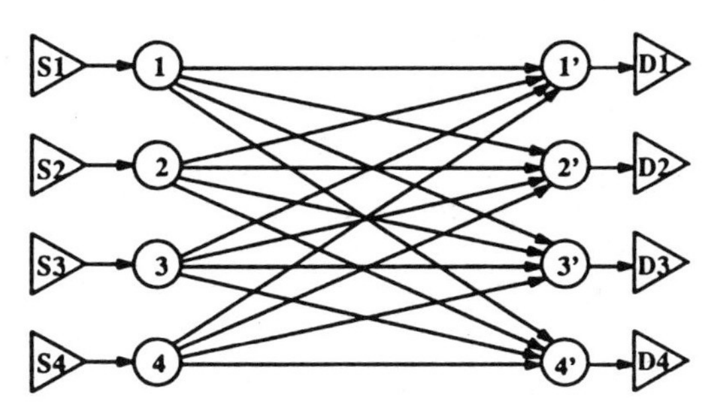

a. $\Sigma_{i=1}^{2} \Sigma_{j=1}^{4} x_{ij} \leqslant u$ where $u < S1 + S2$

b. $x_{11} + x_{12} + x_{13} \leqslant u$ where $u < S1$

3.11 Pandora's Press is a printing company which deals almost exclusively in the publication of mail-order catalogs. Since most of its clients tend to be relatively unstable (in a business longevity sense), Pandora's operates in a job shop manner. Pandora's Press owns and operates five printing presses of varying capabilities in terms of speed, clarity, color reproduction, and paper size and type. At the beginning of each day, Pandora's manager reviews the outstanding printing orders and determines which machines are capable of handling each job in one day. (Each job will be completed in one day and must be run entirely on one machine). He then estimates Pandora's profit (considering varying production, set-up, depreciation, and labor costs as well as revenue) for each feasible machine-to-job assignment. There are five outstanding customer orders today. These, along with the estimated profits of feasible machine assignments, are shown in the table.

Estimated Profit ($1000s)

	Machines				
Customer Orders	1	2	3	4	5
Acme Audio	2	3	–	–	–
Buck's Bikes	–	–	4	–	–
Capital Camping	5	–	–	2	2
Dimension Designs	3	7	–	2	2
Eaton Electronics	–	–	6	–	–

a. Create a netform which will tell Pandora's manager which orders to produce today and which machine to use for each order so as to maximize profits. (It may not be possible to process every order.) Solve your model with a network optimization or LP software package.

b. The manager has decided that it is very important to minimize his machine idle time. Modify your model to insure that (1) as many machines as possible are utilized and then that (2) profit is maximized. (This is meant to be a pre-emptive goal priority statement). Solve the modified model. How much profit is lost as a result of the machine utilization goal?

c. Based on his knowledge of his clients, the manager gives the following priority ranking to the customer orders: (1) Capital Camping; (2) Buck's Bikes; (3) Dimension Designs; (4) Acme Audio; (5) Eaton Electronics. He still wishes his machines' utilization goal to have highest priority. However, beyond that, he now wants to insure that the customer orders will be chosen to be processed in the above order. Finally, profit maximization is the lowest priority goal. Modify your model to accommodate this new set of goals and solve. Did the optimal solution change? Has the total profit changed?

3.12 Nutworks Inc. produces both Narcs and Widgets in plants A and B. Production rates are 2 Narcs per hour and 5 Widgets per hour. The capacity of plant A is 60 h regular time plus 30 h overtime and the capacity of plant B is 80 h regular time and 40 h overtime. Production cost is $100 per hour regular time and $150 per hour overtime. Nutworks has received orders from customer P for 100 Narcs and 250 Widgets and from customer Q for 200 Narcs and 300 Widgets. The shipping cost per unit (Narcs or Widgets) from plant A to customers P and Q is $10 and $5, respectively, and the shipping cost from plant B to customers P and Q is $5 and $15, respectively, both for Narcs and Widgets. The selling price per unit for Narcs is $90 for the first 100 units and $80 for each unit over 100. Widgets sell for $50 for the first 150 units and for $40 for each unit over 150. Nutworks has guaranteed to fill each order within the planning horizon. If, for any reason, an order cannot be filled, then Nutworks agrees to pay a penalty of $20 per Narc or $5 per Widget not delivered.

a. Draw the netform which will maximize total profit for Nutworks.

b. Can the network be modified for different production rates at plants A and B? Explain!

REFERENCES

Charnes, A. and W. Cooper. 1961, *Management Models and Industrial Applications of Linear Programming*, 2 vols., Wiley, New York.

Cohon, J. L. 1978, *Multiobjective Programming and Planning*, Academic Press, New York.

Klingman, D. and N. Phillips. 1984, "Topological and Computational Aspects of Preemptive Multicriteria Military Personnel Assignment Problems," *Management Science*, Vol. 30, No. 11, pp. 1362–1375.

Zeleny, M. 1973, "A Selected Bibliography of Works Related to the Multiple Criteria Decision Making," *Multiple Criteria Decision Making*, USC Press, Columbia, SC.

Zeleny, M. 1974, *Linear Multiobjective Programming*, Springer-Verlag, New York.

Zionts, S. 1974, *Linear and Integer Programming*, Prentice Hall, Englewood Cliffs, NJ.

4

DYNAMIC NETWORK MODELS

The models developed in preceding chapters have all involved "snapshots" of a single time period's duration. Many problems also have a significant dynamic, time-dependent component. Additional techniques to formulate such problems effectively are the focus of this chapter.

4.1 THE INVENTORY CONNECTION

Inventory structures lie at the heart of a wide variety of dynamic models. In an abstract sense, anything that involves some form of continued existence from one period to another may be conceived as an inventory flow, which may be augmented, depleted or transformed, but, subject to all of these, persists until some final disposition. The themes and variations of basic inventory models—at least in the setting of network-related structures—underlie many dynamic models that may at first appear to have little in common with inventories. Consequently, we begin the examination of time-phased models with inventory systems.

4.2 A PROGRESSIVE ILLUSTRATION

To illustrate the basic principles, we start with a very simple sort of inventory problem and modify it by stages. For concreteness, consider the situation where a grain distributor can buy wheat from farms and then sell it directly to local markets or store it for sale in subsequent periods. Relevant data affecting the distributor's decision of how much to buy, sell and maintain in inventory in each of three successive periods are shown in Table 4.1.

The indicated upper limits on the numbers of bushels that can be purchased and sold are based on projected availabilities and demands. The maximum quantity that can be stored in inventory from one period to the next is 800 bushels, and the distributor's storage cost is $0.40 per bushel. Finally, the distributor has 300 bushels already in inventory at the start of period 1, and for a

Table 4.1

	Periods		
	1	2	3
Purchase price/bushel ($)	2.00	3.00	2.50
Selling price/bushel ($)	3.10	4.20	3.50
Max bushels to purchase	1400	1800	1600
Max bushels to sell	1000	2000	2500

safety reverse wants to have at least 200 bushels on hand at the end of period 3. (It may be imagined that period 4 supplies and demands cannot yet be forecast adequately, and the required ending inventory for period 3 is the distributor's way of hedging against this uncertainty. Also, because no cost data are considered after period 3, the storage cost after this period may also be disregarded.)

To model this problem, we first show a network diagram in Figure 4.1 without any costs or bounds attached. An understanding of this "skeletal" diagram is aided by considering how the flow conservation requirement can be interpreted at each of the problem nodes. For example, at node 1 the "flow in = flow out" requirement can be expressed as: the beginning inventory (flow on the horizontal arc into node 1) plus period 1 purchases (flow on the vertical arc into node 1) equals the ending inventory (flow on the horizontal arc out of node 1) plus period 1 sales (flow on the vertical arc out of node 1). Since the ending inventory for period 1 is the same as the beginning inventory for period 2 (and is the flow on the arc from node 1 to node 2), the flow conservation requirement for node 2 can be expressed in exactly the same manner. In short, at each node of the network the flow conservation requirement can be translated into "beginning inventory + purchases = ending inventory + sales."

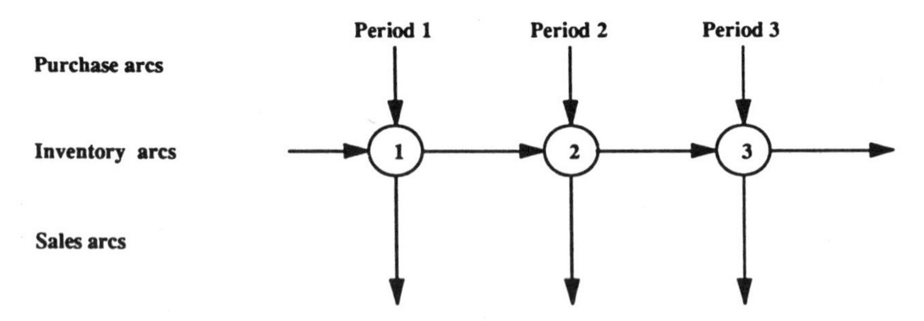

Figure 4.1 Inventory problem.

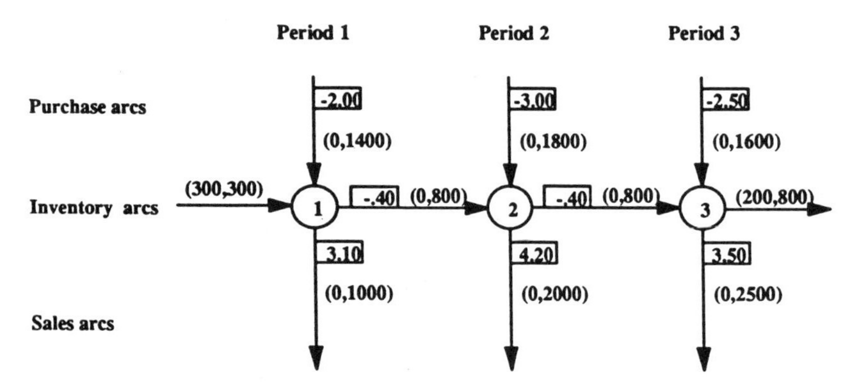

Figure 4.2 Wheat distributor's inventory problem (profit maximization objective).

Question: Using the preceding framework, complete the network diagram for the wheat distributor's problem by attaching the appropriate items of information to the various arcs. Use a profit maximization objective (representing costs as negative values and revenues as positive values).

The answer to the question may be checked by reference to Figure 4.2. Several of the features of Figure 4.2 deserve special comment. Note that the bounds for the initial inventory compel exactly 300 units to enter the system instead of allowing from 0 to 300 units to enter. This is because the 300 bushels of wheat on hand cannot stay in a period that precedes period 1. (Time marches on!) This use of exact bounds (lower bound = upper bound) is a common attribute of time-phased problems, where an initial available quantity of some goods cannot simply be "left behind" but must continue on through the system. (Even if unused, it must show up in a subsequent period's inventory.)

It is also common, as in this example, to impose some lower bound on ending inventory as way of providing a safety factor for operations that continue beyond the time horizon of the model. (In the specific example illustrated, the solution would not be changed by setting both the lower and upper bound on the ending inventory arc to the same value, i.e., 200.) In some cases, it is seen as desirable to put a positive lower bound on each period's inventory, and not just on ending inventory.

Although it may initially seem that applying the distributor's storage cost of \$0.40 per bushel only to ending inventory is in error, this construction actually approximates the total storage cost on average inventory. Under the assumption that the distributor buys and sells wheat at a constant rate during each time period, then the inventory levels change linearly from period to period (see Figure 4.3(a)). Thus if we define Y_t as the ending inventory for time period t, then

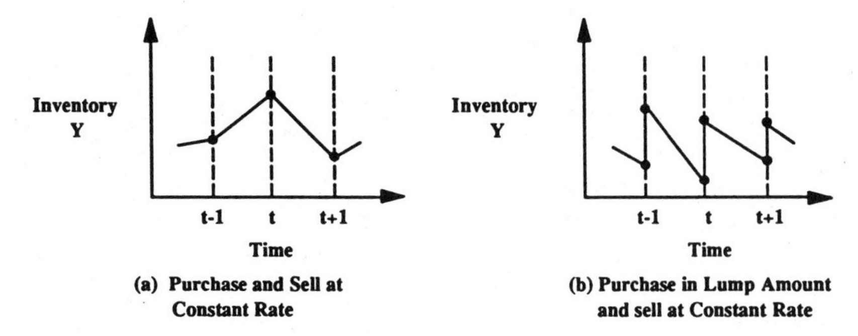

Figure 4.3 Inventory patterns.

the average inventory for period t is $1/2(Y_t + Y_{t-1})$ and for T time periods,

$$\begin{aligned}\text{Total storage cost} &= \$0.40 \sum_{t=1}^{T} 1/2(Y_t + Y_{t-1}) \\ &= \$0.40 \left[\sum_{t=1}^{T-1} Y_t + 1/2(Y_0 + Y_T) \right] \\ &\approx \$0.40 \sum_{t=1}^{T-1} Y_t\end{aligned}$$

The term we have dropped, $1/2(Y_0 + Y_T)$, would have little effect on the optimal solution since Y_0, the initial inventory at the start of period 1, is fixed, and Y_T, the ending inventory for the final period T, is constrained to some safety stock level. (That is, this term is almost a constant, and adding a constant to the objective function will not change the optimal solution.)

Note that under the assumption that the distributor buys and sells wheat at a constant rate during each time period, the maximum and minimum inventory levels will occur at the end or beginning of each time period. Thus applying the inventory upper bound of 800 to the ending inventory arc for period t (which is also the beginning inventory arc for period $t + 1$) is also correct.

However, if flow rates are not constant, then the model must be modified. For example, suppose the distributor buys wheat in a lump amount at the beginning of each period and sells it at a constant rate throughout the period (see Figure 4.3(b)). Then the maximum inventory level for each period would occur immediately after the lump purchase but before sales (and the minimum would occur at the end of the period as before). Figure 4.2 would need to be modified as shown in Figure 4.4.

Models that deal with time-dependent conditions must of course involve some uncertainty. Regardless of the methods for estimating supplies, demands, and prices over time, pinpoint accuracy cannot be expected. It may seem that, in the face of uncertainty, the value of an analytic model would be somewhat

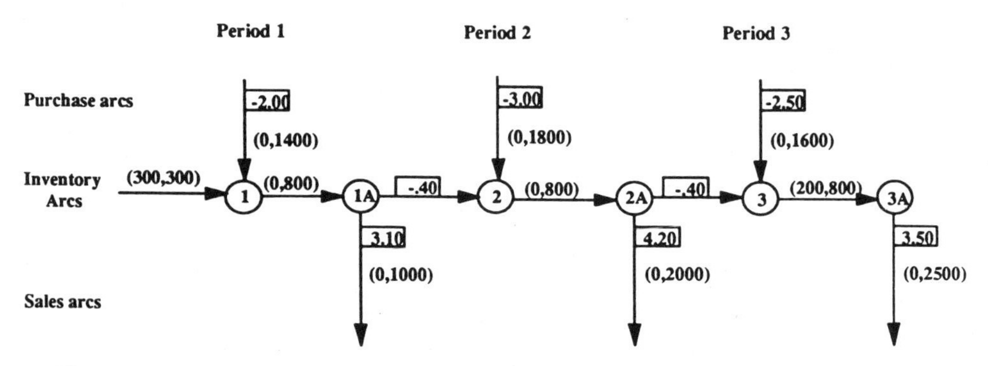

Figure 4.4 Wheat distributor's inventory problem with purchases in a lump amount and sales at a constant rate (profit maximization objective).

impaired. All the same, experience has shown that there is merit in applying such models to uncertain conditions, as attested by the many companies that have saved impressive dollar amounts by using these models. Often, as in other contexts, the key to success is to employ "what if" analyses, exploring major alternatives by re-solving a model with different parameter estimates.

This brings up another important aspect of time-phased models. While the models include decisions that are to be made in the future, the only ones we can implement are today's decisions. Of course, we can tentatively plan to submit a purchase order for a certain volume of goods next month, or to maintain next month's inventory at a certain level, but as next month approaches, we may well come by information that calls for revision of our plans. Projections about June and July prices may look a bit different as we approach June than they did when we were approaching May. Consequently, it makes sense to continually update the model, obtaining new solutions as conditions change, so that the decisions we can implement (those for today) are based on the latest and best information available.

4.3 ADDITIONAL INVENTORY COMPONENTS

Inventory problems often contain one or more elements in addition to those illustrated in the preceding example. In practical settings, these problems may be compounded by goal programming, decreasing returns to scale, and other model structures introduced in previous chapters. In addition, generalized networks and certain more advanced netforms of subsequent chapters provide other model structures that are extremely useful in practical inventory contexts. For simplicity, we focus here on the fundamental inventory model components, without attempting complex amalgamations. Once the principles behind the components are understood, there is no difficulty combining them with the other netform structures to meet the needs of particular settings.

4.4 HANDLING BACKORDERS

In practical applications, it is often not possible, or cost effective, to satisfy current demands (customer orders) out of inventories and purchases during certain periods. Analogously, in financial settings it may not always be the best policy to pay all debts and financial obligations precisely when they come due. In such cases, effective management must allow for options of meeting current demands out of future resources (inventories, purchases, production, cash, etc.). We speak of this process as ***backordering***.

Clearly, it is essential to maintain good control over backordering. For example, effective management of backorders can be a key to improving customer relationships. Many companies are turning to computer-based optimization of networks and netforms for precisely this reason. In addition, in financial contexts, large dollar amounts stand to be saved or lost depending on the coordination of present and future resources.

The way to represent backorder options in networks is fairly straightforward. Viewing a backorder as shipping from a later period's inventory to meet an earlier period's demand, we can model such a shipment by a "back arc" (i.e., an arc that goes backward against the flow of time) as in Figure 4.5. The backorder arcs in Figure 4.5 are those leading diagonally backward from Period 2 to Period 1 (node 2 to node 1B) and from Period 3 to Periods 1 and 2 (node 3 to nodes 1B and 2B). Thus, in this illustration, there are 2 backorder options to meet demand from a single period earlier and 1 backorder option to meet demand from two periods earlier. Note that the only change to the previous form of the inventory model is to insert in each time period a node and arc that make it possible to differentiate backorders from orders met currently. (The

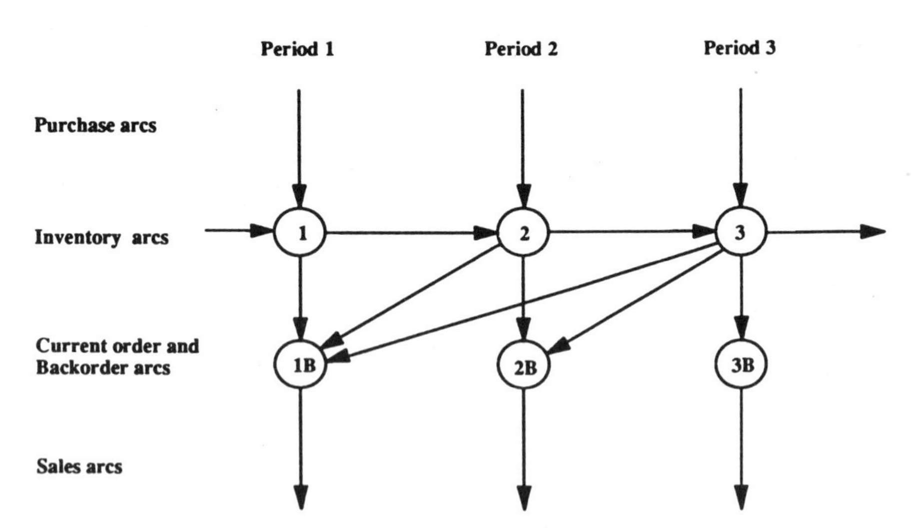

Figure 4.5 Structure of inventory model with backorders.

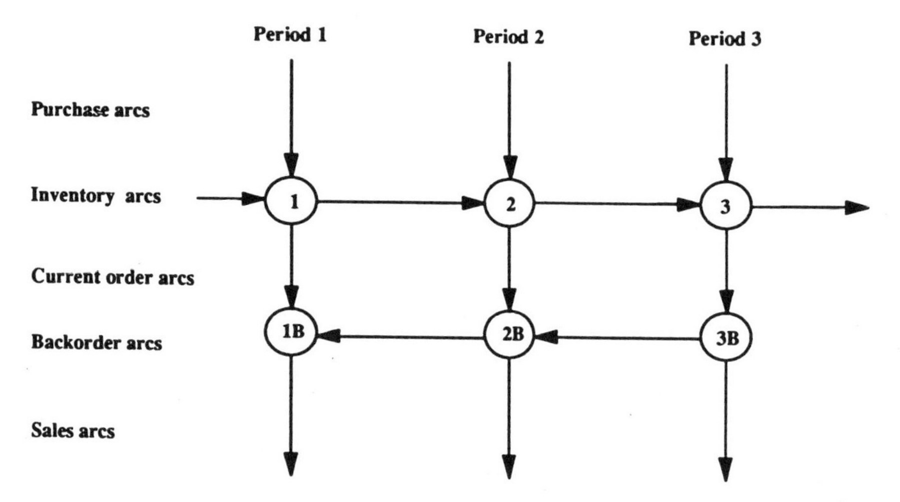

Figure 4.6 Alternative inventory model with backorders.

extra node in period 3 that separates the current order arc from the sales arc, node 3B, is not needed for this function and can optionally be deleted.)

If the distinction as to which later period's inventory is used to meet a current period's backorder is not important, then Figure 4.5 can be simplified (number of arcs reduced) by modeling the backorder arcs similarly to the inventory arcs as shown in Figure 4.6.

Likewise, if we wish to distinguish which period's purchases are used to meet a period's demand in addition to making a similar distinction for backorders, then we can use the diagonal arc concept for both inventory and backorders, as shown in Figure 4.7. (The inventory arcs point diagonally forward in time, while the backorder arcs point diagonally backward in time.) Note that although

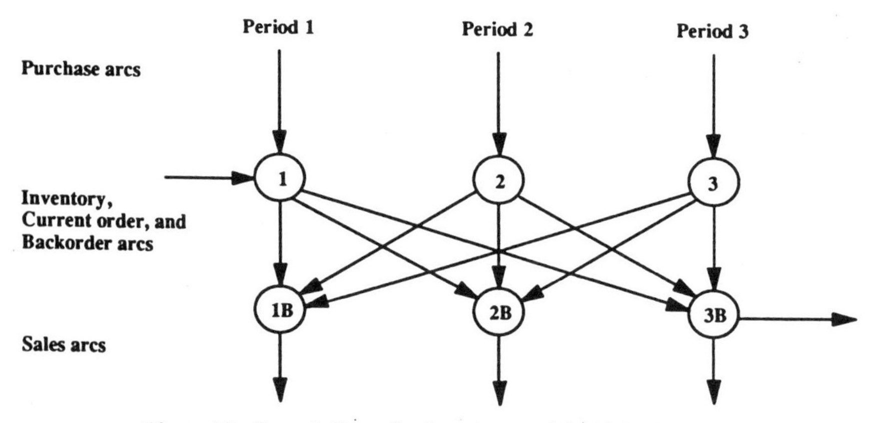

Figure 4.7 Second alternative inventory model with backorders.

Figure 4.7 would provide increased solution detail, it does so at the expense of increasing the number of arcs. For large problems, this might significantly increase the time required for computer solution. However, when viewed sideways, Figure 4.7 has the general structure of a transportation problem where purchases are supplies, sales are demands, and all arcs go from supply points to demand points. Thus a specialized transportation solution algorithm could be used to solve the problem with greater efficiency than a generic network algorithm.

Costs can be attached to backorder arcs to represent concessionary price reductions to customers whose orders are met late, or, in a financial setting, interest charges and penalties incurred for paying off a debt after the regular due date. In addition, companies sometimes wish to enlarge such costs to represent potential losses of customers (or "loss of customer goodwill"). The amount assessed for such loss can be experimentally varied in "what if" analyses; for example, attaching a large cost to see what policies result when backorders are avoided as much as possible. Bounds may also be placed on backorder arcs to limit the quantities backordered in various periods.

Question: Expand the earlier Wheat Distributor example by providing a network diagram that handles the following additional conditions: (a) the distributor can backorder from periods 2 and 3 to meet sales demand of a single period earlier at a cost of $0.20 per unit backordered, up to a limit of 100 units backordered in each period; (b) the distributor can backorder up to 50 units from Period 3 to Period 1 (independent of the amount backordered to Period 2) at a cost of $0.35 per unit.

Going further with the conditions of this question, a situation that calls for combining these inventory model structures with structures from previous chapters is illustrated by the following.

Question: The wheat distributor wants to impose two conditions in addition to those of the preceding exercise: (c) the total amount backordered to Period 1 cannot exceed 120 units; (d) the total amount backordered from Period 3 cannot exceed 115 units.

Figure 4.8 handles the combined conditions of the two preceding questions. Note that it is not possible to handle conditions (a), (b), and (d) using the alternative inventory model structure in Figure 4.6.

4.5 INTEGRATING PRODUCTION AND INVENTORY

The integration of production and inventory is becoming increasingly important as the interacting influences spanning these activities become more pronounced. Failure to link production and inventory decisions leads to suboptimizing—that is, optimizing over different parts of the operation without accounting for their interdependencies—with a significant risk of producing solutions that are far from optimal for the company as a whole.

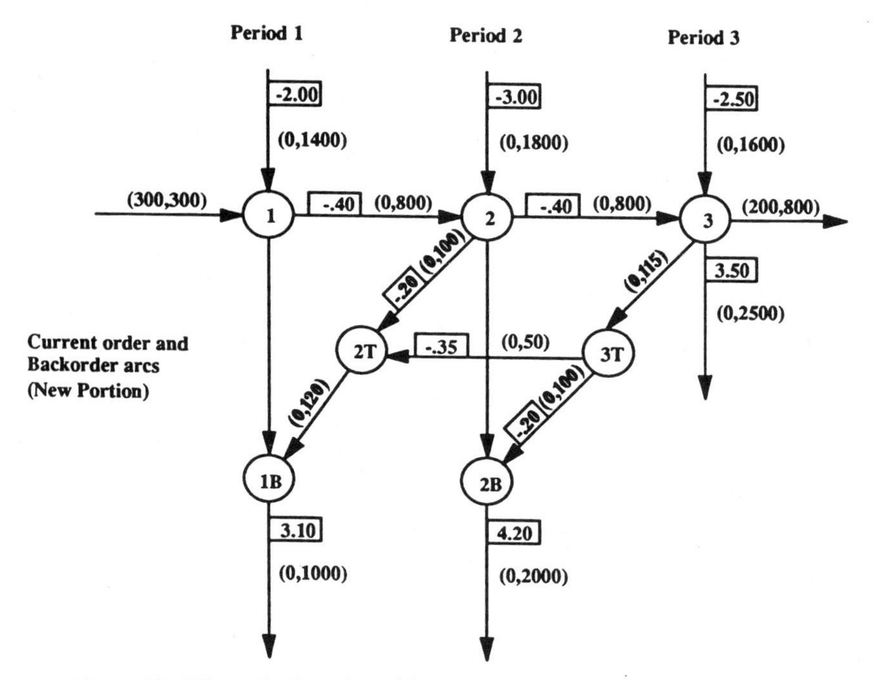

Figure 4.8 Wheat distributor's problem expanded (profit maximization objective).

Indeed, the separate classification of production and inventory is itself misleading, because commonly there are inventories at the production end of operations as well as at the distribution end. There may even be several inventory levels along the way from production to distribution (though sometimes a company will find it appropriate to aggregate some of the levels). We will indicate the form of a "two-level" system, from which extrapolation to additional levels will be apparent.

Consider a firm that purchases raw materials, which it can either maintain on hand in raw materials inventory or process to make a finished product. The finished product, in turn, can be maintained in a finished product inventory or sold. This two-level integration of production and inventory is illustrated in Figure 4.9. (Complications such as backorder arcs are omitted for simplicity.)

Figure 4.9 extends the simple inventory model by simply adding another "layer," nodes 1P, 2P, and 3P. (Of course, general models will typically contain more than 3 periods.) The production arcs in this diagram, which represent the processing of raw materials to yield finished products, have much the same role as the purchase arcs. That is, they bring goods into the finished product section just as the purchase arcs bring goods into the raw materials section. The flow on all of these arcs must be in the same units. For example, if we purchase tons of steel to manufacture steel beams, then all costs and bounds must be translated either to be in terms of tons (using the number of tons per beam to convert the

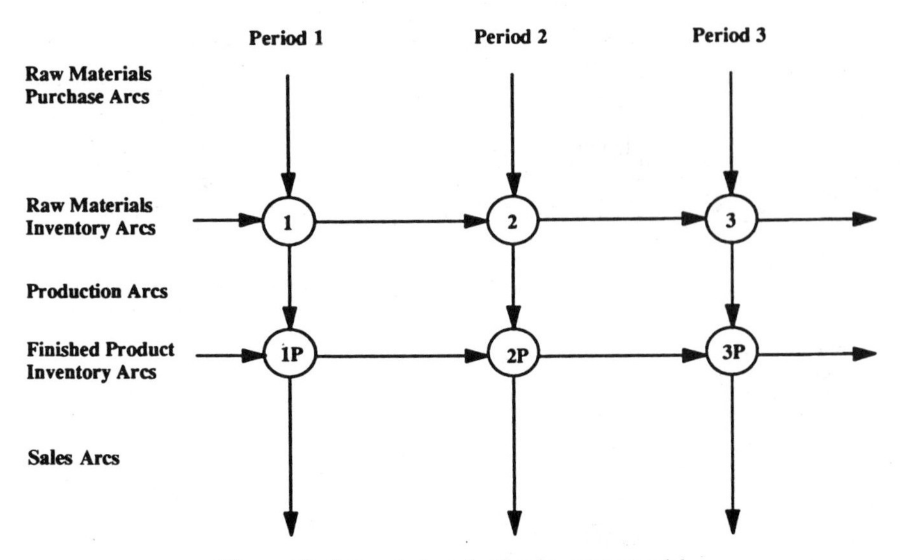

Figure 4.9 Integrated production-inventory model.

raw materials parameters) or in terms of beams (using the number of beams per ton to convert the product parameters). The importance of a model such as this, once appropriate costs, bounds, and revenues are attached, is that it enables the purchasing, production, and inventory decisions to be coordinated simultaneously, without neglecting their interdependencies. Note too that the model can readily be extended to encompass a more complex distribution component. For this the sales arcs need only be replaced by multiple arcs leading to different customers.

4.6 MODELING TIME LAGS

In many situations, purchasing and production are not instantaneous processes, but many involve time lags. For example, a purchase order placed in one period may not result in delivery and receipt of goods until the following period. Similarly, raw materials processed in one period may not result in finished goods until the next. In such cases, purchase and production arcs can simply lead from earlier periods to later periods, according to the extent of their lags. (The interval of time that a single period has been selected to represent has a bearing, of course. If purchase lag times and producing lag times are important, and occur for intervals of different length, it may be useful to subdivide time periods sufficiently to allow both effects to be captured without undue distortion.)

The modeling of time lags is illustrated by the following.

Question: Create a four-period network model for a combined production-inventory system in which purchases and production both typically lag by one period (i.e., are completed one period after initiated). In addition, production may be expedited, so that it is completed in the same period started, by incurring an extra cost. Do not bother to put costs, revenues, or bounds on the arcs (since they have not been specified), but indicate the production arcs that incur extra cost. Indicate all arcs representing flows into or out of Periods 1–4.

The solution is shown in Figure 4.10. (The vertically descending production arcs are those that incur extra cost.) This example points up additional features of time-lagged models. Purchase and production decisions have already been made preceding the first period of the model, and therefore are represented by arcs that bring constant flows into Period 1 nodes, nodes 1 and 1P, just as do the initial inventory arcs. (One could alternately combine these initial incoming flows by reclassifying them all as "initial inventory.") Note that the initial "pre-Period 1" purchase and production arcs, like the initial inventory arcs, are drawn in the form of supply arcs, without tail nodes. (If tail nodes were added, it would be necessary then to add a further initiating "supply arc" into each. Flow cannot occur at a node with no incoming arcs, as a result of the requirement that flow in = flow out.)

Likewise, in Period 4 we must contend with decisions that go "outside" the model, but in this case which are initiated inside. How should these decisions be handled? As with ending inventories, the customary approach is to specify bounds on the arcs leading beyond the last period, representing target conditions desired to be satisfied at the end.

One may at first wonder why the two extra nodes, 5 and 5P, and demand arcs were added after the concluding period. (No extra nodes or supply arcs were added before the starting period.) By placing bounds on the final demand arcs, control can be exercised over the combined final purchase and raw materials inventory flows, and similarly over the combined final production and finished product inventory flows. If it is desired only to limit these final flows separately,

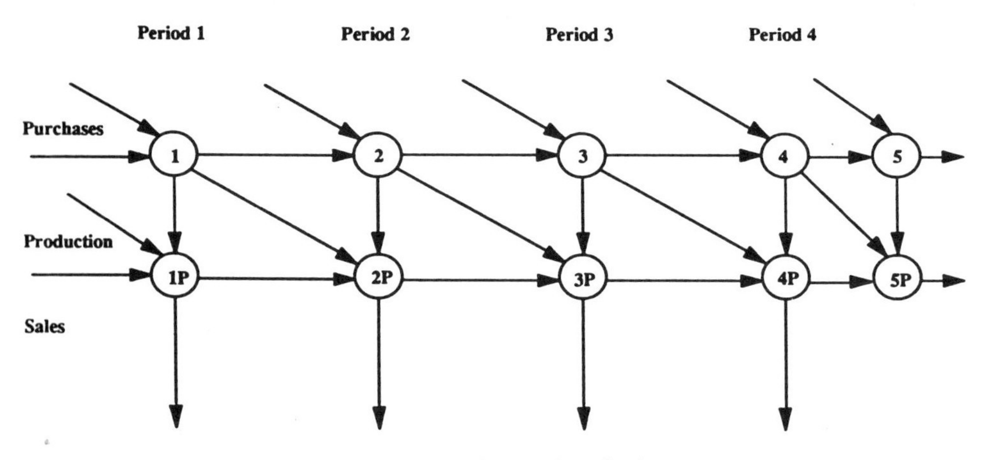

Figure 4.10 Lagged purchase and production example.

and not in combination, then the concluding demand nodes and arcs could be dropped. (This would leave the peculiar situation of the final purchase arc for Period 4 being "stranded," without any nodes, and hence without connection to the rest of the model. However, that would be consistent with the intent of removing the joint restriction on final purchases and inventories.)

These observations show there is some latitude in defining conditions at the end of a time-phased model. At some point, however, the decision must be made to discontinue the connections to the future, and to control the ending part of a time-phased model by imposing bounds. We call such bounds ***post-period*** bounds.

If judgment falters about appropriate post-period bounds to impose, one approach is to begin with a "best guess," and solve the resulting model. Then the bounds that have been imposed on ending flows may be revised by examining the solution flows that apply to previous periods. (For example, intuition may suggest that these bounds should be close to the solution flows for periods close to the end.) The process of revising such bounds may be repeated until the user is satisfied that the relationship between the post-period bounds and flow values in earlier periods is appropriate.

The ambiguity about "connections leading to the future" is often encountered in the real world. The resolution of such ambiguity—just as the resolution of uncertainty in general—must always involve assumptions of varying degrees of arbitrariness. The intelligent approach is to keep track of assumptions that may prove significant, and to explore their consequences using techniques such as solving alternative scenarios.

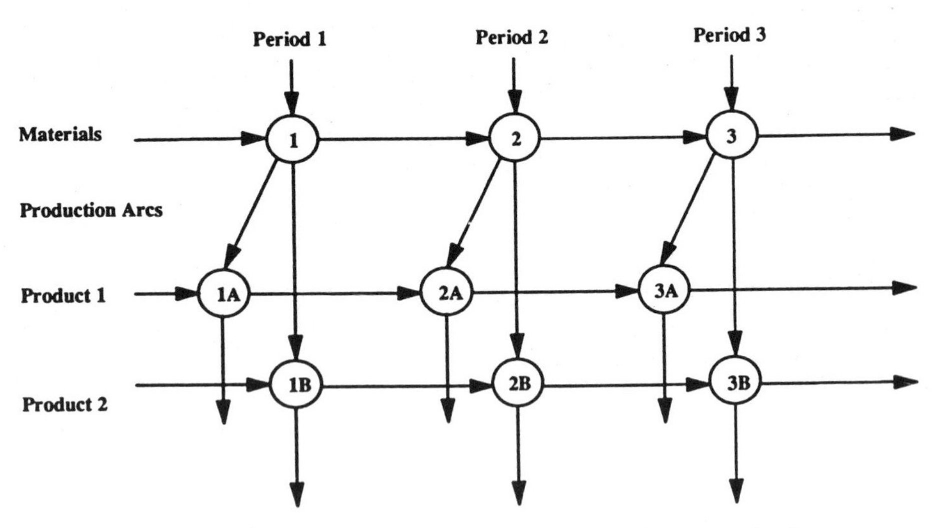

Figure 4.11 Two different products from the same raw materials (by different production processes or by the same process at different plants).

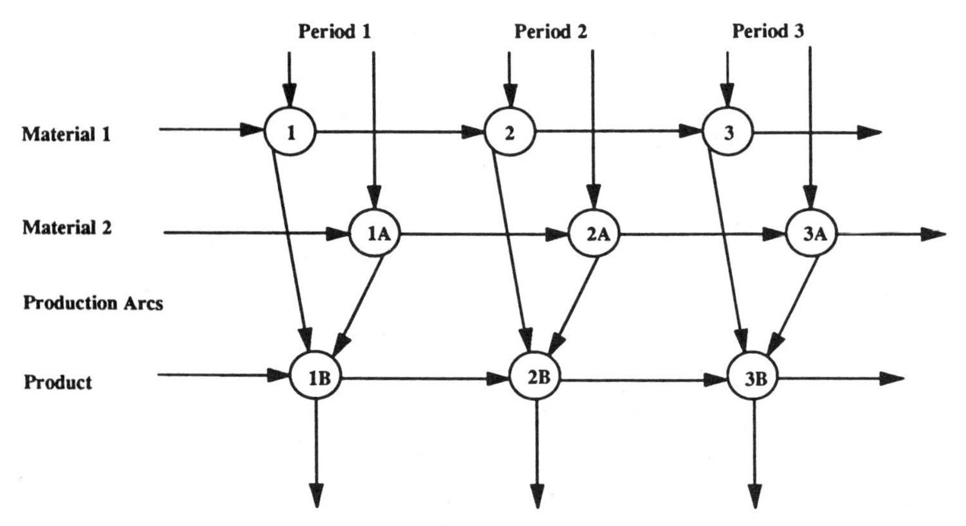

Figure 4.12 Two different raw materials produce the same product (by different processes or by the same process at different plants).

4.7 PARALLEL AND MULTI-PRODUCT PRODUCTION-INVENTORY SYSTEMS

The modeling techniques for single-product production-inventory systems can readily be extended to certain multi-product systems. The most direct extension is for parallel systems, where each component consists of a network exactly of the form previously described. However, the components coincide over some common portion, as where different products may be produced from the same raw materials inventory, or the same product may be produced from different raw materials. (The products or raw materials may or may not be physically differentiable; for example, they may be identical except for being purchased, manufactured or stored at different locations.)

Examples of common structural variations, where two different products are made from the same raw materials, and where two different materials are used to make the same product, are provided by Figures 4.11 and 4.12.

Once again, the units of flow must be consistent. In Figure 4.11, if the two products use different amounts of the raw materials, then all flow parameters would have to be in terms of raw material units (for example, tons). However, in Figure 4.12, scaling to final product units (for example, beams) would be necessary if different amounts of the two materials are used to make the product. Figures 4.11 and 4.12 may of course be expanded to include more than two products or raw materials, and to handle backorders, time lag effects, and so forth.

4.8 JOINT PURCHASE LIMITS

We will show one additional model variant. Suppose a company has two plants, each with its own production-inventory system, and must impose a joint limit on the quantity of materials purchased for the plants in each period. Each plant uses exactly the same materials, purchased from the same source, and the total volume of purchases must be restricted in each period due to a budget constraint. (For example, if each unit purchased in Period 1 cost $10, and the Period 1 purchasing budget is $10,000, then a limit of 1000 units is imposed on total Period 1 purchases for both plants combined.) A partial diagram for the situation is shown in Figure 4.13. A complete version of Figure 4.13 additionally includes production arcs leading downward from the materials sections of the two plants, and so forth.

There is a potential difficulty in this model in the more complex setting where the plants actually utilize different raw materials or purchase them at different prices. For instance, suppose Plant 1 purchases its materials at $10/unit and Plant 2 purchases its materials at $5/unit. Then a budget limit of $10,000 does not directly translate into a "single" joint limit on the total number of units purchased at the plants. One way to handle this difficulty is to ***scale*** all bounds, costs, and revenues involving Plant 2 to refer to 2-unit "bundles." Then, each bundle at Plant 2 costs $10, and in terms of these new units, the unit purchase cost at Plant 1 and Plant 2 is the same. Consequently, a joint bound of 1000 units can now acceptably be applied to control the combined volume of purchases, assuring that the budget limit of $10,000 will be satisfied.

Such scaling is a nuisance, but more than justified where it permits such things as budget constraints to be handled where they might otherwise be excluded. However, it is possible to compound conditions in ways that no amount of scaling will unscramble. Fortunately, generalized network models

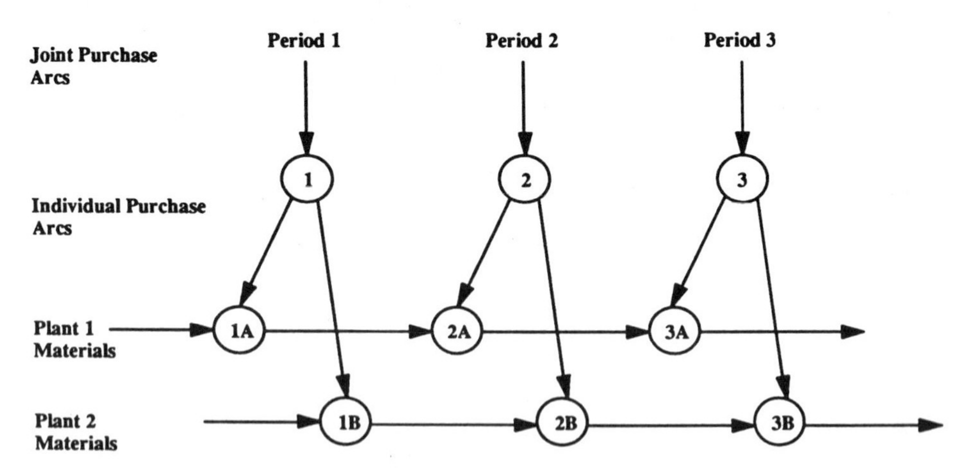

Figure 4.13 Joint purchase limitations: partial diagram (lower portion omitted).

and their extensions, which we examine in the next chapter, make it possible to handle such complications without having to resort to scaling.

4.9 OTHER TIME-PHASED MODELS

The kinds of constructs introduced in the inventory setting carry over to a variety of other time-phased applications that evoke the image of inventories only secondarily, or not at all. We now show how this comes about.

A frequently encountered type of time-phased problem arises in certain transportation and distribution settings, as, for example, in deciding how to route empty trucks or buses. The goal is to select the next locations for picking up new loads, over a span of several days or weeks, in a way that minimizes "empty miles," that is, the miles consumed in order to move empty carriers to their next loading locations. More generally, the objective may be expressed as that of minimizing the cost of re-positioning the fleet. Another related application involves the movement of containers (or railroad cars, etc.) by carriers able to transport quantities of such containers to alternative destinations, again with the goal of minimizing the cost of re-positioning to meet demands over time.

Consider, for example, a situation involving the re-positioning of empty barges among three ports over four time periods. In each period, due to knowledge of the number of barges scheduled to be loaded and unloaded, the net number of empty barges either required to be received from outside, or available to be sent elsewhere, is known for each port. Example information is summarized in Table 4.2. In addition, barges sent from Ports X and Z require 1 time period to reach Port Y, and barges from Port Y likewise require 1 period to reach either Port X or Port Z. Ports X and Z can send barges to each other only by the customary routes that go through Port Y (hence requiring two periods for journeys in either direction between these ports). Disregarding costs (for simplicity), a network diagram for this problem is shown in Figure 4.14.

Table 4.2 Net Barges Required (R) or Available (A)

	Periods			
Ports	1	2	3	4
X	1(A)	2(A)	6(R)	4(R)
Y	3(A)	4(R)	3(A)	4(R)
Z	8(A)	3(R)	5(A)	2(A)

It should be noted that the initial net availabilities at each port, the arcs into nodes X1, Y1, and Z1, will reflect the outcome of re-positioning decisions prior to Period 1 as well as the net effect of loading and unloading barges in this initial period. Correspondingly, the horizontal demand arcs extending from Period 4 nodes, nodes X4, Y4, and Z4, will carry the net availabilities to be transmitted to subsequent periods. In this respect the starting net availabilities of Period 1 and the ending net availabilities of Period 4 have a different interpretation than the net availabilities and requirements depicted by vertical arcs in Periods 2, 3, and 4.

A port that winds up with a surplus of empty barges at the end of some period need not ship them all to other ports, but can carry some of them over to the next period. This is handled by the horizontal arcs that connect a port in one period "to itself" in the next. These arcs have exactly the same function as the horizontal inventory carrying arcs of earlier examples. (They may, for example, have costs and bounds applicable to barges retained in port.) The diagonal arcs express the remaining shipping possibilities described in the example.

Finally, note that the net availabilities and requirements of the vertical arcs are expressed as exact supplies and demands (with lower bounds and upper bounds equal). This is done for the same reason that initial availabilities are expressed as exact supplies (rather than supplies with 0 lower bounds). As occurs for starting inventory in the standard inventory models, no available supply quantity can be "left unused," because if not shipped it must still be carried over at its originating port from one period to the next. Similarly, no amount beyond a required demand quantity can be extracted from the system, because any excess must likewise be carried forward into the future.

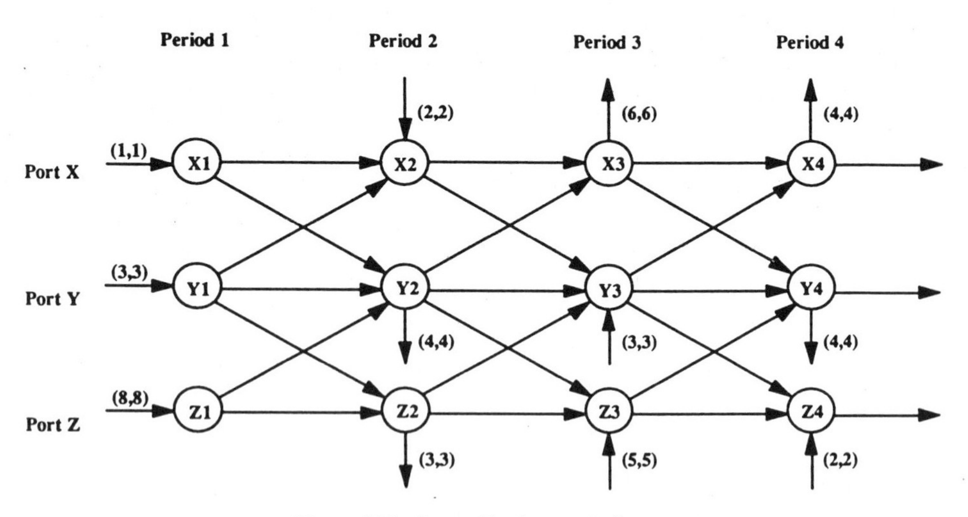

Figure 4.14 Re-positioning empty barges.

4.10 DYNAMIC MODELS AS LAYERED TIME SLICES

In the preceding examples the progression of time was depicted by a succession of single period "slices," arranged from left to right. It is quite common also to represent time's passage by arranging such slices from top to bottom, particularly in the case where each layer represents a more complex network structure. One may imagine drawing the network for each period on its own sheet of paper, and then stacking the sheets, with Period 1 on top, Period 2 next, and so forth. Vertically descending inventory arcs connect each period to its successor. (Backorder arcs are vertically ascending.) In this manner, time-phased models can readily be created for problems whose single-period snapshots are quite involved.

To illustrate the process, suppose a company has two plants, two warehouses, and three customers, and can make shipments in each time period as shown in Figure 4.15. Each such period diagram will have its own supplies and demands (amounts that can be produced at the plants and amounts ordered by the customers). In addition, if Warehouse 1 and 2 can carry forward inventory from one period to the next, this can be represented by downward arcs from the warehouse nodes in one period to the same nodes in the period following. Similarly, downward initial inventory arcs can enter nodes 1 and 2 in the first period diagram (with exact supplies), and leave these same nodes in the final period diagram.

A two-period example incorporating the preceding diagram is shown in Figure 4.16. Each period appears on its own plane, with warehouse inventory arcs passing vertically between and beyond the planes. Costs and bounds are again omitted for convenience.

Just as in the simple inventory models, where purchases and production may not be instantaneous processes, the more complex dynamic models may likewise entail "time lagged" interconnections. Such models follow the same principles illustrated in simple contexts.

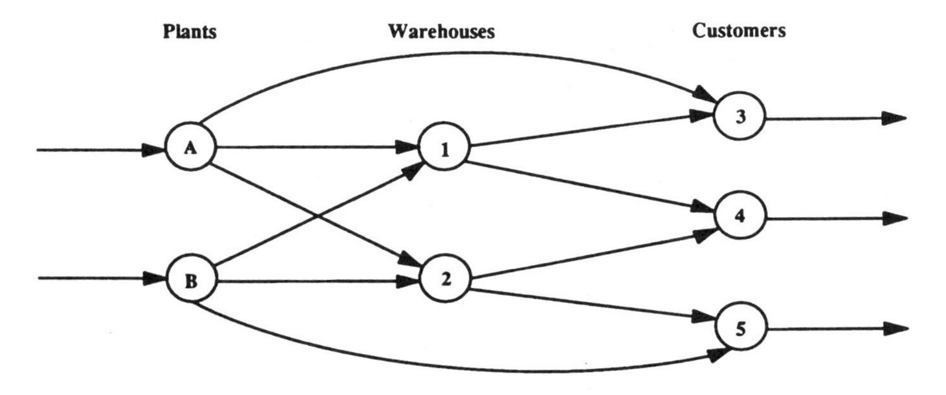

Figure 4.15 Distribution model.

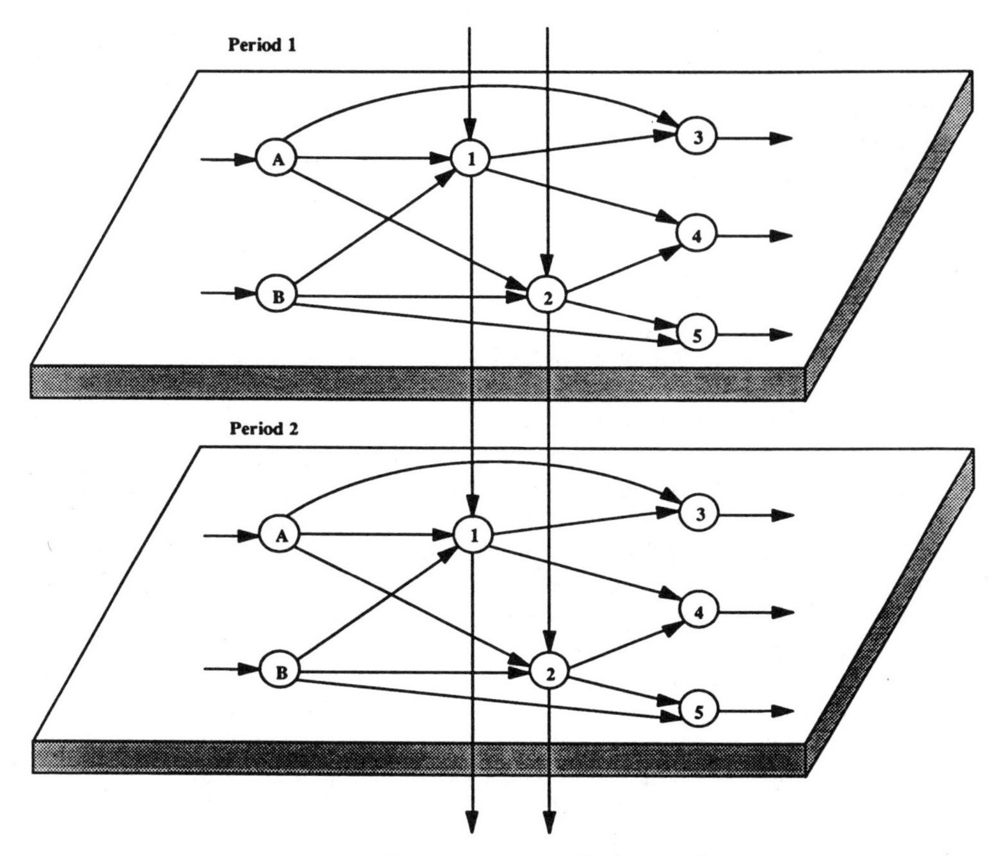

Figure 4.16 Two-period distribution model.

APPLICATION: AGRICO CHEMICAL COMPANY

The Agrico Chemical Company is a subsidiary of The Williams Companies, one of the nation's 350 largest industrial companies. Agrico is one of the largest manufacturers of chemical fertilizers in the United States. Led by progressive management, Agrico had grown from a relatively small firm to a company with sales exceeding $500 million in less than a decade.

In the mid-1970s, however, sharply escalating distribution costs, together with the highly seasonal demand pattern of chemical fertilizers, was creating a complex problem that standard techniques involving charts and cost figures

Reprinted by permission. An Integrated Production, Distribution and Inventory Planning System, F. Glover, G. Jones, D. Karney, D. Klingman, and J. Mote. *Interfaces*, Vol. 9, No. 5, pp. 21–35, 1979.

could not handle. The complexity of the situation affected the three major segments of the company's operation: production, distribution, and inventory. In response to their problems, management decided to develop an integrated computer-based planning system.

To aid in their problem solution, Agrico enlisted an outside team of management science consultants. The consultants brought on board the knowledge of advanced network methodology to deal with the multiple-time-period distribution problem. The Agrico distribution problem was indeed large in scale, as it consisted of 4 production plants, 78 distribution centers, and approximately 2000 customers.

An integral part of the production, distribution, and inventory (PDI) system is a dynamic trans-shipment network model. A simplified three-month "snapshot" of the overall model structure is provided in Figure 4.17. Costs, supplies, demands, and bounds are omitted for simplicity.

A state-of-the-art trans-shipment algorithm enabled the production, distribution, and inventory problem, with approximately 6000 equations and 35,000 decision variables, to be solved in less than 1 min of CPU time on an AMDAHL V-6 computer. The efficiency of the algorithm enabled the PDI system to be used extensively to evaluate the benefit/cost impact of alternative capital investments and has lead to multi-million dollar savings [Glover et al. 1979].

CASE: CITGO PETROLEUM CORPORATION, PART A

Citgo Petroleum is the nation's largest independent "downstream" marketer of petroleum products, selling over 5.2 billion gallons of product per year. This translated into a 1984 dollar sales figure of $4.5 billion, enough to rank Citgo 132nd on the Forbes 500 Largest Industrials. When Southland Corporation acquired Citgo from Cities Service in 1983, it did not purchase Cities Service's crude oil reserves. Since neither Southland nor Citgo owned any oil reserves, Citgo then had to make money on its downstream marketing activities alone. To accomplish this, Southland wanted to develop a system to determine on a weekly basis which marketing activities are likely to be profitable, since product gross margins are very dynamic in the petroleum industry. Also, due to increases in oil prices and interest rates, the cost of financing Citgo's working capital for refined product inventory, trade receivables, and accounts receivables had substantially increased since 1973. Thus a system was needed to help management control working capital as well as enhance the coordination of their supply, distribution, and marketing activities.

Citgo's marketing operations span the eastern two-thirds of the United States. The corporation has five distribution center terminals and 27 product storage terminals which are connected to a distribution network of pipelines, tankers, and barges (Figure 4.18). In addition, Citgo sells product from nine leased terminals and over 350 exchange terminals through agreements with

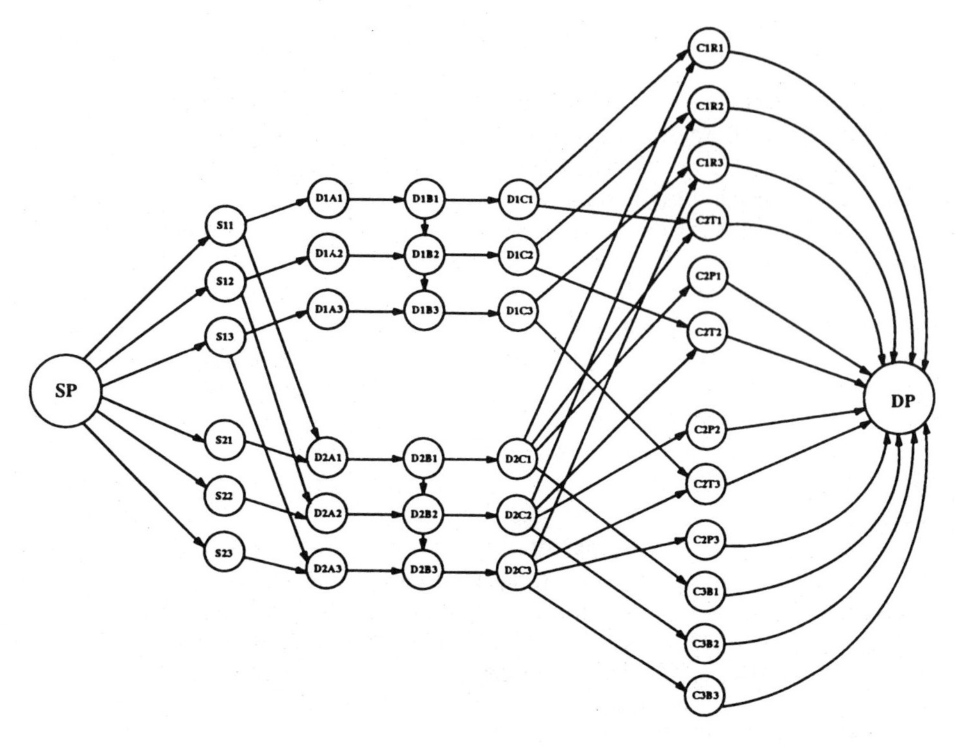

ARC LEGEND

SP to Sit	Supply Policy Arc for Supply Site i in Period t
Sit to DjAt	Transportation Arc from Supply Site i to Distribution Center j in Period t
DjAt to DjBt	Unloading Dock Arc for Distribution Center j in Period t
DjBt to DjCt	Loading Dock Arc for Distribution Center j in Period t
DjBt to DjBt + 1	Inventory Arc for Distribution Center j during Period t
DjCt to Ckmt	Transportation Arc from Distribution Center j to Client k by Mode m in Period t
Ckmt to DP	Demand Policy Arc for Client k by Mode m in Period t

NODE LEGEND

SP	Supply Policy
Sit	Supply Site i in Period t
DjAt, DjBt, DjCt	Distribution Center j in Period t A. unloading dock B. inventory C. loading dock
Ckmt	Client k by Mode m in Period t m = { R Rail, T Truck, P Pipeline, B Barge }
DP	Demand Policy

Figure 4.17 Agrico PDI network model. Reprinted by permission. An Integrated Production, Distribution and Inventory Planning System, F. Glover, G. Jones, D. Karney, D. Klingman, and J. Mote. *Interfaces*, Vol. 9, No. 5, pp. 21–35, 1979.

Reprinted by permission. An Optimization Based Integrated Short-Term Refined Petroleum Product Planning System, D. Klingman, N. Phillips, D. Steiger, R. Wirth, R. Padman, and R. Krishnan, *The Institute of Management Sciences*, Vol. 33, No. 7, 1987, pp. 813–830.

other petroleum marketers. The addition of Southland's 7-Eleven stores increased this marketing and distribution network to all of the 48 contiguous United States.

Products marketed by Citgo include four grades of motor fuel (regular, unleaded, super unleaded, and diesel) as well as No. 2 fuel oil, turbine fuel, naphtha, and several blended motor fuels. Since the motor fuels and No. 2 fuel oil make up 85% of total sales volume, only these products were considered in the system. Citgo acquires its supply of product through the following options: refining at the Citgo refinery in Lake Charles, Louisiana, spot purchases on one

Figure 4.18 Citgo product distribution network.

of five major spot markets, and product exchanges and trades with other industry refiners. This case is a simplified version of the real problem [Klingman et al. 1986, 1987] and concerns only one product. Exchanges and trades will be explained in Part B of the case. Part C adds other modifications which make the model more realistic.

For the purposes of this case, you may assume the following: Citgo's forecasted Gulf spot market price is 80 cents per gallon for each of the next 5 weeks. Citgo traders have determined that the company can purchase up to 20 gallons each week at this spot price. The Lake Charles Refinery can process up to 200 gallons per week. Company policy dictates that at least 50 gallons must be refined each week. The product from the Lake Charles refinery and the product purchased on the Gulf spot market are sent to the Lake Charles Distribution Center. From the Lake Charles Distribution Center, the company has four options:

1. It may send product to the Explorer pipeline.
2. It may send product to the Colonial pipeline.
3. It may send product to a Gulf port, where it is loaded onto tankers.
4. It may sell the product on the Gulf spot market.

The Gulf port, at which the tankers are loaded, can handle up to 100 gallons per week. Citgo has preliminarily scheduled 10 and 20 gallons per week for the Explorer and Colonial pipelines for each of the next 5 weeks, respectively. These pipelines have a maximum capacity of 50 gallons/week for Explorer and 60 gallons/week for Colonial.

From the pipelines and tankers the product is shipped to terminals. The Colonial pipeline serves the following cities: Meridian, Birmingham, Atlanta, Spartanburg, Charlotte, and Richmond. The Explorer pipeline serves Dallas and St. Louis, and the Lake Charles tanker serves New York. The cost of transportation is currently 0.5 cents per gallon per 100 miles. This cost is the same for all modes of transportation. There is also a handling cost associated with each terminal. Each terminal has three types of demand: retail demand, wholesale demand, and exchange partner demand. Only the Lake Charles Distribution Center may put product into the pipelines; terminals may only draw product from the pipelines.

Demand forecasts are available for both retail and exchange partner lifting; however, wholesale demand is forecasted as a price/volume demand function. A typical terminal's price/volume function and total revenue function are shown in Figures 4.19 and 4.20. The segment of the price/volume function corresponding to "reasonable" industry pricing alternatives is approximately linear, as indicated by the dashed line in Figure 4.19. Since this linear approximation has negative slope, the resulting total revenue function is concave as shown in Figure 4.20. We use a piecewise-linear approximation to the total revenue function as indicated by the dashed line in Figure 4.20. Five discrete points on

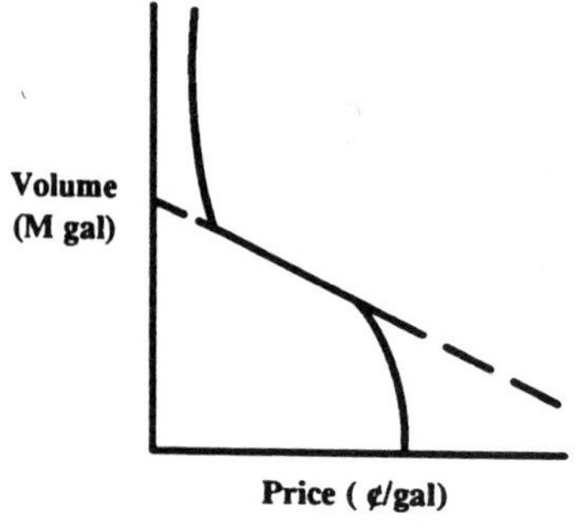

Figure 4.19 Price/volume function.

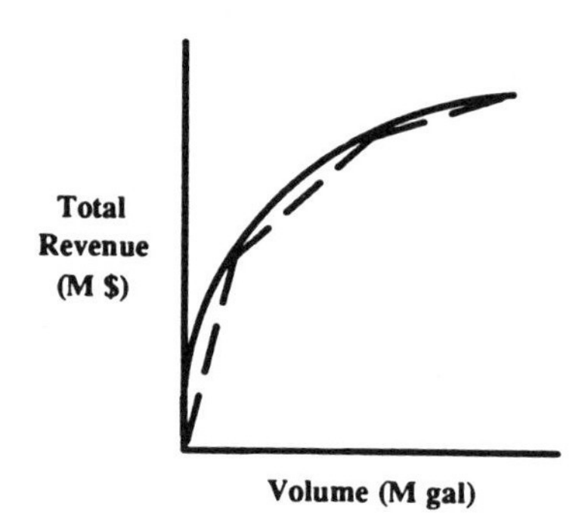

Figure 4.20 Total revenue function.

the price/volume function are given in the tables for the St. Louis terminal. (The prices apply to the total volume sold, not just the incremental amount. Thus to find the slope of the linear segments of the total revenue function, you must calculate change in revenue for each demand level. The demands represent the volume where the slope changes (i.e., the "break points" in the piecewise-linear approximation to the total revenue function). The wholesale demand for each of the other terminals is expressed in the tables as a single point estimate.

Southland has dictated that Citgo must meet *all* retail demand at a specified price known as OPIS Low. Southland allows Citgo to determine how much of the wholesale demand should be met.

Timing

Time is an important consideration for Citgo. Transportation of product takes weeks in some cases. This is important due to the time value of money. Since Citgo does not receive money for its product immediately after production, product in transit is, in effect, inventory that must be financed. Citgo estimates that the transportation time value of money is equal to 0.5 cents per gallon per 100 miles.

Citgo's decisions concerning how much product to refine and purchase are made on a weekly basis. It also forecasts demand in terms of gallons per week; however, product that leaves the Lake Charles Distribution Center through the Colonial pipeline cannot reach all of the terminals on the pipeline within 1 week. Product that leaves Lake Charles in week 1 can reach Birmingham by the end of week 1, Charlotte by the end of week 2, and Richmond by the end of week 3. A schedule for the Colonial pipeline and other necessary information is given in the tables that follow.

Holding cost is 1.5 cents per week. This figure includes the time value of money. During the week before week 1, Lake Charles sent 45 gallons to the Colonial pipeline. The Meridian and Birmingham terminals drew a total of 15 of these gallons. Two weeks before week 1, Lake Charles sent 50 gallons to the Colonial pipeline. After the Charlotte terminal made its draw (one week before week 1), there remained 10 gallons from the 50-gallon shipment.

Colonial Pipeline Schedule

Week	Destination
1	Lake Charles Distribution Center Meridian Birmingham
2	Atlanta Spartanburg Charlotte
3	Richmond

Terminal Inventories

Terminal	Initial Storage Inventory (gallons)
Lake Charles	10
Meridian	1
Birmingham	3
Atlanta	5
Spartanburg	1
Charlotte	3
Richmond	4
Dallas	8
St. Louis	7
New York	10

Distance

	Miles
Lake Charles–Meridian	200
Meridian–Birmingham	100
Birmingham–Atlanta	100
Atlanta–Spartanburg	150
Spartanburg–Charlotte	50
Charlotte–Richmond	150
Lake Charles–Dallas	250
Dallas–St. Louis	600
Lake Charles–New York	1000

Forecasted Price/Demand for Colonial Pipeline: Week 1

Terminal	Retail Demand (gallons)	Wholesale Demand (gallons)	Wholesale Price (cents)
Meridian	1	1	84.5
Birmingham	7	5	85.5
Atlanta	11	9	86.8
Spartanburg	1	1	87.1
Charlotte	5	5	87.6
Richmond	6	4	88.0

Forecasted Price/Demand for Explorer Pipeline: Week 1

Terminal	Retail Demand (gallons)	Wholesale Demand (gallons)	Wholesale Price (cents)
Dallas	12	15	86.5
St. Louis	8	8	87.1
		9	87.0
		10	86.9
		11	86.8
		12	86.7

Forecasted Price/Demand for Tanker: Week 1

Terminal	Retail Demand (gallons)	Wholesale Demand (gallons)	Wholesale Price (cents)
New York	25	20	92.0

Terminal Handling Costs

Terminal	Handling Cost (cents)
Meridian	0.5
Birmingham	0.5
Atlanta	1.0
Spartanburg	0.5
Charlotte	0.5
Richmond	1.0
Dallas	1.0
St. Louis	1.0
New York	1.5

Terminal Inventories: Three-Period Model

Terminal	Initial Storage Inventory (gallons)	Minimum Inventory (gallons)	Maximum Inventory (gallons)
Lake Charles	10	2	15
Meridian	1	1	2
Birmingham	3	3	6
Atlanta	5	4	9
Spartanburg	1	1	2
Charlotte	3	2	5
Richmond	4	2	5

Forecasted Retail and Wholesale Demand for Colonial Pipeline: Three-Period Model

Terminal	Retail	Wholesale	
	Demand (gallons)	Demand (gallons)	Price (cents)
	Week 1		
Meridian	1	1	84.5
Birmingham	7	5	85.5
Atlanta	11	9	86.8
Spartanburg	1	1	87.1
Charlotte	5	5	87.6
Richmond	6	4	88.0
	Week 2		
Meridian	1	1	84.5
Birmingham	6	6	85.5
Atlanta	10	8	86.8
Spartanburg	1	1	87.1
Charlotte	5	4	87.6
Richmond	5	5	88.0
	Weeks 3—5		
Meridian	1	1	84.5
Birmingham	7	6	85.5
Atlanta	10	9	86.8
Spartanburg	1	1	87.1
Charlotte	5	5	87.6
Richmond	4	4	88.0

Case Questions

1. Using the data provided, but assuming a one-week transit time to all terminals, develop a one-period (week) network model for Citgo Petroleum to decide where to distribute its product. For the one-period model, you should ignore holding cost and the time value of money, and let all supply, demands, and bounds correspond to week 1 data. (Hint: Create a node for each transportation mode. Connect each terminal node to the transportation node that serves it.)
2. Citgo uses an internal transfer price between refining and marketing equal to the Gulf Coast spot price. Why might you use use a slightly different price within the model? What price would you recommend for the cost associated with the transfer of product from the Lake Charles Refinery to the Lake Charles Distribution Center within the model? Why?
3. The Gulf Coast spot price represents a buy/sell price. Why might you use a slightly different price for buying as opposed to selling product on the Gulf spot market within the model? What price would you recommend?
4. As mentioned earlier, company policy dictates a minimum amount of refining, and the company has preliminarily scheduled certain amounts for the pipelines. How would you alter the network model you developed to determine if these levels are reasonable?
5. Develop a three-period network model for the Colonial pipeline. The first two periods should be weeks 1 and 2, and the third period should be weeks 3–5. The last time period is long enough to allow product to be shipped anywhere in the distribution network. The first two shorter periods allow for more detailed planning to be done in the near-term where forecasted data are more accurate, and decisions with shorter lead times are more pressing. (Hint: Create three Colonial pipeline nodes for each time period, each representing the segment of the pipeline that product can traverse in a week. Refer to the Colonial Pipeline Schedule table to determine which terminal nodes to connect to which pipeline node. The pipeline nodes will then be connected in a manner to reflect the time lag in product distribution. For example, product initially in the pipeline at the Charlotte terminal in week 1 will not reach the pipeline at the Richmond terminal until the next time period.)

CASE: CITGO PETROLEUM CORPORATION, PART B

Delivery and Receipt Agreements

Besides receiving product from the spot market and its own refinery, Citgo has one other source of product, other oil companies. Oil companies will often buy or sell product from each other. An agreement to buy product is known as a receipt agreement, and an agreement to sell product is known as a delivery agreement. In these agreements the companies must agree upon the following factors:

1. The location from which the product will be sent or received.
2. The number of gallons to be sold.
3. The price per gallon.
4. A time by which the delivery of product must be completed.

It is customary in these agreements for the company that is purchasing the product to pay for the entire amount "up front". This can bring the time value of money into effect: if the company that is selling the product delays its delivery, it has effectively raised the cost to the buyer (profit to the seller). Citgo estimates the time value of money to be $0.015 per gallon per week.

EXCHANGE AGREEMENTS

Suppose Citgo had a surplus of product at Richmond, and a shortage in Waco and Austin. Further suppose that Shell Oil had a surplus of the same product in Waco and Austin, and a shortage in Richmond. If Citgo and Shell tried to correct their imbalances each using public distribution networks, the correction could take several weeks, and would incur transportation costs. If, instead, the companies agreed to trade product, Shell giving Citgo product in Waco and Austin in exchange for Citgo's product in Richmond, the correction would be virtually immediate, and both companies would incur significantly lower transportation costs. Since transactions of this type benefit both parties, they are used extensively by petroleum companies, and are known as exchange agreements. In an exchange agreement, five things must be known:

1. The exchange partner's forecasted demand at each Citgo terminal covered in the agreement.
2. Citgo's forecasted retail demand at each exchange partner's terminal covered in the agreement.
3. A base location ("bulk" payback point).
4. A "location differential" charge and a terminal handling charge which balances the difference in the transportation costs incurred by each partner to deliver product from the base location (e.g., if, in the example above, the companies agree that the location differential should be $0.04 per gallon at Waco, and the terminal handling cost at Waco is $0.005 per gallon, then it will cost Citgo $0.045 for every gallon they lift at Waco).
5. A maximum amount by which the contract can get out of balance. (Imbalances occur when one partner takes more product than the other partner in a given week. One way to correct these imbalances is to pay back the amount owed in bulk to the distribution network at the base location. If the partner does not wish to correct the imbalance immediately, it may carry the imbalance into the next period.)

Case Questions

1. Citgo has entered into a delivery agreement in which they have sold 7 gallons of product at $0.88 per gallon to Mobil at the city of Charlotte. Citgo will receive Mobil's payment for all 7 gallons at the beginning of week 1. The contract specifies that all 7 gallons must have been delivered by the end of week 2. The contract also states that at least 2 gallons must be sent in week 1, at least 1 gallon must be sent in week 2, and at most 4 gallons can be sent in week 2. Show how this agreement could be included in the network model developed in Question 5 of Part A. Use the transportation costs and transportation time value of money from Part A. Ignore the terminal handling costs. (Hint: Create a node for the delivery agreement. Connect this node to the pipeline network at the same spot where the Charlotte terminal node is connected. Ignore the sales price since the total revenue for this sale is fixed, that is, it is a constant.)
2. Citgo has entered into a receipt agreement in which they have purchased 9 gallons of product at the city of Richmond for $0.90 per gallon. Under the terms of the contract, Citgo must take the product in the Colonial pipeline, and must pay for all 9 gallons at the beginning of week 1. The contract requires that all 9 gallons must have been received by the end of week 3. The contract also specifies that the minimum receipt in any week is 1 gallon, and the maximum in week 1 is 2 gallons. Show how this agreement could be included in the network model developed in Question 5 of Part A.
3. Citgo has entered into an exchange agreement with Shell Oil for product at Shell's Macon and Albany terminals in return for product at Citgo's Meridian and Birmingham terminals. Forecasted exchange demand is given in the table below.

Forecasted Exchange Demand (gallons)

	Week	Meridian	Birmingham	Macon	Albany
Citgo	1			3	4
Shell	1	1	6		
Citgo	2			2	4
Shell	2	1	6		
Citgo	3			3	5
Shell	3	1	6		

The demand shown for Macon and Albany is Citgo's forecasted retail demand. Citgo may also wish to satisfy wholesale demand at these terminals. Citgo has decided to let the model determine how much wholesale demand should be met at these terminals, at a price of \$0.88 per gallon. The companies agree that they will not allow the contract to get out of balance by more than 3 gallons. If it does, a bulk payback will be made to the pipeline which serves the base location, the city of Atlanta. The companies agree that Citgo should pay a location differential of \$0.01 per gallon at Macon, and \$0.02 per gallon at Albany. Terminal handling costs are \$0.005 per gallon at each terminal. Show how this agreement could be included in the model developed in Question 5 of Part A. (Hint: You should construct an exchange agreement node for each time period, which is connected to the distribution network in the same spot as the base location and whose supply is equal to the total Shell forecasted demand at Citgo's terminals for that time period. This supply can be used to satisfy Citgo's forecasted demand at the exchange terminal nodes of Macon and Albany.)

4. In the above models, the time value of money is treated as a constant. What are the advantages and disadvantages of this approach?

CASE: CITGO PETROLEUM CORPORATION, PART C

Use the information about Citgo in Parts A and B to answer the following questions.

Case Questions

1. The real Citgo supply/distribution/marketing (SDM) model has several distribution centers at which product may be transferred from one transportation carrier (tanker, pipeline, etc.) to another. Transportation costs on the arcs represent only the costs from the nearest upstream distribution center, rather than from the product source. Explain why this is necessary.
2. Given an SDM model similar to that of Citgo, present a partial model which, if shortages occur, will allow meeting retail demand in later time periods to be less important than meeting retail demand in early time periods. (Assume retail demand need not be met, and there are no backorders.) Explain computation of costs, etc.
3. Explain the difference in the way time is handled in the Agrico PDI model versus the Citgo SDM model.
4. Given an SDM model similar to that of Citgo, present a partial model which allows exchange contracts
   a. To make fixed bulk payments at locations in the distribution network other than the base location.
   b. To receive fixed bulk payments at locations in the distribution network other than the base location.

   Explain computation of costs, bounds, etc.

5. Explain how the Citgo SDM model may be used to represent unlike product trades.
6. Present a partial model which is equivalent to the Citgo SDM model but where exchange terminal nodes (belonging to other oil companies and part of an exchange agreement) are eliminated to reduce problem size.

EXERCISES

4.1 A metal casting company can purchase its coal needs over three time periods from either of two suppliers. Because prices of the suppliers vary from period to period, the company may find it advantageous to buy more coal than it needs in a given period, and store the unused amount for later use. The cost of coal from the two suppliers, and maximum amount that can be purchased from each, are shown in the following table.

	Cost of Coal/ton ($)			Maximum No. of Tons Available		
	Period 1	Period 2	Period 3	Period 1	Period 2	Period 3
Supplier 1	900	1300	1000	70	80	60
Supplier 2	1200	1100	1200	90	70	80

Finally, assume the company requires 130 tons in Period 1, 120 tons in Period 2, and 150 tons in Period 3, with a required ending inventory of 5 tons (exactly).

a. Provide a network diagram for the problem as stated (for the objective of minimizing cost).

b. How does the diagram change if the company has 20 tons of coal on hand at the beginning of Period 1, the maximum amount of coal that can be stored in any period is 15 tons, and the cost of storing coal from one period to the next is $35/ton? (Attaching a storage cost to ending inventory makes no difference to the problem solution. Why?)

4.2 The Hunky Dory Boat Company buys boats from an independent contractor, customizes them in an operation that requires only a couple of days, then either sells them directly or puts them into inventory. Due to the speed of customizing and other special conditions of operation, purchases, assembly, and sales can all be assumed to take place at the start of each period. Hunky Dory decides to develop a 2-period model.

Beginning inventory of customized boats is 5 and ending inventory should be 5 also (without cost or profit attached). Additional relevant data appear in the table below:

	Period 1	Period 2
Sales upper limit	35	40
Sales price ($)	9000	9400
Purchase upper limit	50	20
Purchase price ($)	3500	4200
Customizing cost ($)	800	800

Holding customized boats in inventory for a period actually produces a revenue, because the inventorying consists of docking the boats, and there are always people willing to rent the boats (so far, without hurting sales prices), yielding a net gain of $60/boat after cleanup and other expense. Backorder cost for one period is $40/boat.

a. Prepare a cost minimizing network model with customizing nodes C1 and C2, inventory nodes I1 and I2, and sales nodes S1 and S2. Do not create arcs that require "composite" costs or profits, but make the network so that each cost or profit occurs on its own arc.

b. Identify which nodes could be removed, causing appropriate pairs of consecutive arcs to be replaced by single arcs (combining costs and bounds of the consecutive arcs).

c. Show the change that occurs if the company is limited to purchasing at most 60 boats in the two periods combined.

4.3 The following inventory model diagram will not always successfully model the situation where backorders can extend for either one or two periods. Create specific backorder costs and bounds for the "back arcs" that can be used to illustrate ways in which this construction might fail. (Indicate all the different ways of failing you can think of.)

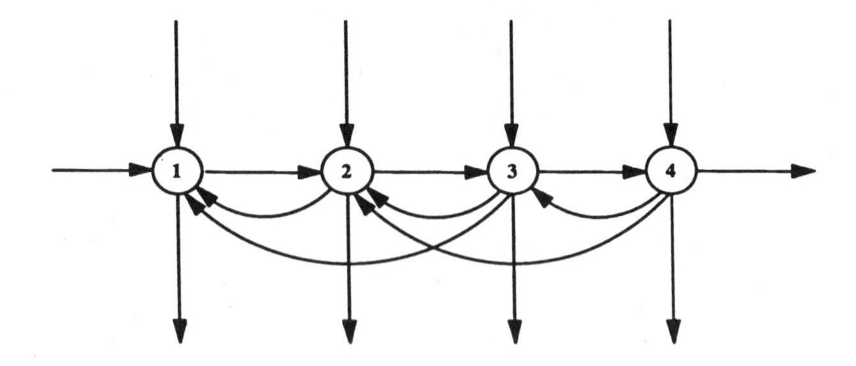

4.4 Consider the example problem given in the text for shipping empty barges between 3 ports in 4 time periods (illustrated by Figure 4.14, "Repositioning empty barges"). Extend this example by incorporating the following conditions. (Assume a cost minimizing objective.)

a. The cost of sending empty barges between Ports X and Y is $50/barge (each way), and between Ports Y and Z is $65/barge. The costs of retaining barges in Ports X, Y, and Z are respectively $4, $3, and $5 per period.

b. An alternative channel exists making it possible to send barges one way from Port Z to Port X at a cost of $90/barge. The trip on this channel requires 2 periods to complete, and at most two barges can initiate a trip in any given period. Assume 1 barge was sent on this channel in each of the two periods that preceded Period 1, and that the net availabilities of Periods 1 and 2 have not been adjusted to account for this. Further assume that the company decides to handle its "connections to the future" by stipulating that exactly 1 barge should be sent (i.e., initiate a trip) on this channel in each of Periods 3 and 4, and that availabilities in these periods likewise have not been adjusted to reflect this decision. (Incorporate this condition two ways, one where connections to periods outside the model's time frame are shown as explicit arcs, and the other where such connections are absorbed by adjusting net availabilities.)

4.5 The Pick Axe Mining Company mines and sells a metal used in manufacturing special strength alloys. Because of different weather conditions (some of the mining is surface mining) and variations in prices of fuel to run its machines, the amount and cost of ore that can be extracted vary from period to period. The company believes it has enough information, however, to predict costs and extraction limits fairly accurately for five periods into the future, and is also reasonably confident of its demand predictions over that span of time. The estimates of available ore, cost of mining, demand for the ore, and sales prices are as follows. The company's objective is to minimize cost.

	Periods				
	1	2	3	4	5
Units of ore available[a]	300	230	290	410	420
Cost/unit to mine ($)	40	45	55	40	45
Units expected demand[b]	250	310	230	390	460
Selling price/unit ($)	80	75	85	80	75

[a]Any amount up to this quantity can be mined.
[b]Any amount up to this quantity can be sold.

Assume that up to 40 units of ore can be stored from one period to the next, at a storage cost of $5/unit. Also assume a beginning inventory of 20 units of ore, and a required ending inventory of 20 units (exactly).

a. Provide a network diagram for the problem as stated. Show how the network would change for (b)–(e) below.

b. Assume up to 20 additional units can be mined in each of Periods 2 and 4, at a cost of $55/unit and $50/unit, respectively.

c. Assume further that up to 25 additional units of demand could be made available in Period 4 if the unit selling price for these additional units was dropped $5 below the price of the first 390 units.

d. Suppose the company is able, in addition to (b) and (c), to backorder demand for one period; that is, to meet part of the demand in an earlier period out of ore mined in the immediately following period. Assume a backorder cost of $5/unit.

e. In addition to (d), assume that ore mined in Period 5 can go also to meet demand of two periods earlier, at a backorder cost of $7/unit.

4.6 Suppose in Exercise 4.5 that the Pick Axe Mining Company must first process the ore it extracts for one time period before this ore is available to meet demand. Upon mining, unprocessed ore can either be sent immediately to be processed (so that it will be able to meet demand in the next period), or it can be held in storage until a later period, to be processed at that time. All storage information of Exercise 4.5 refers to inventories of processed ore. Information concerning unprocessed ore is as follows.

	Periods				
Unprocessed Ore	1	2	3	4	5
Unit cost to store ($)	4	4	5	5	5
Limit on amount stored	40	40	40	40	40
Unit cost to process ($)	12	17	11	20	18
Limit on amount processed	320	320	320	320	320

Assume an initial inventory of 150 units of processed ore and 20 units of unprocessed ore. Assume also an ending inventory requirement of exactly 20 units of unprocessed ore. Finally, assume a combined requirement of exactly 300 units of processed ore from Period 5's final processing decision and processed ore inventory (which together provide processed ore available to meet demand in Period 6). Prepare a netform for each of the stipulations (a)–(e) of Exercise 4.5.

4.7 If the problems of Exercises 4.5 and 4.6 could be solved very rapidly (e.g., in a manner of seconds while sitting at a computer terminal), and if the costs and bounds could be changed conveniently, what steps might an analyst employ to make better decisions under conditions of uncertainty?

4.8 Suppose a firm having two plants and three demand points is planning a production schedule for three periods. Assume over Periods 1, 2 and 3 that Plant A has available supplies of 4, 7 and 7, respectively and that Plant B has available supplies of 8, 10 and 11, respectively. Assume over Periods 1, 2, and 3 that Demand Point X has demands 3, 10 and 10, respectively; similarly, Demand Point Y has demands 4, 2, and 6, respectively; and Demand Point Z has demands 2, 6, and 4, respectively. Let c_{ij} be the cost of shipping a unit from Plant i to Demand Point j in any period. Assume supplies and demands are fixed.

a. Construct a transportation model to minimize total shipping costs.

b. Is there a feasible solution to this numerical example? Justify your conclusion. Give a general rule for discerning whether such a problem has a feasible solution.

c. Explain how to modify the formulation when the total supply exceeds total demand.

d. Suppose each unit of inventory held at the end of a period incurs the cost h. Revise your model. Will adding this cost change the optimal solution in part (a)? In part (c)?

e. Construct a trans-shipment model to minimize total shipping plus inventory costs. Is there an advantage to using a trans-shipment model rather than a pure transportation formulation for the model?

4.9 In the next four time periods, a metal fabrication company must meet demand commitments for 12, 9, 18, and 22 units of its product, respectively. The company's production capacity is 10 units per period, assuming that its work force is employed for one shift a day. The associated direct cost is $100 per unit. By using overtime, the company also can produce 6 more units each time period at a direct unit cost of $150. Because the demand commitments fluctuate considerably, the company expects that it may have to build inventory in some periods to meet demand in subsequent periods. Each unit of inventory at the end of each period incurs a holding charge of $10. Formulate a network model to minimize total costs.

4.10 Radiotopes has in stock four uranium power sources. The quality of these items deteriorates with time as measured in monthly periods. Currently two are new and two are 1 month old. The company has contracted to sell the stock as follows: they must deliver one item 1 month from now, one item 2 months from now, one item 4 months from now, and one item 6 months from now. The revenue they receive for each item is a function of its age at the time of delivery; this function is denoted by $R(A)$, A being the relevant age. Formulate a netform model that can enable Radiotopes to determine which item they should supply at each delivery date so that they can maximize total revenue.

4.11 Given a PDI model similar to that of Agrico, present a partial model (cost minimizing) which will allow demand to be filled by backorders when necessary. Assume that there is no time limit on backorders (i.e., current demand can be satisfied in the current or any subsequent period); however, there is a cost per unit, C_B, for each period that delivery is delayed. Show costs and three time periods.

4.12 Present a partial PDI model which will allow demand to be satisfied by backorders kept for at most two periods (i.e., current demand can be met only in the current period or the subsequent two periods). Again, the cost per unit per period is C_B. Show costs and three time periods.

4.13 Present a partial PDI model (with no backorders) in which customer demands are prioritized when shortages occur. Show one time period and three customers where Customer A's demand must be met before Customer B's demand which must be met before Customer C's demand.

4.14 For the PDI model, management has decided that a buffer stock (minimum inventory) of 100 units is desirable. Also, current storage capacity is 800 units. Storage costs are $10 per unit for the first 500 units and $12 per unit thereafter. The logistics manager (who has some network modeling experience) has proposed the following cost minimizing construct to model the inventory policy:

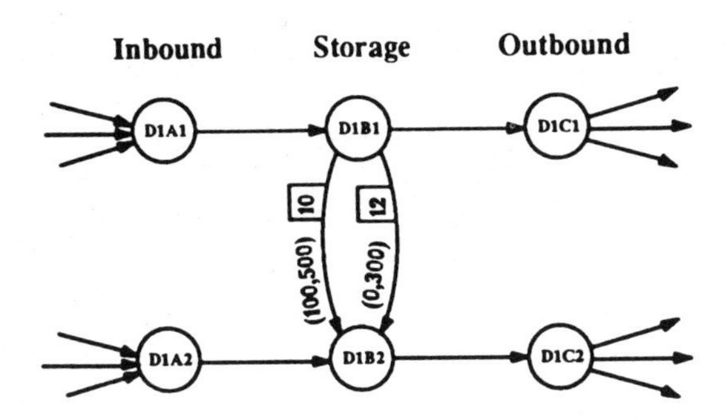

You realize that these rigid inventory requirements could possibly force the PDI model solution to be infeasible. Propose an alternative inventory construct which will capture management's policies but will also allow them to be violated if necessary to get a feasible solution (i.e., inventory levels can drop below 100 or rise above 800 but your model will impose severe penalties for these occurrences).

4.15 A manufacturer of small electronic calculators is working on setting up his production plans for the next 6 months. One product is particularly puzzling to him. The orders on hand for the coming season are:

Month	Orders
Jan	100
Feb	150
March	200
April	100
May	200
June	150

The product will be discontinued after satisfying June demand. Hence there is no need to keep any inventory after June. The production cost, using regular manpower, is $10/unit. Producing the calculator on overtime costs an additional $2/unit. Inventory carrying costs are $0.50/unit per month. The regular shift production is limited to 100 units/month and overtime production is limited to an additional 75 units/month. Draw a netform model which will yield the minimum cost production schedule.

***4.16** A caterer, who is in charge of serving meals for the next 5 days, is faced with the problem of deciding on the daily supply of fresh napkins. His requirements for the 5 days are known to be 100, 300, 200, 150, and 100, respectively. His alternatives are:

a. Buy new napkins at the price of 50 cents apiece.

b. Send soiled napkins to the laundry where they can receive either 48-h service at the cost of 5 cents a piece, or 24-h service at the cost of 10 cents apiece.

Either soiled or clean napkins may be kept in inventory from day to day. Formulate a network model to minimize cost. (Hint: Define a clean napkin node and a dirty napkin node for each day. The napkin requirements then become bounds on the arc from the clean to the dirty node each day.)

***4.17** At Dallas Airport, Transgulf Airlines needs varying numbers of ground personnel according to the time of day. The minimum requirements are as follows.

Time of Day	Personnel Needed	Number of Personnel Starting Beginning of Period
12:00–4:00 a.m.	75	x_1
4:00–8:00 a.m.	250	x_2
8:00–12:00 p.m.	300	x_3
12:00–4:00 p.m.	50	x_4
4:00–8:00 p.m.	350	x_5
8:00–12:00 a.m.	150	x_6

Assume that personnel can report for work at the start of any of the above six time periods and will remain on the job for eight consecutive hours. The linear programming model to minimize the total number of personnel hired is shown below.

$$\text{Minimize} \quad \sum_{i=1}^{6} x_i$$

$$\begin{array}{llllllll}
\text{Subject to} & x_1 & & & & & + x_6 & \geqslant 75 \\
& x_1 + x_2 & & & & & & \geqslant 250 \\
& & x_2 + x_3 & & & & & \geqslant 300 \\
& & & x_3 + x_4 & & & & \geqslant 50 \\
& & & & x_4 + x_5 & & & \geqslant 350 \\
& & & & & x_5 & + x_6 & \geqslant 150
\end{array}$$

$$x_1, x_2, \ldots, x_6 \geqslant 0$$

a. Convert the algebraic statement to a balanced form and draw a netform of the problem (Hint: Include surplus variables, multiply some rows by -1 to get the proper format, and then create a redundant constraint.)

b. Can you think of a different, more intuitive network model which you could draw without writing the LP and transforming it?

c. If the number of time periods were odd, could this problem still be modeled as a network?

***4.18** Around 435 BC, Sparta decided to draft reserve troops to supplement its regular army. New warriors could be enlisted for 1, 2, or 3 years. Let x_{1T}, x_{2T}, and x_{3T} be the number of warriors enlisted in Year T for 1, 2, and 3 years, respectively. The associated unit costs are c_{1T}, c_{2T}, and c_{3T}. In each Year T, the minimal total reserve warrior strength was set at R_T; R_T varied from year to year. As a Spartan general, you could find an optimal enlistment policy for the ensuing 5 years by solving the problem as a linear programming model. For simplicity, let $T = 1$ denote the year 435 BC. The linear programming model to minimize total cost is shown below.

$$\text{Minimize} \quad \sum_{T=1}^{5} \sum_{i=1}^{3} c_{iT} x_{iT}$$

Subject to

$$\begin{array}{lll}
\text{Year 1} & x_{11} + x_{21} + x_{31} & \geqslant 100 \\
\text{Year 2} & x_{21} + x_{31} + x_{12} + x_{22} + x_{32} & \geqslant 120 \\
\text{Year 3} & x_{31} \qquad x_{22} + x_{32} + x_{13} + x_{23} + x_{33} & \geqslant 110 \\
\text{Year 4} & x_{32} + x_{23} + x_{33} + x_{14} + x_{24} & \geqslant 140 \\
\text{Year 5} & x_{33} + x_{24} + x_{15} & \geqslant 130
\end{array}$$

$$x_{iT} \geqslant 0, \quad i = 1,2,3, \ T = 1,2,\ldots, 5$$

a. Can the problem as it is currently stated without performing any mathematical operation be expressed as a netform? Explain.

b. Reconsider the problem above now letting x_{ij} = number of warriors enlisted at the beginning of period i and discharged at the beginning of period j. At the beginning of each time period, the number of warriors enlisted plus the number available from previous multi-period enlistments must be greater than or equal to the total number required during the period. This is the conceptualization used in the original formulation. An alternative conceptualization could be one which focuses on marginal requirements rather than total requirements. At the beginning of the first year, the marginal requirement and total requirement are the same since we assume that there are initially no reserve troops. (That is, the total requirement is 100, and the amount required to get from zero initially to a total of 100, the marginal requirements, is also 100.) Therefore, the decision is how many to enlist in the first year for each possible tour of duty (i.e., to discharge at the beginning of period 2, 3, or 4), x_{12}, x_{13}, x_{14}. Also, since the total requirement is "at least" 100, more than 100 could be enlisted. The amount enlisted in excess of the total requirement for period 1 can be referred to as "slack" and designated S_1. Therefore, the sum of those enlisted minus the slack must equal 100. The total requirement in period 2 is at least 120. Therefore, the Army must grow by 20 if no slack was created in period 1, and in general by $20 - S_1$. Also, some of those enlisted in period 1 will be discharged at the beginning of period 2 and must therefore be replaced. Consequently, the marginal requirement at the beginning of period 2 (i.e., $\Sigma_{j>2} x_{2j}$) must equal the sum of: (1) the amount by which the Army must grow $(20-S_1)$; (2) the amount who will be discharged $(\Sigma_{i<2} x_{i2})$; and (3) any slack created this period (S_2). In general,

$$\sum_{j>T} x_{Tj} = (R_T - R_{T-1}) - S_{T-1} + \sum_{i<T} x_{iT} + S_T, \quad \text{for } T = 1, \ldots, 6$$

Rearranging yields

$$\sum_{i<T} x_{iT} - \sum_{j>T} x_{Tj} - S_{T-1} + S_T = -(R_T - R_{T-1}) \quad \text{for } T = 1, \ldots, 6$$

where $S_0 = S_6 = R_0 = R_6 = 0$ (note that variable x_{56} exists). Write the six constraints of this formulation in detached coefficient form.

c. Can the formulation in part (b) be expressed as a network? If yes, draw its netform and explain the intuitive interpretation of its elements in relation to the original problem statement.

d. Using the x_{ij}'s as defined in part (b), re-write the original formulation in detached coefficient form and convert the original constraints to the new constraints using elementary row operations. (Hint: It may be enlightening to use the $-(R_T - R_{T-1})$ symbols for the right-hand side of the new formulation and the R_T symbols for the right-hand side of the old formulation.

***4.19** A company operates moving vans between cities X and Y and cities A, B, and C. The schedule for the next 10 days is shown in the following table.

Dates	Origin City	Destination City
3	X	A
4	Y	A
6	X	C
6	Y	A
9	X	B
9	Y	A
10	X	A
10	X	C
10	Y	B

After a van goes from an origin city to a destination city, it may return to either origin city. Vans may be kept in inventory at the origin cities but not the destination cities. The total time (in days) to drive from city to city, in either direction, is shown in the second table:

	A	B	C
X	2	3	2
Y	1	2	1

Suppose three vans are at City X and three at City Y on Day 1. Let the expense of returning a van to city X from cities A, B, and C be p_A, p_B, and p_C, respectively, and similarly let the cost of returning a van to City Y be q_A, q_B, and q_C, respectively. The moving company wants to find a routing for vans from cities A, B, and C back to cities X and Y that minimizes total cost.

a. Construct a netform model that places three vans at each city (X and Y) at the end of the time horizon. Hint: Let nodes correspond to (date t, city i) for all possible departure/arrival dates t and cities i = X, Y, A, B, or C. For example, since on Date 6 a van must leave city X and go to city C, taking 2 days of travel time, then one van must travel between nodes 6X and 8C. That is, create an arc between nodes 6X and 8C which has bounds of (1, 1). Note that the shipping times make it impossible to send a van from some of the nodes in sufficient time to meet the requirements at some other nodes. For example, node 8C cannot meet the requirement at node 9X because it takes 2 days to go from city C to city X.

b. Suppose the company wants to find the minimum number of vans required to meet the schedule and then to minimize shipping costs. How would you modify your network in part (a) to incorporate these two pre-emptive objectives?

4.20 A bottler and distributor for the Chug-a-lug Beer Company buys beer wholesale from the main plant, then bottles it and sells it to groceries and other outlets. The distributor wants to develop a 3-period planning model for its operations, taking the following information into account. (For this model, purchases and sales are assumed to take place at the start of each period. All supply and demand units are in bottles of beer; all prices are in dollars.)

	Periods		
	1	2	3
Sales demand (upper limit)	8,000	11,000	6,000
Sales price ($)	1.40	1.90	2.00
Purchase supply (upper limit)	5,000	9,000	10,000
Purchase price ($)	0.50	0.65	0.75
Bottling cost ($)	0.10	0.10	0.10
Back order costs ($)			
From Period 2	0.10	n/a	n/a
From Period 3	0.30	0.10	n/a

(The $0.30 entry in the bottom left of the table says it costs $0.30 to back-order from Period 3 to Period 1; n/a means "not applicable.") Back-orders can come out of all bottled beer available at the start of a period, whether from inventory or freshly bottled. The company maintains no inventory of beer except in bottled form. At most, 8000 bottles can be held in inventory from one period to the next, but there is a revenue of $0.12 for each bottle in inventory for the period, paid by the warehouse owner who believes the presence of the bottles, due to Chug-a-lug Beer's distinctive aroma, keeps the rodents out of the warehouse. There is a limit for all three periods combined that no more beer can be purchased than would fill 20,000 bottles. Beginning inventory on hand (no cost or revenue attached) is 400 bottles of beer. Ending inventory, following the start of Period 3, is to be exactly 50 bottles, which will be given to the warehouse owner free as a goodwill gesture. The distributor plans to skip town at that point. (Empty bottles can be obtained at once from a glassworks across the street, so they have no inventory, and their cost is included in the bottling cost already mentioned. Also, there is no time lag in the bottling process.) Formulate this problem as a revenue maximization network problem. Create a combined purchase node, labeled P; three bottling nodes, labeled B1, B2, and B3; three inventory nodes, labeled I1, I2, and I3; and three sales nodes, labeled S1, S2, and S3. (Do not eliminate any node or include any extras.) In general, construct the network so that costs or revenues that appear on arcs are those cited, and not "composite" values.

REFERENCES

Glover, F., G. Jones, D. Karney, D. Klingman, and J. Mote. 1979, "An Integrated Production, Distribution, and Inventory Planning System," *Interfaces*, Vol. 9, No. 5, pp. 21–35.

Klingman, D., N. Phillips, D. Steiger, R. Wirth, and W. Young. 1986, "The Challenges and Success Factors in Implementing an Integrated Petroleum Products Planning System for Citgo," *Interfaces*, Vol. 16, No. 3, pp. 1–19.

Klingman, D., N. Phillips, D. Steiger, R. Wirth, R. Padman, and R. Krishnan. 1987, "An Optimization Based Integrated Short-Term Refined Petroleum Product Planning System," *Management Science*, Vol. 33, No. 7, pp. 813–830.

5

GENERALIZED NETWORKS

The preceding chapters have presented a wide array of situations capable of being modeled by pure networks. At the same time, there are many important practical problems that cannot adequately be captured by a pure network representation. This chapter introduces generalized networks, which expand our modeling capabilities.

5.1 GENERALIZED NETWORKS: A PRACTICAL STEP BEYOND PURE NETWORKS

The most conspicuous limitation of pure networks is that all problem activities must be represented by arcs. A pure network arc does not allow flow to change its magnitude as it moves from the tail node to the head node. Yet there are a variety of settings in which the process of moving from one junction (state, location, time period, etc.) to another does induce a flow change. For example, the activity of investing money typically does not yield the same quantity of money at the end of a period as at the beginning (at least not if the investment is prudent), but yields an amount which is augmented by the rate of return. Conversely, money expressed in terms of purchasing power may decrease in value from one period to the next. Similarly, perishable goods held in inventory, water flowing in irrigation channels, electrical power carried on transmission lines, crops planted and harvested, livestock raised for market, all involve quantities that naturally diminish or grow.

Moreover, there is another way that quantities change during transfer that is even more common in practical application, involving the transformation of flow from one unit of measure to another. Examples include processing raw materials to produce a finished good, converting water power into electricity, exchanging currency of one type for another, distilling a liquid to produce a gas, and so forth. A useful subclass consists of transformations that produce conceptual change rather than a physical change, called ***re-expression***. For example, a job may be re-expressed in terms of the number of hours it requires to be processed, a bus may be re-expressed in terms of the number of passengers it can carry, a truck may be re-expressed in terms of the number of trips it can

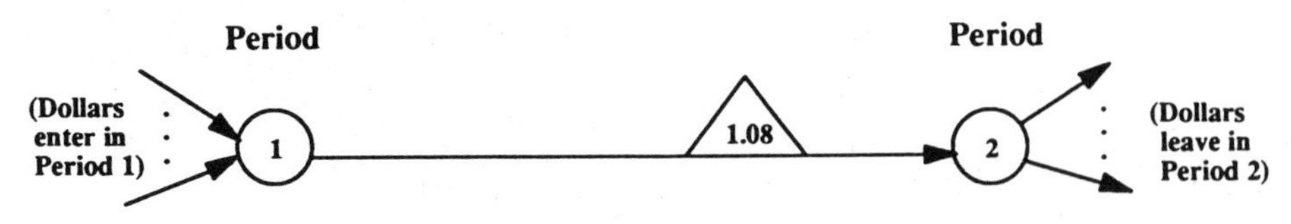

Figure 5.1 Generalized arc for investing at 8% interest.

make, a machine may be re-expressed in terms of the volume of goods it can process, and so forth.

Each of these ways of changing flow requires a means to describe how the magnitude of the flow alters as a result of the activity being performed, that is, a ***coefficient of change***. In the network context, such a coefficient of change is called a ***multiplier***, because in fact it is multiplied by the flow to change the flow magnitude. A multiplier is represented pictorially by attaching it to an arc within a triangle, placed next to the node that receives the changed flow magnitude. An arc containing such a multiplier is called a ***generalized arc***.

To illustrate, suppose that every dollar invested in Period 1 earns 8% interest by Period 2, hence returns 1.08 times its original quantity. The generalized arc representing this investment activity, with multiplier attached, is shown in Figure 5.1. The multiplier indicates that for any amount of flow (dollars of investment) carried on the arc, 1.08 times this amount is delivered to the node for Period 2.

Consider next an activity in which airplanes carry passengers from a City A to a City Z, and each plane carries 130 passengers. This situation can be represented by Figure 5.2. Again, the multiplier indicates that the amount of flow on the arc (number of planes) delivers 130 times this amount (number of passengers) to the node for City Z. (We may imagine that the arcs into node A and out of node Z connect to a larger network structure, not shown.)

Generalized arcs, like ordinary arcs, can also have attached costs (or profits) and bounds. Here a distinction needs to be made. *The flow to which costs and bounds apply is the original arc flow, before it is transformed by a multiplier.* A multiplier is therefore positioned next to the node that receives the changed flow, and other quantities precede it. Thus, Figure 5.3 represents the example where

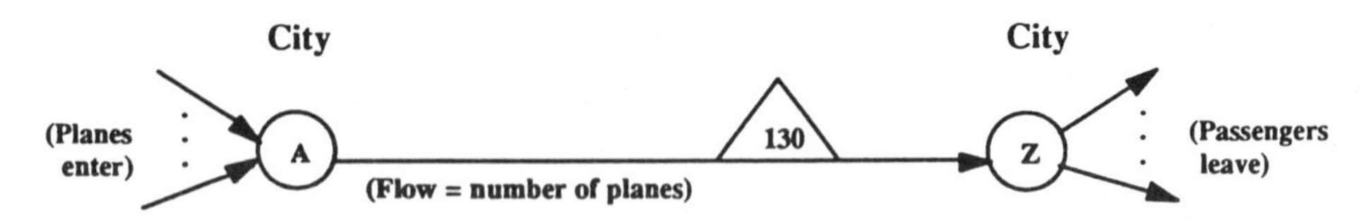

Figure 5.2 Generalized arc for planes carrying passengers.

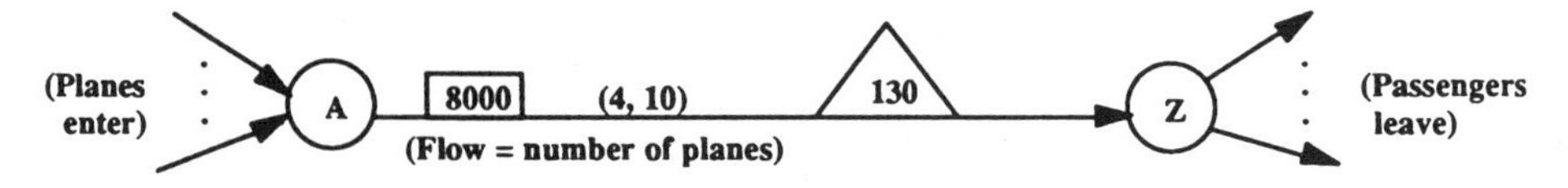

Figure 5.3

from 4 to 10 planes are required to fly from City A to City Z, and each flight costs $8000.

We may expand the example by supposing it is possible to fly to City Z not only from City A but also from City B. Further, suppose at least 260 and at most 650 passengers are required to reach City Z from both A and B combined. To handle this situation, we create an additional arc, whose flow is expressed in numbers of passengers, to which the bounds can be applied directly. This arc issues from a new node which is introduced to receive the combined flows from City A and City B (using the technique for modeling combined flow introduced in Chapter 3). Upon stipulating a flight cost of $8000 from City A and of $6000 from City B, the appropriate diagram is shown in Figure 5.4.

Networks containing constructions such as these, involving generalized arcs, are called ***generalized networks***. They are sometimes also called ***networks with gains and losses***, conveying the fact that multipliers can cause flows to increase or decrease (for multiplier values greater than 1 and less than 1, respectively). The term "generalized" is appropriate, because an ordinary arc is just a generalized arc in which the multiplier is implicitly 1. (Exactly 1 unit of flow leaves an ordinary arc for every unit that enters.)

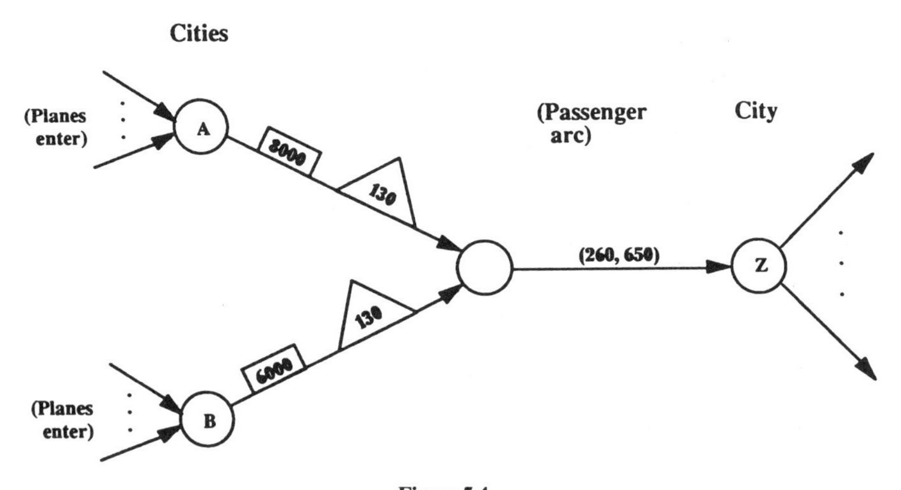

Figure 5.4

5.2 GENERALIZED NETWORKS IN PRODUCTION AND INVENTORY APPLICATIONS

Generalized networks are especially prevalent in production and inventory applications such as those described in Chapter 4. To illustrate, consider a simple inventory setting involving perishable goods, in which 4% of those held from one period to the next are lost. (Only 0.96 of those on hand at the start of a period make it to the end.) A three-period diagram, showing only the multipliers applicable to goods in inventory, is shown in Figure 5.5. No multiplier is placed on the initial inventory arc, reflecting the fact that initial inventory is normally viewed as the amount on hand after deterioration has already occurred. Similarly, no multiplier is attached to the ending inventory arc, because we are concerned only with the amount that leaves the final period.

In more complex models involving both inventory and production processes, multipliers become useful to handle such things as conversions of raw materials to finished goods. To illustrate, consider a paper manufacturing company that processes wood pulp to make paper. After purchasing, each pound of wood pulp either goes into a raw materials inventory or is processed to yield 0.2 reams of paper. Wood pulp costs $0.10 per pound, and paper sells for $3.00 per ream. Capacity is available to store 500 pounds of pulp and 100 reams of paper per period. Holding costs are 10% of the value of the item per period. The manufacturer can sell at least 200 reams and at most 800 reams of paper per period. An appropriate 3-period production-inventory diagram appears in Figure 5.6.

The 0.2 multiplier on the production arcs, converting pounds of pulp inventory into reams of paper, is a simple addition to the type of production-inventory structures of Chapter 4, but has far-reaching consequences for expanding the range and realism of such models.

One can readily compound this sort of model with different multipliers representing conversion factors of different processes, leading to different final

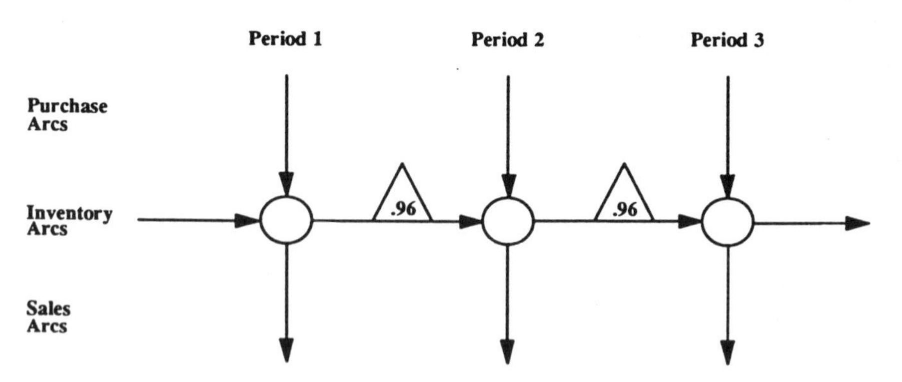

Figure 5.5 Perishable goods.

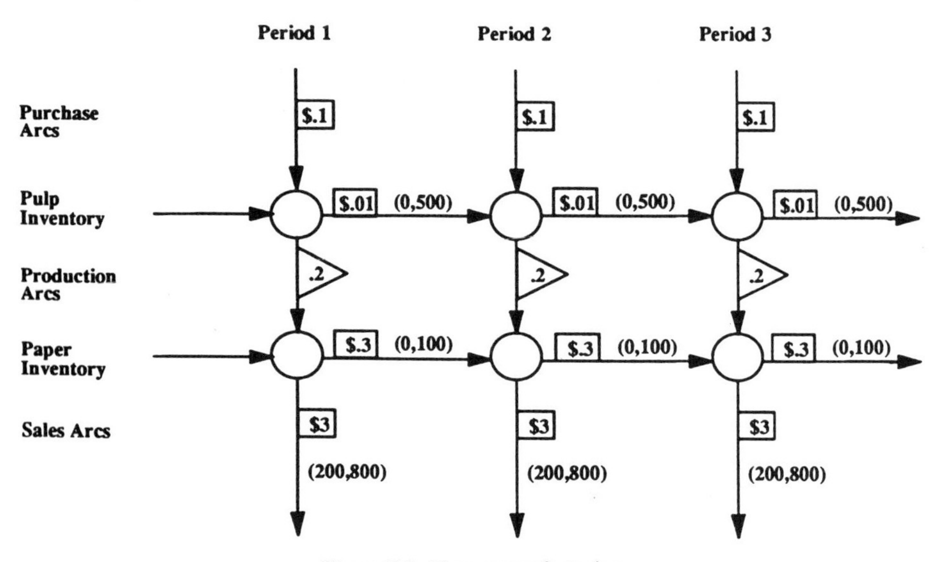

Figure 5.6 Paper manufacturing.

products. Similarly, multipliers representing inventory deterioration, as in the earlier illustration, can be appended.

5.3 CASH FLOW MODELS

Cash flow models provide a particularly fertile field for generalized network applications. Because they involve multiple time periods, they may be viewed as a form of inventory model (in a broad sense), and indeed bear a strong resemblance to the types of inventory models previously discussed.

For a simple illustration, Figure 5.7 represents a 3-period example in which

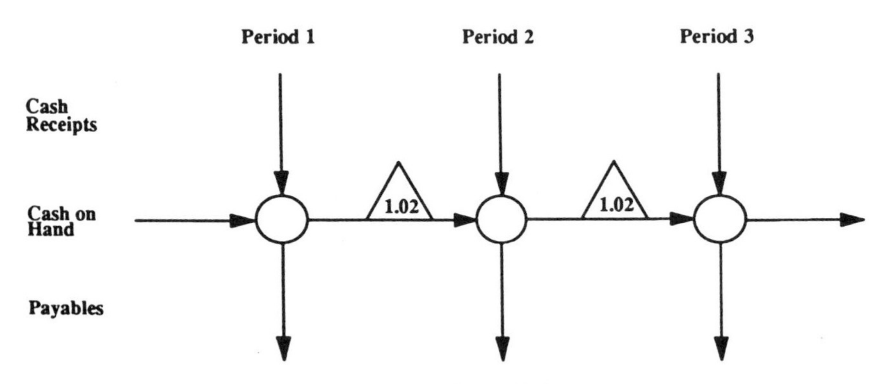

Figure 5.7 Cash flow with interest.

cash receipts (from sales and receivables, for example) feed into cash on hand. The latter may be used to meet payables, or carried forward at an interest rate of 2% per period. Viewing receipts as supplies and payables as demands, note that the 2% interest rate gives a multiplier of 1.02 for each dollar carried forward.

Now we extend the model in Figure 5.8 to show additional useful features of cash flow analysis that can be handled by generalized networks. Consider the common situation where payables need not be paid at once, but may be paid out of future funds. In particular, suppose that a 1-period delay in meeting payables means that the company must pay \$1.03 for each \$1 owed, and a 2-period delay requires payment of \$1.07 for each dollar owed. Note that this is equivalent to saying that each \$1 paid retires \$1/1.03 of debt from one period earlier and \$1/1.07 of debt from two periods earlier.

This illustration shows that arcs to meet payables out of future funds correspond to the "back arcs" used in earlier inventory models to handle backorders. The versatility of these arcs is greatly increased by the addition of multipliers, both here and in the inventory setting. (For example, as shown by Exercise 5.3 at the end of the chapter, multipliers make it possible to represent loss of sales as a result of using backorders.)

Cash flow models can readily be expanded to include different categories of receipts and payables. To illustrate, suppose the preceding example deals with accounts payable, and that the company also must handle notes payable. Figure 5.9 represents the situation where \$1.04 must be paid to retire each \$1 of notes payable from a period earlier, and no debt is allowed to go more than one period without payment (A = accounts and N = notes).

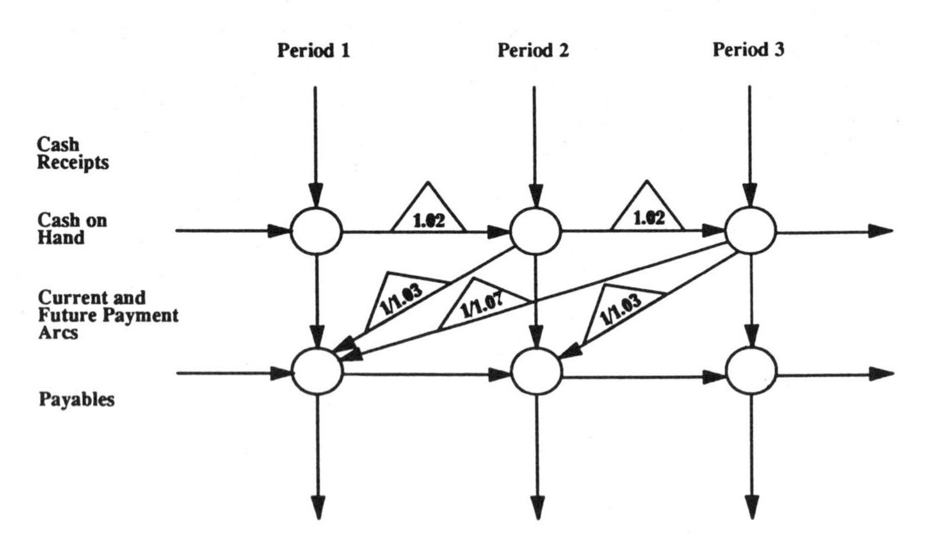

Figure 5.8 Cash flow with interest and delayed payables.

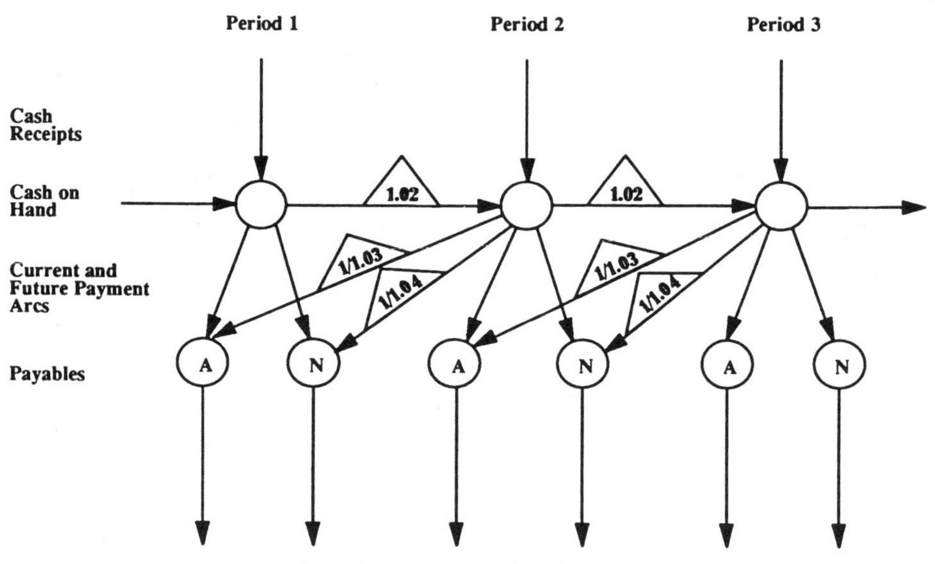

Figure 5.9 Two categories of payables.

Question: Develop a complete 4-period cash flow model with the following characteristics.

	Period 1	Period 2	Period 3	Period 4
Cash receipts ($)	800	900	700	800
Accounts payable ($)	600	700	500	500

Beginning cash on hand is $200, and cash carried forward from period to period may be allocated in any desired split between a bank account that pays 1% rate of interest per period and bonds that pay 3% rate of interest every two periods. However, the bonds can only be purchased in Periods 1 and 2, and cannot be cashed in until two periods after purchase. Finally, the payment of payables can be postponed at most one period, in which case $1.02 must be paid to retire $1 owed. The objective is to maximize cash on hand at the end of the model horizon (the amount carried beyond Period 4).

Figure 5.10 handles the conditions of the preceding question. It may be noted in Figure 5.10 that the opportunity to invest in bonds is modeled by arcs that extend for two time periods (since the bonds cannot be cashed after only a single period). The objective of maximizing the cash on hand at the end is handled by the unit profit figure attached to the rightmost arc. (It may be noted that the accounts payable portion of the model for Period 4, consisting of two consecutive arcs, can be replaced by a single arc.)

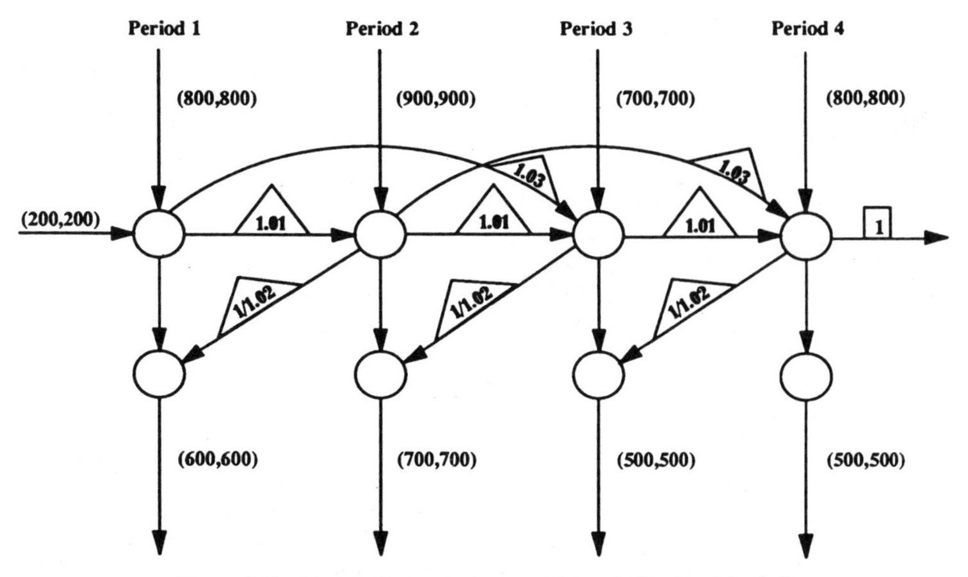

Figure 5.10 Four-period cash flow model (maximization objective).

Managing the timing of investments and payments, as, for example, to maximize return, is a valuable aspect of this type of cash flow model. A more sophisticated version can easily be constructed that allows the investor to withdraw cash in each period, and permits future cash to be valued less highly than present cash. Lower bounds may also be imposed to withdraw a certain minimum amount of cash in each period.

To illustrate these considerations in the context of the preceding problem, suppose that each \$1 withdrawn is valued as \$1 in Period 1, \$0.99 in Period 2, \$0.98 in Period 3, and \$0.97 in Period 4. Further, suppose a minimum of \$50 is

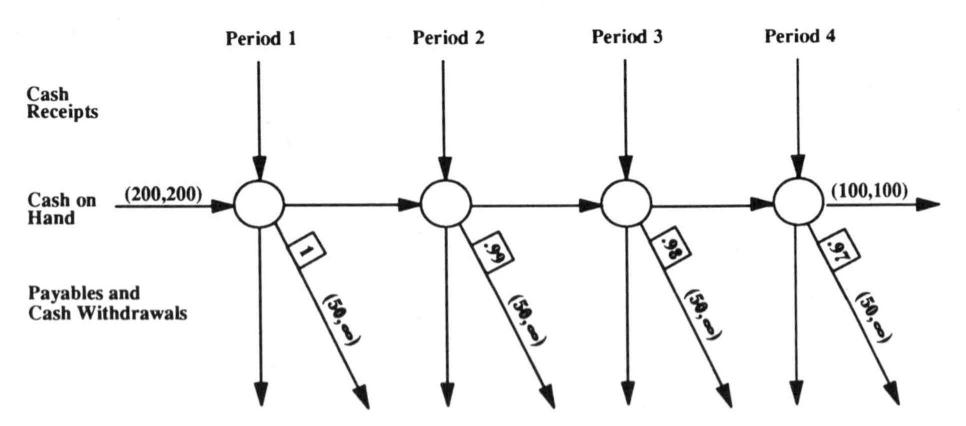

Figure 5.11 Simplified model with periodic cash withdrawals (maximization objective).

to be withdrawn each period, and $100 must be left on hand at the end of Period 4. The general form of the model to maximize return is depicted in Figure 5.11. (Arcs and data not relevant to the concerns of this illustration have been dropped or simplified to emphasize the role of the cash withdrawal arcs.)

The cash withdrawal arcs may be identified in this illustration as those leading downward diagonally to the right. Note that they add negligible complexity to the model. If the preceding simplified diagram is expanded back to include all the elements of the full cash flow example, the cash withdrawal arcs retain exactly the positions shown, with the net effect of adding a single demand arc per period.

5.4 CONSTRUCTION DIFFERENTIATING POTENTIAL AND ACTUAL

The ability of generalized networks to re-express the context of an activity by means of arc multipliers makes it possible to create a useful distinction between potential and actual. This distinction, and its value as a modeling technique, may be illustrated as follows. We expand on the example in Figure 5.4 in which airplanes carry passengers from two cities, A and B, to a City Z. Figure 5.4, which includes the costs of flights, multipliers representing the number of passengers carried by each plane, and bounds on the combined number of passengers to be received at City Z, is repeated below.

Reflection suggests that this diagram is not an entirely accurate portrayal of real-world conditions. Instead of saying that a plane carries 130 passengers, it would be more reasonable to say that it merely has the capacity to do so; or, in other words, that it carries a ***potential*** of 130 passengers. Of this potential

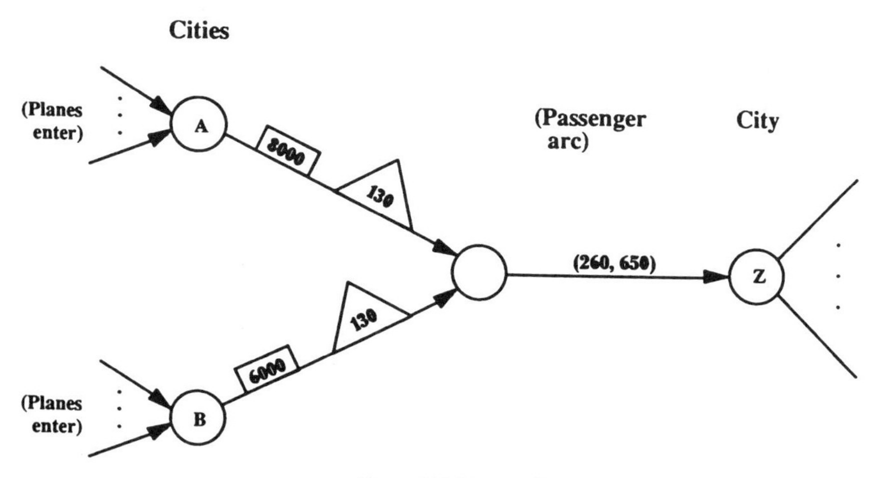

Figure 5.4 (*Repeated*).

number, part will consist of passengers actually carried, and the remainder will consist of passengers not carried (i.e., empty seats). A generalized network model can differentiate potential and actual by the simple device of introducing a new node to receive the potential quantity, followed by two arcs, one for realized potential (quantity actually carried) and one for unrealized potential (empty seats).

In the present example, to differentiate potential and actual passengers carried, it is only necessary to add a single arc to handle empty seats, because a node and arc to receive combined passenger flows from City A and City B already exists. The resulting diagram appears in Figure 5.12. It should be noted that the arc for empty seats has no destination node. (By contrast, the arcs out of node Z, which are shown only incompletely, may lead to other nodes of the network.) The arc for empty seats is a surplus arc (as defined in Chapter 2), since it carries surplus—in this case, unrealized potential—out of the system. Because there is frequently a cost associated with potential that is not used, it is not uncommon to attach costs to surplus arcs in this context.

Question: Extend Figure 5.12 to handle the stipulation that there is a penalty cost for each empty seat on planes from City A and City B of \$10 and \$15, respectively, in addition to the fixed flight costs.

Figure 5.13 achieves the goal of the question. Again, note the arcs for empty seats (passengers not carried) are surplus arcs, and have no head nodes. This means of differentiating potential and actual flows can be used advantageously in a number of production and distribution settings, as demonstrated in exercises at the end of the chapter.

5.5 THE COMPLICATION OF DISCRETENESS

The preceding example brings up an important consideration that must be squarely faced in the generalized network context—whether flows must take discrete (i.e., integer, or whole number) values, or whether they may acceptably be fractional. In the pure network setting, optimal solutions always exist that automatically make all flows integer-valued (if all data are integer-valued and any feasible solutions exist at all). Moreover, these are exactly the solutions that standard solution methods find. In the generalized network setting, however, the best solutions may well occur by assigning some flows noninteger values. Thus, it could happen that a problem resembling that of Figure 5.13 would achieve a cost minimizing solution by sending $12\frac{1}{3}$ planes from City A to City Z. It may or may not be possible to reconcile this with practical considerations, depending on the situation. If not, then the requirement of discreteness must be explicitly enforced.

Discrete-valued models constitute a highly important domain in their own right, and are given special attention in Chapter 6. They make it possible to handle not only the obvious types of indivisibilities (as where only integer

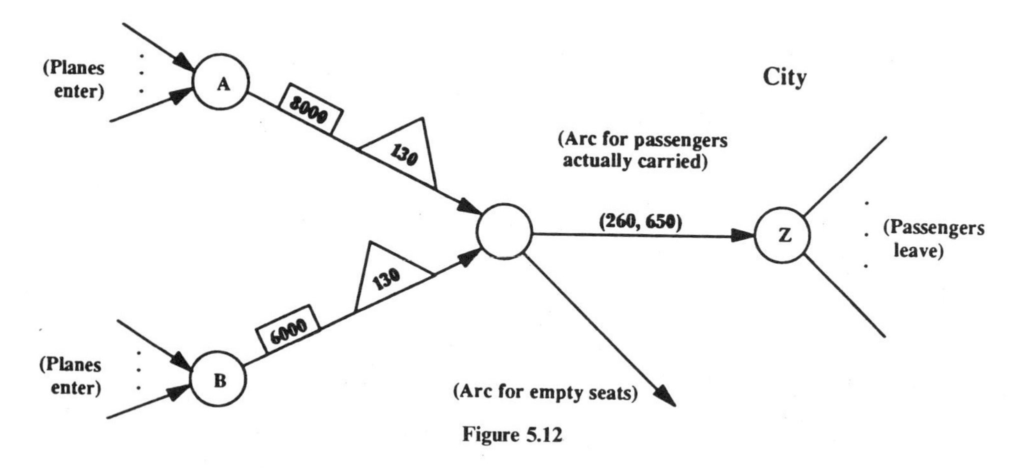

Figure 5.12

numbers of plants and warehouses can be built), but a variety of nonlinearities as well. It is important to keep in mind when dealing with generalized networks that discrete and continuous (i.e., nondiscrete) models are different, and require different solution methods. One manifestation of this difference is that discrete models typically require considerably more computer time to solve.

Because of the relative ease of dealing with continuous models, practitioners often give a fair amount of thought to whether they can live with a continuous solution when integer-valued flows are required. No hard and fast rules exist. A preliminary study can undertake to determine the "damage" to objective function and problem constraints from using doctored (usually rounded) continuous solutions in place of optimal integer solutions. This is a prudent measure if the discrete element is suspected to be critical, because examples exist

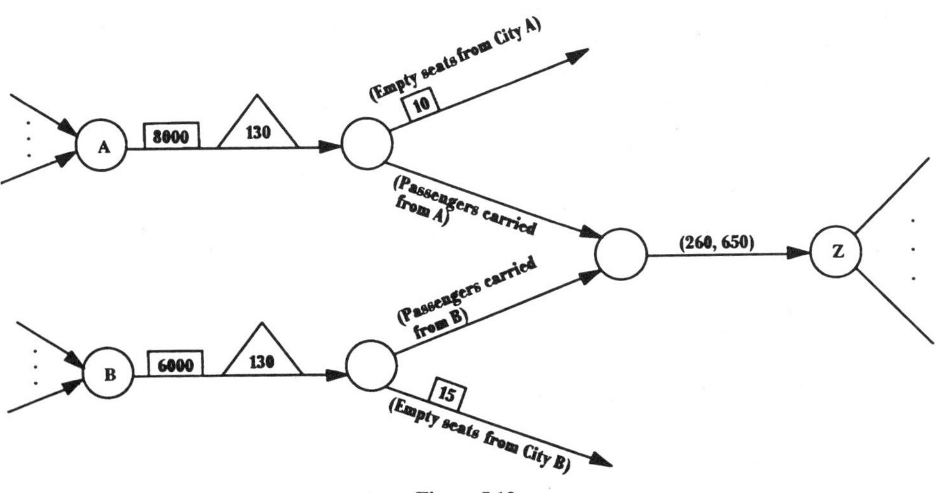

Figure 5.13

where rounded solutions are far from the corresponding optimal integer solutions and so can result in substantial losses of money, instead of putting operations solidly in the black. At the same time, there are situations where the use of refined discrete optimization techniques is an unnecessary luxury.

Short of conducting a sensitivity study (or being immersed deeply enough in a particular problem context to have special insights into the matter), there are a few simple indicators that bear on the acceptability of rounding continuous solutions to integers. In general, it is likely to be safer to round to integer values where: (1) only a small proportion of the flows that are supposed to be discrete receive fractional values in a continuous solution; (2) flows that receive fractional values are not for crucial activities; (3) the amount by which a fractional flow differs from an integer is a small part of its total value (e.g., it is usually safer to round 999.7 to 1000 than it is to round 1.4 to 1). Unfortunately, none of these indicators is foolproof.

The complication of discreteness has been highlighted in advance precisely because it is such an important issue, and many practitioners fail to appreciate its significance (to their peril). At the same time, it should be stressed that many of the modeling techniques developed here for continuous generalized networks carry over directly to discrete models, as will become clear in the next chapter.

5.6 DESIGNING GENERALIZED NETWORKS FROM ALTERNATIVE PERSPECTIVES

Generalized network models can often be designed from two different perspectives. The following example makes this concrete.

Two tasks can be performed by either John or Marc, both of whom can be hired for up to 40 h. John can complete Task 1 in 28 h and Task 2 in 27 h, charging $8/h for his services. Marc can complete Task 1 in 26 h and Task 2 in 22 h, charging $10/h. John and Marc can work fewer hours on either task, completing the fraction of the task corresponding to the ratio of time worked to the time required to do the entire task. (Equivalently, a fraction of a task done by a particular person consumes that same fraction of the total hours he would devote to the full task. Thus, to do half of Task 1 would take John 14 h and Marc 13 h.) The problem is to determine how many hours to hire John and Marc in order to complete the two tasks at minimum total cost.

For formulation purposes, the two different perspectives that present themselves are that of assigning tasks to people and that of assigning people to tasks. This potential for dual visualization arises in many other generalized network (and pure network) problems, and it is useful to explore its consequences.

Consider first the perspective of assigning tasks to people, that is, where tasks assume the role of origins and people assume the role of destinations. The appropriate generalized network diagram is shown in Figure 5.14. To verify that the diagram faithfully represents the preceding problem, note first that the supply of exactly 1 unit for each task expresses the requirement that each task

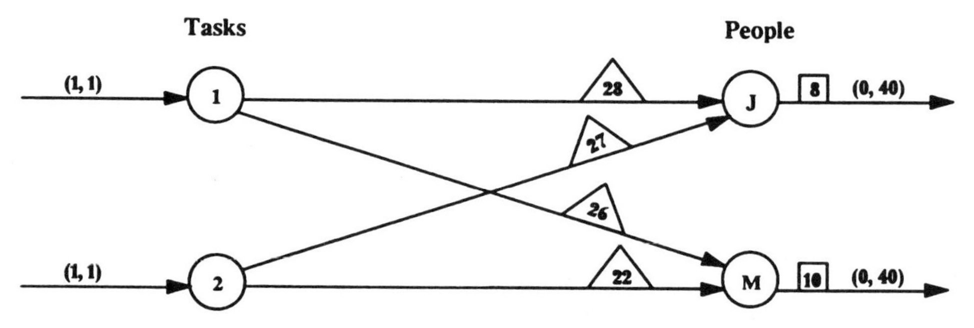

Figure 5.14 Perspective 1: assigning tasks to people.

must be completed. Next, the multipliers on the arcs out of node 1 indicate that assigning Task 1 totally to John (node J) will consume 28 h of his time, while assigning it totally to Marc (node M) will consume 26 h of his time. If the task is split half and half between them (corresponding to a flow of $\frac{1}{2}$ on each of the arcs out of node 1), then $\frac{1}{2} \times 28 = 14$ h of John's time will be consumed and $\frac{1}{2} \times 26 = 13$ h of Marc's time will be consumed. Similar remarks apply to the arcs out of node 2. Finally, the costs and bounds on the arcs out of nodes J and M correspond to the hourly wage rates and total available hours for John and Marc.

For comparison, consider now the reverse orientation, in which people take the role of origins and tasks take the role of destinations. Figure 5.15 not only reverses the orientations of all arcs in the earlier diagram, but inverts the values of the multipliers. To verify its correctness, note that the arcs out of node J, whose flow is measured in hours of John's time, specify that each hour John devotes to Task 1 will complete 1/28 of that task, and each hour he devotes to Task 2 will complete 1/27 of that task, as appropriate. In short, reversing orientation can modify the units by which flows on arcs are measured. Specifically, each arc that changes the unit of measure (as from jobs to hours in Figure 5.14) will change this measure in exactly the reverse way (from hours to

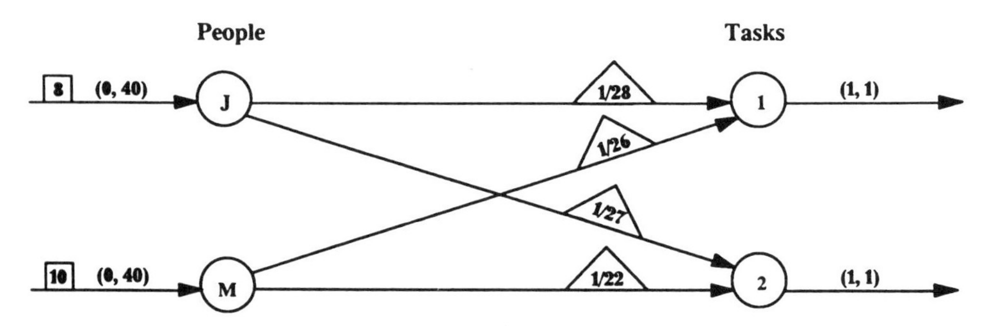

Figure 5.15 Perspective 2: assigning people to tasks.

jobs) when oriented in the opposite direction. This can have implications not only for multipliers but also for costs and bounds, and, as a result, for the convenience of certain elaborations. This is illustrated by the following question.

Question: Give a diagram for both Perspectives 1 and 2 in the situation where Marc also charges \$2 for each hour he works on Task 1 and \$3 for each hour he works on Task 2. (These are surcharges, in addition to his \$10 hourly rate.) However, he may not work more than 20 hours on either task by itself.

Perspective 2 turns out to be more convenient for handling the conditions expressed in the question, and we present its representation in Figure 5.16. The interpretation of the costs and bounds in Figure 5.16 is straightforward. Now consider how the appropriate costs and bounds can be created for Perspective 1. First, observe that Marc's surcharge of \$2/hour for Task 1 converts into a total surcharge of \$52 to do the entire task (provided he could devote all 26 h he requires to complete it). Thus, \$52 is the proper cost to go on the arc from Task 1 to Marc. If only half the task gets assigned to Marc, the surcharge is $\frac{1}{2} \times \$52 = \$26$. This agrees with what we know from the alternative perspective since Marc will complete half the task in 13 h, yielding \$26 at his surcharge rate of \$2/h.

Next, the upper bound of 20 h that Marc can devote to Task 1 implies that he can complete at most 20/26 (or 10/13) of the task, given that it takes him 26 h to do it all. Thus 10/13 is the appropriate upper bound to go on the arc from Task 1 to Marc.

In a similar manner, it may be seen that \$3/h converts to a surcharge of \$66/task when Marc works on Task 2, and the upper bound of 20 h converts to a bound of 20/22 (or 10/11) limiting the portion of Task 2 completed. These observations lead to Figure 5.17.

There are two important morals to this illustration. The first is that the roles of "supplies" and "demands" can be interchanged in generalized networks just as in pure networks. The second is that one perspective can be more convenient that the other, depending on the way a problem is stated and the conditions to be incorporated.

In general, it is valuable to become adept at looking at problems from

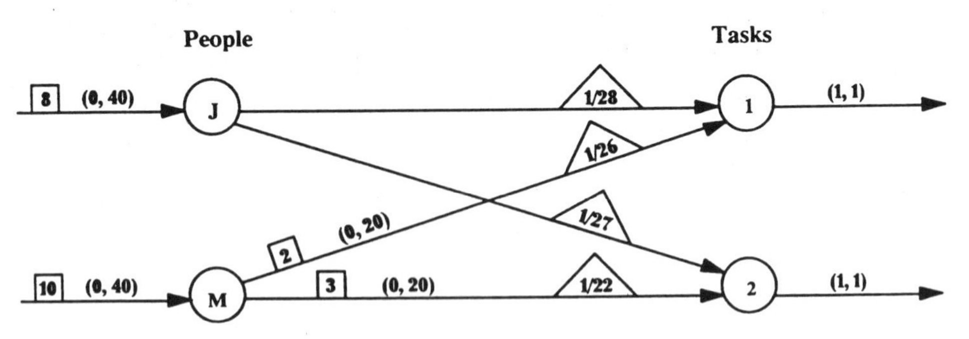

Figure 5.16 Perspective 2.

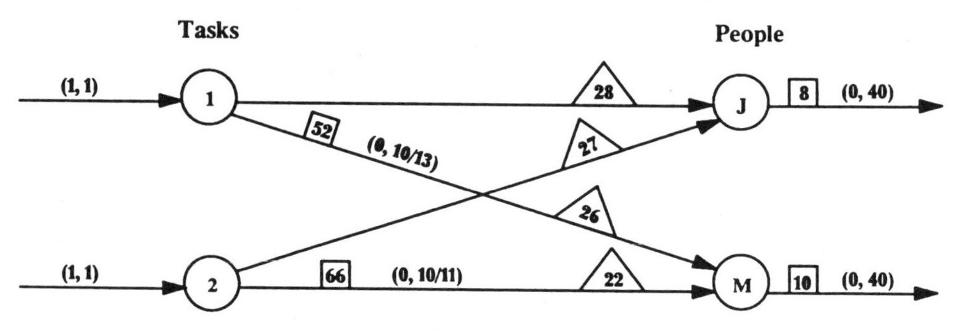

Figure 5.17 Perspective 1.

alternative perspectives. Prior to the point where such shifts in perspective become second nature, the key to identifying appropriate costs, bounds, and multipliers is to keep close track of the units in which arc flows are expressed.

5.7 CHOICES TO USE MORE OR FEWER NODES AND ARCS

Examples where problems can be modeled in varying degrees of compactness—as, for instance, where one may choose to retain or remove nodes that have a single entering and a single leaving arc—have already been encountered in this and previous chapters. From the standpoint of the modeler, the preferred representation is the one that communicates the nature of the problem more clearly.

However, generalized networks typically invite more alternative representations than pure networks. To illustrate, consider a manufacturing situation, in which workers from two different departments, A and B, can be assigned to a project that involves working on either Machine 1 or Machine 2. At most 3 workers are available from each department, and each worker assigned to the project can contribute 40 h of work. Also, exactly 130 h must be devoted to Machine 1 and exactly 80 h must be devoted to Machine 2.

We may imagine that these stipulations refer to a portion of a larger problem, whose complete form we temporarily disregard. We will also for the moment disregard costs, and examine different alternatives for modeling the portion described.

The terminology of the problem suggests the image of assigning people to a project. Then, each person assigned can be converted to 40 h of available work, which may in turn be applied to either of the two machines. Figure 5.18 shows a relatively direct representation that accords with this image. Since we have not stipulated that this model should be discrete, we are implicitly allowing a fractional number of workers to be assigned. In the present case we will suppose it is admissible for workers to devote part time to the project, and hence accept the natural interpretation of $\frac{1}{2}$ a worker as "a worker that works half time."

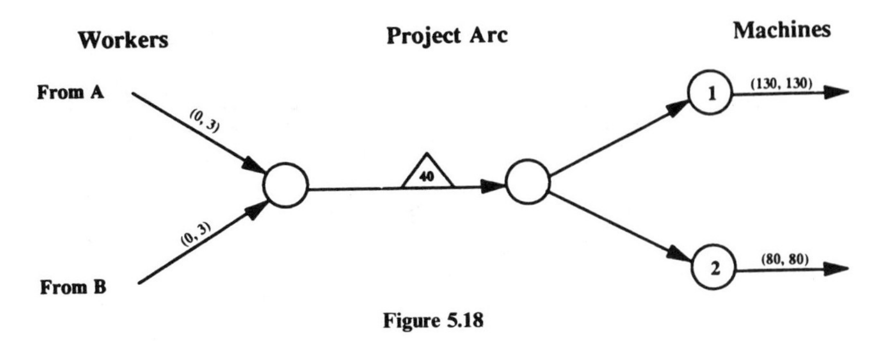

Figure 5.18

Consideration of Figure 5.18 shows that the project arc can be viewed as superfluous, since each arc that leads to a machine can convert from workers to hours. This gives rise to the construction in Figure 5.19.

Now suppose (for a moment) that each worker assigned to the project is less skilled on Machine 1 than Machine 2, so that the effective number of hours that can be given to Machine 1 is 30 rather than 40 (i.e., every 4 h only "counts" as 3). This can be handled in Figure 5.19 by replacing the multiplier into Machine 1 by 30. However, this condition can be modeled a bit more intelligibly using Figure 5.18. Specifically, putting a multiplier of $\frac{3}{4}$ on the arc into Machine 1 in Figure 5.18 effectively communicates the fact that hours devoted to this machine are utilized with only 75% efficiency.

Disregarding this latter type of elaboration, still another representation of the original problem is possible by immediately converting workers into hours, as shown in Figure 5.20.

If we now consider, as before, the alternative situation where the workers are less skilled on Machine 1 than Machine 2, we do not have to resort to Figure 5.18 for a convenient representation. The 75% efficiency for hours devoted to Machine 1 can be modeled in Figure 5.20 by simply attaching a multiplier of $\frac{3}{4}$ on the arc into the node for Machine 1.

On the other hand, consider yet another alternative situation where workers from Department A are less skilled all around than those from B, so that each

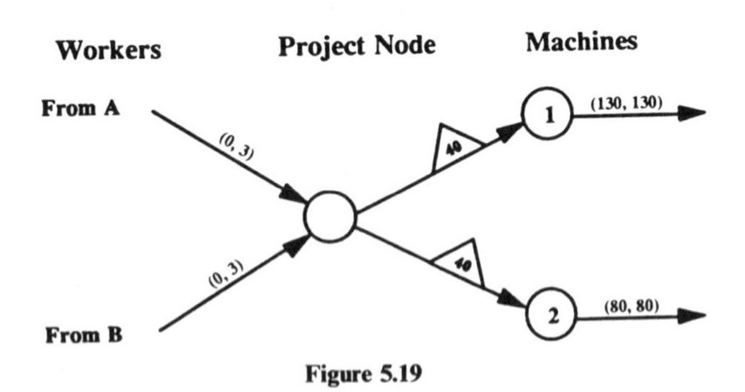

Figure 5.19

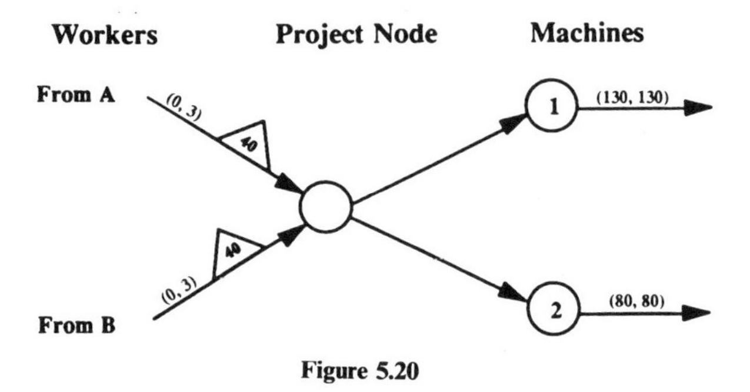

Figure 5.20

40 h contributed to the project only counts as 20 (regardless of the machine). This can be modeled using Figure 5.20 by replacing the multiplier on the arc from Department A by 20. In each of Figures 5.18 and 5.19, the condition is handled by attaching a multiplier of $\frac{1}{2}$ to the arc from Department A, indicating that workers from A are only 50% efficient. Thus, in this case, Figures 5.18 and 5.19 yield a slightly more intelligible representation.

Finally, an additional set of possibilities remains for modeling the original problem, in which the project node is eliminated entirely. We show one of them in Figure 5.21. The image of assigning workers to a common project is lost in Figure 5.21, but in fact greater flexibility is achieved. For example, if workers from A are 90% efficient on Machine 1 (each 40 h is worth 36) and workers from B are 80% efficient on Machine 2 (each 40 h is worth 32), then this can be modeled by changing the two appropriate multipliers of Figure 5.21 to 36 and 32. However, no changes of the multipliers in any of Figures 5.18–5.20 will handle this.

Question: Using the same structure as Figure 5.21, show a way to attach multipliers to the arcs so that the 90% efficiency of workers from Department A on Machine 1 and the 80% efficiency of workers from Department B on Machine 2 appear as multipliers of 0.9 and 0.8 on appropriate arcs.

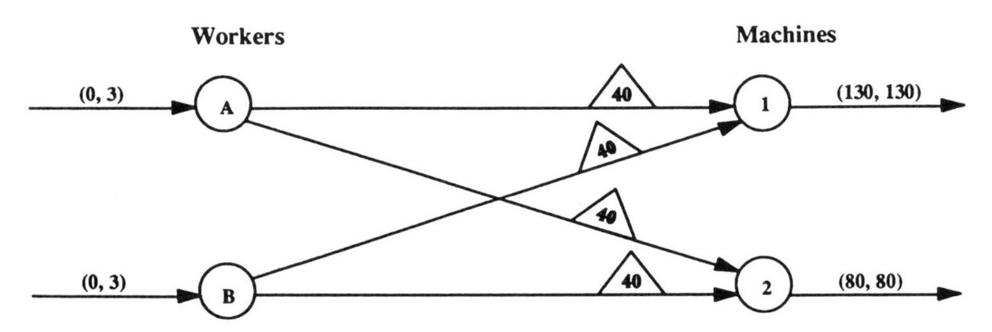

Figure 5.21

When options of the sort represented by Figure 5.21 exist, the sensible approach is to use whatever representation seems most natural to the problem context. (This can be especially helpful when the modeler returns to the diagram "cold.") Adding footnotes or other explanatory notation is a good idea, particularly in more complex settings. In general, it is more important to get a representation that works and is reasonably understandable than to be concerned whether the result is appealing in some abstract sense.

5.8 NEGATIVE MULTIPLIERS

We now consider a somewhat novel type of construction which further expands the concept of what can happen to flow on an arc. An arc with a negative multiplier extracts flow from the nodes at both ends of the arc. For example, in Figure 5.22, the multiplier of -2 extracts two units of flow from node B for every unit of flow which leaves node A. However, if zero flow leaves node A on the arc, then no flow can be extracted from node B via this arc. (This is consistent with the fact that an arc with a positive multiplier cannot contribute flow to its head node unless the arc carries flow.)

The inclusion of an arrowhead on a negative multiplier arc is somewhat misleading though, since it tends to suggest that flow will be going to the arrowhead end. To avoid this implication, the linking arc may be shown without an arrowhead as in Figure 5.23.

There is an additional reason why removing the arrowhead supports the idea that flow is extracted. The tail end of the arc, which has no arrowhead, has the function of extracting flow from node A. Thus it may be construed that we have implicitly been dealing with -1 multipliers attached to the tail ends of arcs, without showing them.

One of the major applications of negative multipliers is to model proportional slacks. For example, suppose that each unit of product X consists of one type A component and three type B components. Initially we have 100 type A components and 300 type B components. If we view these supplies as fixed, then adding a restriction that says three type B components must be used for

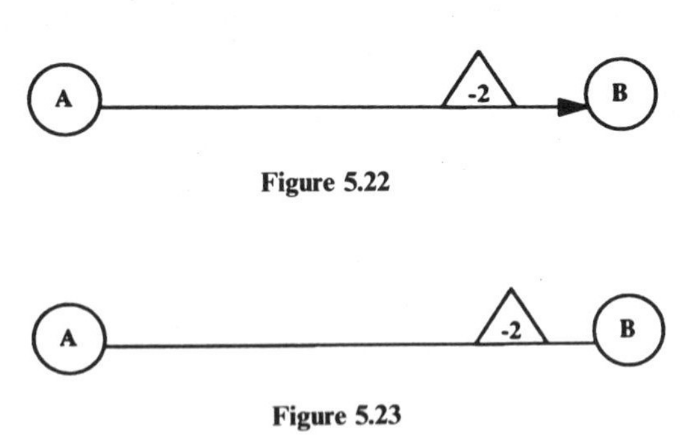

Figure 5.22

Figure 5.23

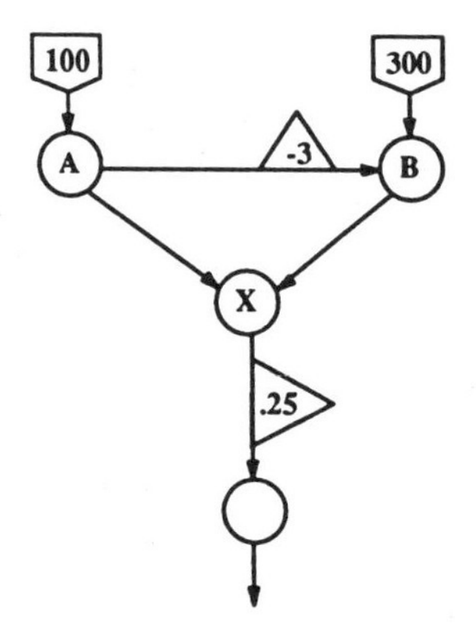

Figure 5.24 Proportional slack.

every one type A component used is equivalent to saying that three type B components must be left over for every type A component left over. The latter interpretation is modeled in Figure 5.24. The arc between nodes A and B functions as a slack arc. For every unit of flow (slack) that leaves node A on this arc, three units of flow (slack) are extracted from node B. Since the supplies are fixed, then the flows to node X from nodes A and B will also satisfy this one-to-three ratio.

Proportional slacks can also be used to insure that resources are consumed at equal rates. To illustrate, suppose that plant A has 25 units of resource and plant B has 35 units of resource available to supply three customers who each want at least 10 units. Management wants to distribute the resources at equal rates. That is, for every unit of resource shipped from plant A, 35/25 units should be shipped from plant B. As in the previous example, for fixed supplies, this is equivalent to requiring 35/25 units of slack at node B for every unit of slack at node A. The model is shown in Figure 5.25. Chapter 6 presents further uses of negative multipliers in discrete flow networks.

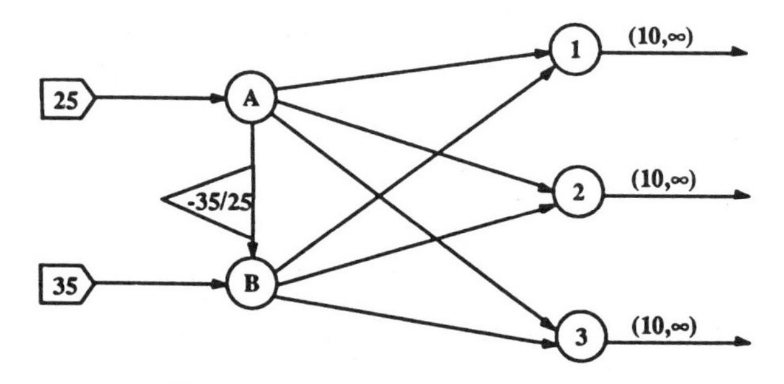

Figure 5.25 Use resources at equal rates.

5.9 ALGEBRAIC STATEMENT OF GENERALIZED NETWORK MODEL

A generalized network problem can be characterized by a coefficient matrix which has at most two nonzero entries in each column. As with pure network problems, each row of the coefficient matrix is associated with a node and each column with an arc. That is, a node corresponds to a problem equation and an arc corresponds to a variable. The coefficient matrix can be transformed using elementary row and column operations (if the variables are bounded) so that at least one entry in any column with two nonzero entries is -1. In this way, a directed arc is created that leads from the node associated with the -1 to the node associated with the other nonzero entry, which is the arc multiplier. (If both entries are -1, the arc may be directed either way.) A single nonzero entry in a column is represented by an arc that touches only one node.

These ideas are illustrated by the following generalized network problem.

$$\text{Minimize} \quad 1X_{12} + 5X_{13} + 3X_{23} + 1X_{24} - 4X_{32} - 9X_{34}$$

Subject to:

$$\begin{array}{lllllllll}
① & X_1 - 1X_{12} - 1X_{13} & & & & & & & =0 \\
② & 2X_{12} & & -1X_{23} & -1X_{24} + 1/3X_{32} & & & & =0 \\
③ & & 1/2X_{13} + 1X_{23} & & & -1X_{32} & -1X_{34} & & =0 \\
④ & & & & -1/5X_{24} & & +3X_{34} - X_4 & & =0
\end{array}$$

$$0 \leqslant X_1 \leqslant 5, \quad 0 \leqslant X_{12} \leqslant 3, \quad 0 \leqslant X_{13} \leqslant 4, \quad 0 \leqslant X_{23} \leqslant 6$$

$$0 \leqslant X_{24} \leqslant 5, \quad 0 \leqslant X_{32} \leqslant 3, \quad 0 \leqslant X_{34} \leqslant 7, \quad 10 \leqslant X_4 \leqslant 10$$

The associated network is shown in Figure 5.26. Note that in the "Total Inflow − Total Outflow = 0" equations, the inflow variables are multiplied by the arc multipliers.

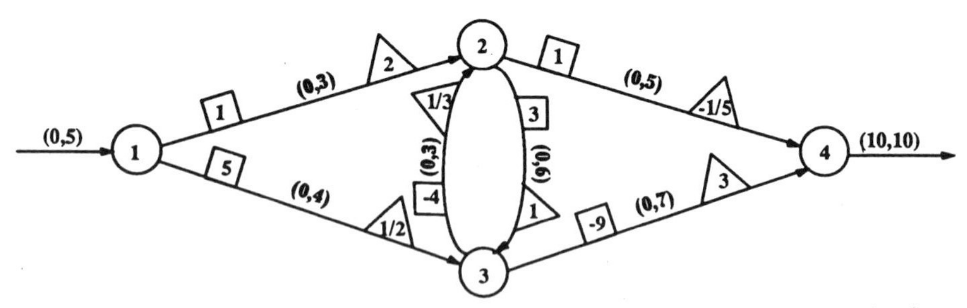

Figure 5.26 Generalized network. Reprinted by permission. Improved Computer-Based Planning Techniques, Part 1, F. Glover, J. Hultz, and D. Klingman, *Interfaces*, Vol. 8, No. 4, pp. 16–26, August, 1978.

Charnes and Cooper [1961] observed that some generalized network problems may be scaled to pure network problems by multiplying the rows and columns by appropriate factors. Glover and Klingman [1973] showed that any generalized network problem whose detached coefficient matrix does not have full row rank is equivalent to a pure network problem. Furthermore, they developed an efficient procedure to simultaneously determine the rank of the matrix and transform (if possible) the generalized network problem into a pure problem. An example is shown in Figure 5.27 where R_i is the row adjustment factor for row i and K_j is the column adjustment factor for column j. For example, to scale the column associated with the variable x_{12} to the pure network format, multiply row 1 by 1, row 2 by 2, and column 1 by 2. (The determination of these adjustment factors is not straightforward. Refer to Glover and Klingman [1973] for details.)

Many times, the scaling of a generalized network to a pure network may be derived intuitively by examining the network itself. This is especially true of problems whose multipliers are used solely to convert raw materials to finished goods. For example, in the paper manufacturing problem in Figure 5.6 we could omit the multipliers by thinking of all flow in terms of pounds of wood pulp. That is, we could scale the costs and bounds on the paper inventory and sales arcs to be in terms of pounds of wood pulp rather than reams of paper, as shown in Figure 5.28. For instance, the bounds on the paper inventory arcs in Figure 5.6 are 0–100 reams of paper with an inventory cost of $.3 per ream. Since each

Generalized Network Problem

	x_{12}	x_{13}	x_{23}	x_{24}	x_{25}	x_{36}	x_{34}	x_{43}	x_{45}	x_{46}	RHS	R_i
Costs	3/2	4	9	4	1	16	1	2/3	16	5		
	-1/2	-4									-4	1
	1/4		-3	-4	-1/2						-10	2
		2	3			-2	-1/2	1/3			0	2
				4			1/2	-1/3	-4	-5	0	2
					3/8				3		6	8/3
						2				5	4	2
K_j	2	1/4	1/6	1/8	1	1/4	1	3/2	1/8	1/10		

Scaled Pure Network Problem

	y_{12}	y_{13}	y_{23}	y_{24}	y_{25}	y_{36}	y_{34}	y_{43}	y_{45}	y_{46}	RHS
Costs	3	1	3/2	1/2	1	4	1	1	2	1/2	
	-1	-1									-4
	1		-1	-1	-1						-20
		1	1			-1	-1	1			0
				1			1	-1	-1	-1	0
					1				1		16
						1				1	8

Figure 5.27 Reprinted by permission. On the Equivalence of Some Generalized Network Problems to Pure Network Problems, F. Glover and D. Klingman, *Mathematical Programming*, Vol. 4, No. 3, pp. 269–278, 1973.

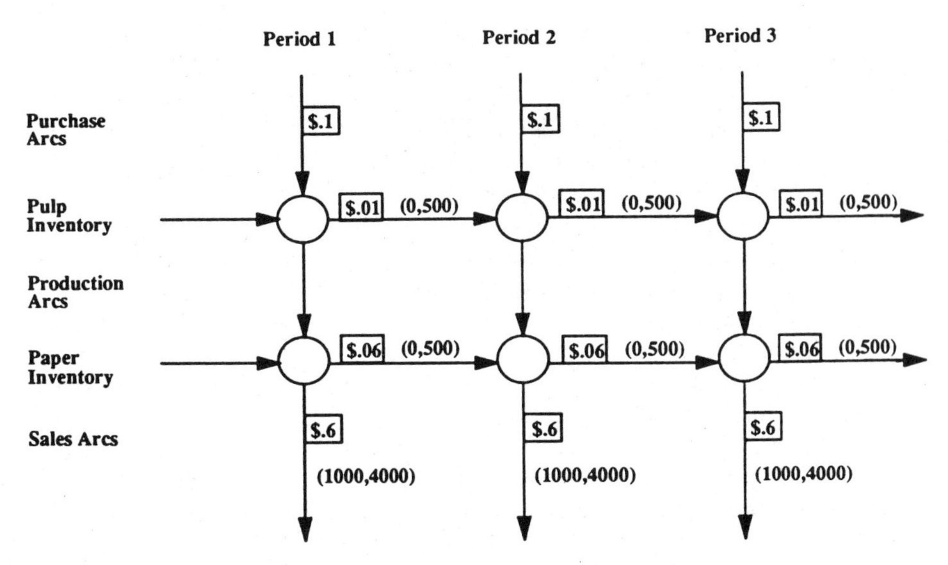

Figure 5.28 Scaled paper manufacturing problem.

pound of wood pulp yields 0.2 reams of paper, then 100 reams of paper would be equivalent to 500 pounds of wood pulp. The cost to inventory the amount of paper equivalent to a pound of wood pulp would be 0.2 (reams/pound) $\times$ 0.3 ($/ream) = 0.06 ($/pound). Note that this cost is not the same as the cost to inventory a pound of pulp (on the pulp inventory arcs), since what we are carrying on the paper inventory arcs is actually finished paper (which has in essence increased the value of the pulp).

5.10 THE GENERALIZED NETWORK DOMAIN: A HISTORICAL NOTE

Generalized networks enlarge the range of application of pure networks by a very substantial increment. Applications involving generalized networks have appeared in the literature with increasing frequency in recent years, coming now to dominate applications involving pure networks. This marks an evolutionary trend that is briefly worth tracing.

As a model type, generalized networks were known in the late 1950s, but received scant attention for nearly a decade and a half. Then, following the discovery that pure networks could be solved with unprecedented efficiency by specialized methods, researchers began to ask whether the same might be true for generalized networks. The question was resolved affirmatively in the early 1970s when computer solution routines were developed that proved able to solve generalized network problems fifty times faster than the best commercial linear programming solution packages (see Grigoriadis [1980] for a survey).

To grasp the practical implications of this solution advance, note that an analyst conducting "what if" investigations could solve a different problem scenario every day for nearly 2 months and still consume no more computer time than required to solve just one scenario using the best procedures previously available. The importance of this for problems where conditions change from day to day is clearly enormous.

The number of problems formulated as generalized networks abruptly increased as a result of this impetus. As it gradually became apparent that generalized networks not only could be solved efficiently, but also shared modeling advantages in common with pure networks (due to their pictorial aspect), the number of practical generalized network applications grew still further.

There is often a lag between theory and practice. In the realm of generalized networks, however, the lag has been of another kind. While industry and government have been markedly expanding their use of generalized network models, few textbooks make more than a passing reference to them, and most make no reference at all. Obviously this is a state of affairs that will not last. Practitioners in the fields of management science and decision support who have an understanding of generalized network models are a solid step ahead of their less knowledgeable peers in dealing with the broad range of problems that can be expressed and solved (to great advantage) within this framework.

APPLICATION: MACHINE SCHEDULING MODEL

A large paper manufacturer has 25 machines which are used to produce various types of paper such as 100% cotton bond. Each week they receive approximately 5800 customer orders, each for a single type of paper. Each machine cannot make every type of paper. The network shown in Figure 5.29 models the problem of scheduling customer orders on machines for a single week.

The supply a_i represents the capacity of machine i in hours. The demand b_j is the pounds of a particular paper product requested in customer order j. The cost coefficients, c_{ij}, are cost per hour of production for customer order j on machine

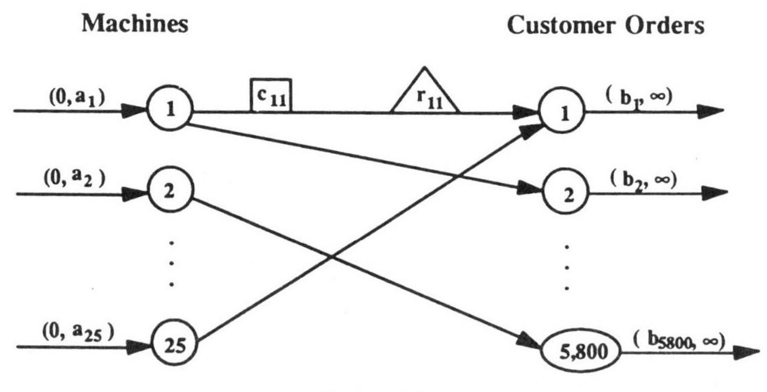

Figure 5.29

i. Finally, the multiplier r_{ij} is machine i's production rate in pounds per hour of customer order j's product type.

The manufacturer wanted each customer's order for a particular paper product to be processed by a single machine in order to assure uniformity of the product. This would necessitate requiring that the generalized network model have an integer solution, a topic discussed in Chapter 6. However, due to the structure of this particular model, at most 25 of the 5800 customer orders would be split between machines. (Each node corresponds to a constraint, and there is a basic variable for each constraint. Thus at most 5825 machine/customer arcs will have nonzero flow. See the discussion of basic solutions in Appendix A.) The manufacturer decided that these split orders could just be postponed until the next week's scheduling run, thus avoiding the decreased solution efficiency of requiring integer solutions. For a typical problem with 25 machines, 5800 customer orders, and 22,000 arcs, solution times on an IBM 370/168 computer were 2 min for a generalized network code (NETG) compared to 37 min for a linear programming code (MPSX-370).

CASE: W. R. GRACE COMPANY, PART A

The W. R. Grace Company case describes a real company and the problem situation which it faced [Klingman et al. 1988]. You should approach this case from the point of view of an independent consultant who is going to submit a proposal for the project. The proposal will consist of a model description and a solution approach. Proposals will be evaluated using two equally weighted criteria: (1) model validity (the capability of your model to find the correct solution) and (2) practicality of approach. (A valid model which requires 6 months of computer time to solve may not be viewed as highly practical. You are not limited to the employment of network models. However, their proven solution efficiency could greatly enhance the practicality of the approach. Beyond that, the actual structure of different network models will also affect their practicality.) There is more than one correct model. However, every model should be capable of finding the "best" solution.

Report Considerations

Your report need not be elaborate or lengthy. However, it should include at least the following elements:

1. A graphical representation of your model (even if it is not strictly a network model) with clearly defined notation. By a graphical representation, we mean a diagram which presents the problem conceptually but does not necessarily contain every variable and constraint.

Reprinted with permission from *Operations Research Society of America*, Vol. 36, No. 6, pp. 811–822, 1988.

2. A short verbal description of the model which clarifies the graphical representation (2 pages maximum).
3. A statement of how all cost parameters are determined and if your model is minimizing or maximizing.
4. Answers to all questions presented with the case description.

Remember, this is a proposal which you hope will win a contract award. Neatness and readability are prerequisite requirements.

INTRODUCTION

The W. R. Grace Company is involved in the production, distribution, and sale of a number of phosphate-based chemical products. The company's phosphate product line has its primary processing and production facilities located in Bartow, Florida. Phosphate rock is mined and processed at Bartow to produce phosphoric acid (P_2O_5), which is the primary resource used in the production of the phosphate products.

Besides the phosphoric acid plant, W. R. Grace has three other large production facilities located at Bartow. For obscure reasons, these three plants are referred to within the company as "300X", "300Y", and "FERT".

The company also has regional plants and warehouses located in different parts of the country which handle all or part of the product line. There are twelve regional warehouses, nine regional granular mixed fertilizer plants, and five blend mixed fertilizer plants.

The company's product line consists of diammonia phosphate (DAP), monoammonia phosphate (MAP), granular triple sulphate (GTS), ronapile triple super (RTS), and mixed fertilizers A and B (GMFA and GMFB).

Production and storage facilities exist at the regional granular mixed fertilizer plants as well as at the plants at Bartow. Phosphoric acid is shipped from Bartow to the regional granular mixed fertilizer plants to be used in the production process. In order to allow for the seasonal fluctuations in demand, finished products can also be shipped to the regional plants from the plants at Bartow.

The blend mixed fertilizer plants and the regional warehouses obtain finished products from the Bartow plants as well as from the regional granular mixed fertilizer plants.

The complex decisions of how to allocate the available phosphoric acid for each product and to each manufacturing plant, as well as how to distribute products across the country, were being made based on the judgment of the individual regional and production managers. In order to coordinate these production, distribution, and inventory decisions over a multi-period planning horizon and thus create an operational plan which would be optimal for the entire organization, W. R. Grace wanted to develop a computer-based planning system using management science techniques. A management consulting firm was brought in for this purpose.

Bartow

Of the three finished goods production facilities in Bartow, Plant 300X uses phosphoric acid to make DAP and GTS; Plant 300Y produces only GTS, and Plant FERT produces MAP and DAP. The production technology is such that a ton of phosphoric acid yields 2.04 tons of DAP, or 2.75 tons of GTS, or 1.83 tons of MAP. RTS, an intermediate product used in the manufacture of other products, is also produced at the Bartow phosphoric acid plant. The yield ratio for RTS is 2.86 tons per ton of phosphoric acid.

All the finished products produced at Bartow as well as the phosphoric acid can be used to meet external demand at Bartow or stored for use in future periods. DAP, MAP, RTS, and P_2O_5 can be shipped to meet demand at the regional warehouses, and DAP can be sent to the blend mixed fertilizer plants.

External demand at Bartow consists of export and regional and national sales. In addition, the company often "loans" P_2O_5 as well as finished products to other companies which repay W. R. Grace in future periods. These exchanges help the companies smooth out some of the seasonal fluctuations in demand.

Regional Granular Mixed Fertilizer Plants

At the regional plants, phosphoric acid is used to make DAP, MAP, GMFA, and GMFB. A ton of phosphoric acid yields a ton of GMFA or GMFB. Yield ratios for all products are the same at all production facilities. The regional plants also have storage facilities for the RTS received from Bartow.

The products can be used to meet demand at the regional plants, stored, or shipped to the regional warehouses. DAP can also be sent to the blend mixed fertilizer plants.

Regional Blend Mixed Fertilizer Plants

DAP, received from the plants at Bartow and from the regional granular mixed fertilizer plants, is utilized in blending processes at these plants. No storage facilities exist.

Regional Warehouses

Facilities exist for the storage and sale of all finished products except RTS at the regional warehouses.

Supply

The Bartow facilities produce phosphoric acid at the rate of 1350 tons per month.

Demand

The management of W. R. Grace is currently planning production for the next 12 monthly periods. Reliable sales forecasts are available for each plant and warehouse, though not at the individual customer level.

For illustration purposes we assume there exists just one regional granular mixed fertilizer plant, one regional warehouse, and one blend mixed fertilizer plant. Demand figures for these facilities are presented in Table 5.1. Assume these demands must be met exactly.

OBJECTIVE

The shipments of W. R. Grace's products are all made by rail, with a fixed fee of $0.03 per ton-mile. The distances between origins and destinations are summarized in Table 5.2. Management wishes to minimize shipping costs.

Table 5.1 Demand (tons)

Period 1	Bartow Plants	Regional Gran. Mixed Fertilizer Plants	Blend Mixed Fertilizer Plants	Regional Warehouses
P_2O_5	200	-	-	-
DAP	350	250	200	300
MAP	150	100	-	100
GTS	350	-	-	350
RTS	50	100	-	-
GMFA	-	50	-	50
GMFB	-	50	-	50
Period 2				
P_2O_5	180	-	-	-
DAP	370	235	185	320
MAP	145	120	-	100
GTS	345	-	-	340
RTS	60	110	-	-
GMFA	-	40	-	45
GMFB	-	60	-	50

Table 5.2 Rail Distance (miles)

	Bartow Plants	Regional Gran. Mixed Fertilizer Plants	Blend Mixed Fertilizer Plants	Regional Warehouses
Bartow Plants	-	500	600	700
Regional Gran. Mixed Fertilizer Plants	500	-	200	300
Blend Mixed Fertilizer Plants	600	200	-	250
Regional Warehouses	700	300	250	-

Case Questions

1. Formulate a generalized network model of the problem.
2. Scale your generalized network model to a pure network format by representing all costs and demands in P_2O_5 equivalent units.
3. Given a solution for the pure network, explain how you would find a solution for the original problem.
4. What is a "necessary" condition for a balanced pure network model to have a feasible solution? Does the model you created satisfy this condition? (Justify your answer).
5. What are the advantages of using the pure network form of the model? What are the disadvantages?

CASE: W. R. GRACE COMPANY, PART B

After the generalized network of Part A had been developed, the consulting group had another meeting with management. In the discussions that followed, it became apparent that a number of production, storage and transportation restrictions had not been considered.

At Bartow, the monthly production capacity of Plant FERT is limited to either 900 tons of DAP or 540 tons of MAP, or a proportional combination of the two products. The production capacity of Plant 300Y is 500 tons of GTS per month, and Plant 300X can produce 1500 tons of DAP or GTS, or a proportional combination of the two products. In addition, management policy stipulates that all of the Bartow plants must be operated at full capacity.

The time available for production at the regional granular mixed fertilizer plant is 1200 hours per month. The time required to produce GMFA, GMFB, DAP, and MAP from a ton of phosphoric acid is 2 h, 1 h, 1 h, and 2 h, respectively. In order to produce GMFA and GMFB from a ton of phosphoric acid, the blending process requires 0.7 and 0.8 tons of RTS, respectively. So the total amount of GMFA and GMFB that can be produced depends on the availability of RTS, or alternatively, the demand for RTS depends on the amounts of GMFA and GMFB produced.

The storage capacity of the regional granular mixed fertilizer plant is 6500 tons per month and that for the warehouse is 8000 tons per month.

Management wants to maximize profits while running the Bartow plants at full capacity. It is not necessary to meet all demands if doing so is not profitable. Sales prices for the products are summarized in Table 5.3 and are assumed to be the same all over the country. Production costs are the same at all plants.

Case Questions

In your solutions to the following problems, any nonnetwork constraints should be stated algebraically. Be sure to define all variables.

1. Modify your model of Part A (adding linear side constraints if necessary) to accommodate the production capacity for Plants 300X, 300Y and FERT.
2. Introduce a blending side constraint for the production of GMFA and GMFB from RTS.
3. The algebraic statement of your network model, together with the side constraints above, make up an LP model for the entire problem. The network portion of this model is referred to as an embedded network.

Table 5.3 Prices ($/ton)

P_2O_5	DAP	MAP	GTS	RTS	GMFA	GMFB
10	30	25	35	32	45	38

Assuming that your solution approach is to first solve the embedded network model, suggest ways to improve (strengthen) the embedded network model in a manner which would make a solution to it a better, but still relaxed, starting solution for the entire LP problem including the constraints corresponding to Problems 1 and 2 above. (Relaxation requires that any solution to the entire LP will be a feasible solution to the embedded network, but not vice versa. Hint: Approximate the production capacity constraints by bounds on the arcs leading into each plant node.)

4. Introduce side constraints for production time and storage capacity.
5. How would you change the model for profit maximization? Assume now that demands need not be met.

EXERCISES

5.1 An investor has $200,000 to allocate between two types of investments. The first investment matures in one year and yields 80 cents for each $1 invested. The second matures in 2 years and yields $1.10 for each $1 invested. (Yields are amounts in addition to the quantity invested.) Both types of investments are available at the start of any given year but money invested may not be withdrawn until an investment matures. Construct a model to determine how the investor should allocate his money to maximize his return at the end of three years.

5.2 A company has a fleet of buses and a fleet of ships for carrying passengers on scenic trips. To power its vehicles the company can purchase two types of fuels. The mileage the buses and ships can get from the fuels are as follows:

	Miles/Gallon	
	Buses	Ships
Fuel 1	16	8
Fuel 2	19	6

For the scheduled trips, the buses must travel exactly 200,000 miles and the ships exactly 30,000 miles. Fuel 1 costs \$1.20/gallon and Fuel 2 costs \$1.25/gallon (\$1.00 buys 0.833 gallons of Fuel 1 and 0.8 gallons of Fuel 2). In addition, the source for Fuel 1 can supply only 2000 gallons (\$2400 worth) of fuel at its current price, and after that must charge \$1.22/gallon (each \$1.00 buys 0.82 gallons). The available fuel budget (upper limit) is \$20,000. Develop a model to determine how much fuel of each type to purchase for each use, to minimize total dollars spent. View the available fuel budget as a supply.

5.3 (Lost Sales from Backorders) Suppose that sales demand is exactly 200 units in Period 1, but that an attempt to meet any part of this demand by backordering from Period 2 will result in losing 2% of that portion of demand. (For example, backordering to meet 100 units of demand will lose 2 units, or in other words, 98 units backordered will offset 100 units of demand.) Assume that sales revenue is \$50/unit, and that a \$3 penalty is attached to each unit backordered (independent of the loss of revenue from reduced sales) to reflect a deterioration in customer goodwill. Identify which of the following partial diagrams, (a), (b), or (c), captures this situation and explain why. (Purchase arcs and inventory carrying arcs are not shown.)

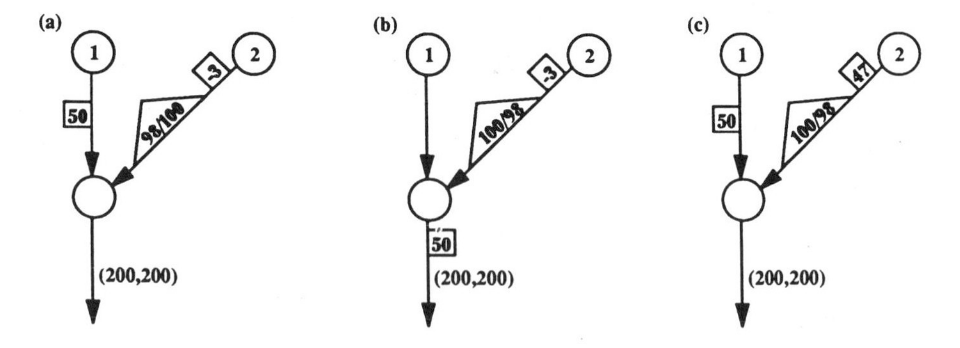

5.4 Refer to Figures 5.14 and 5.15. Indicate what modifications would be required to handle the following conditions (treat each condition independently):

a. In addition to his regular hourly rates, John charges \$1.00 for each hour he works on Task 1 and $1.50 for each hour he works on Task 2.

b. Of Marc's available hours, up to 15 may be spent on odd jobs that bring in a revenue (to Marc's employer) of $12 for each hour he devotes to them.

c. John is able to complete no more than half of Task 1 and no more than two-thirds of Task 2.

d. A third task must also be completed, which John can do in 8 h and Marc can do in 12 h. (The rule of "fractional completion for fractional time spend" still applies.) However, Marc can devote no more than 30 h to Tasks 1 and 2 combined.

5.5 (Integrated Production/Distribution Model 1) A company has two plants that make regular (R) and high fidelity (HF) cassettes. For the period under consideration, the number of available hours of production, the hourly production cost, and the number of cassettes that can be made in 1 h (if devoted to making one kind of cassette only) are shown in the following table.

			No. of Cassettes (of 1 type only) Made per Hour	
	Available Hours	Hourly Cost ($)	R	HF
Plant 1	80	15.00	19	14
Plant 2	120	13.00	16	12

The tapes, once produced, are sent on to two company distribution centers. Shipping costs per cassette appear below:

Shipping Cost per Cassette

	Distribution Centers	
	1	2
Plant 1	0.30	0.18
Plant 2	0.15	0.25

Each distributor has minimum demands for cassettes, to be certain of meeting the needs of regular customers. These demands and the revenue for cassettes (above distribution center handling costs) are as follows:

Minimum Demands and Unit Revenue

Distribution Center	Cassette Type			
	R	R	HF	HF
1	650	$1.80	200	$2.10
2	400	$1.85	400	$2.30

Finally, each center can handle a maximum of 1600 cassettes of both types combined. Identify a profit maximizing generalized network for this problem, starting from the Plants (as "suppliers") and leading ultimately to the Distribution Centers. Label nodes P1, P2 (for plants), D1, D2 (for distribution centers), and R and HF (for regular and hi fi), creating four each of the R and HF nodes.

5.6 (Integrated Production/Distribution Model 2) A candy manufacturer operates two stores that make fudge and divinity and then ship these products to two principal distributors. Store 1 has 160 h of available labor per week, and Store 2 has 200 h of available labor per week. The workers and facilities of the two stores make it possible to produce candy at the rate indicated in the following table.

Pounds of Candy Produced for Each Hour of Labor

	Fudge	Divinity
Store 1	12	9
Store 2	11	10

Shipping costs and the demands of distributors for candy are indicated in the following two tables.

Dollar Shipping Costs Per Pound of Candy

From	To	
	Distributor 1	Distributor 2
Store 1	0.18	0.24
Store 2	0.33	0.20

Weekly Demands for Candy in Pounds (Lower Bound, Upper Bound)

	Fudge	Divinity
Distributor 1	1500, 2500	
Distributor 2	1000, 3000	1000, 3000

The company gets $1.40/lb for fudge shipped to Distributor 1, and $1.30/lb for fudge and $1.50/lb for divinity shipped to Distributor 2. (Distributor 1 does not require divinity.) Assuming labor costs are fixed at each store regardless of hours worked, devise a generalized network for maximizing weekly net revenue. Create only five nodes, labeled S1, S2 (for stores) and D1F, D2F, D2D for distribution center demand for fudge and divinity. Is there a disadvantage in using such a minimal number of nodes?

5.7 The following problem is a simplified version of one routinely solved by the Transport and Road Research Laboratory of England. (The problem is re-solved for new data in each successive month.) Sand, crushed rock, and shale can be extracted at the locations of Aleshire, Bromley, and Chadwick, subject to the restriction that no more than 100 tons/month of any given material can be extracted at any location. In addition, the upper limit on all materials combined that can be extracted during the month is 290 tons for Aleshire, 270 tons for Bromley, and 250 tons for Chadwick. The extracted materials are then to be shipped from their points of origin to destinations Dorset and Exeter, where they are used to produce concrete and sub-base (for road construction). Any of the three materials can be used to make sub-base with complete substitutability (e.g., in principle, 9 tons of sub-base can be made from 9 tons of any of the materials, or from 3 tons of each, etc.). Concrete can be made from either sand or crushed rock (again, with complete substitutability). However, the principle of complete substitutability operates in reality subject to an important qualification. Due to imperfect processing, 1% of the sand and 2% of the shale are lost in making sub-base, while 2% of the sand and 1% of the crushed rock are lost in making concrete. The demands and costs of the system are summarized as follows:

Demands (Tons/Month)

	Concrete	Sub-Base
D	160	170
E	180	140

Distribution Cost ($/ton)

	Destination	
Source	D	E
A	25	40
B	35	25
C	15	30

Extraction Cost ($/ton)

Sand	Shale	Crushed rock
40	50	45

Processing Cost ($/ton of product produced)

	Product	
Destination	Concrete	Sub-Base
D	6	10
E	7	9

Prepare a generalized network for minimizing total cost.

5.8 (Inventories of Perishable Goods) A Chemical Company produces a high-grade liquid propellant used in certain specialized aerospace applications. The manufacturing process requires 1500 gallons of "raw fuel," obtained from a supplier, to produce 675 gallons of the propellant. Because several stages of processing are required, raw fuel must be obtained a full period ahead of the time when the propellant is to be delivered to meet demand. In addition, due to the presence of certain additives, the propellant deteriorates over time, so that 2% of the volume kept on hand from any period to the next becomes a useless residue, unavailable for sale. Supplies of raw fuel and demands for propellant over the next 6 periods, expressed in gallons, are given in the following table, along with associated costs and prices.

	Periods					
	1	2	3	4	5	6
Maximum available raw fuel	8600	10000	9200	9000	8400	8600
Minimum required propellant	2000	2500	2800	3300	3000	2800
Maximum required propellant	3000	3000	5000	5000	4000	4000
Cost/gallon of raw fuel ($)	2.50	2.50	3.00	3.50	3.50	3.00
Price/gallon of propellant ($)	38.00	40.00	42.00	42.00	42.00	42.00

The cost of processing each gallon of raw fuel to make propellant is \$11.00, and the cost to store a gallon of propellant for one period is \$1.30, with an upper limit of 1500 gallons on this storage capacity. (Residue is removed at the end of each period so that the volume is correspondingly reduced. Storage cost includes an allocation to cover this cost of removal.)

Assuming no facilities are available to store raw fuel, but that the company begins with 2500 gallons of propellant on hand in Period 1 and requires an inventory of this same amount for Period 7 (from Period 6 raw fuel purchases plus inventory carried forward), prepare a generalized network to maximize revenues.

5.9 An investor has money-making activities A and B available at the beginning of each of the next 5 years. Each dollar invested in A at the beginning of one year returns \$1.40 (a profit of \$0.40) 2 years later (in time for immediate reinvestment). Each dollar invested in B at the beginning of one year returns \$1.70, 3 years later. In addition, money-making activities C and D will each be available at one time in the future. Each dollar invested in C at the beginning of the second year returns \$2.00, 4 years later. Each dollar invested in D at the beginning of the fifth year returns \$1.30, 1 year later. Money held as cash from year to year earns no profit. The investor begins with \$10,000. He wishes to know which investment plan maximizes the amount of money he can accumulate by the beginning of the sixth year.

a. Construct a generalized network model to find the optimal investment plan.

b. Write the algebraic statement of your network model in part (a). Is it possible to create a balanced generalized network by adding up all the constraints and multiplying both sides by -1 to get a redundant constraint?

5.10 On January 1 a distributorship for a manufacturer of heavy machinery estimates its monthly sales for the first quarter to be as follows:

	January	February	March
Expected sales (in units of product)	144	132	158

The current inventory is 65 units, and inventory purchases can be made on the 1st or 15th of each month, requiring an approximate one-half month lag time before delivery (at which time they are paid for). Sales are expected to cluster around the middle and end of each month, equally divided between them. Delivery on up to 30 units of sales in any half-month period may be postponed to the following half-month period at a cost of $10/unit, and delivery on up to 10 additional units may be postponed for a full month at a cost of $15/unit. The cost for holding a unit in inventory is set at $4/unit for every unit carried between successive half-month periods. Company policy requires an inventory of 80 units (with no back-logged deliveries) to be available for the start of the second quarter. Each inventory purchase is limited to a maximum of 81 units.

a. Prepare a netform to determine the amount to be ordered in each time period to minimize total cost.

b. Suppose that late servicing of demand results in order cancellations as follows: 20% cancellation of outstanding orders which are one-half month late; 40% cancellation of outstanding orders which are a full month late. Modify your model to account for these lost sales. (Hint: Use multipliers to convert the backordered quantities so that they will offset the original demand node requirements.)

5.11 A certain corporation has three branch plants with excess production capacity. All three plants have the capability for producing a certain product, and management has decided to use some of the excess production capacity in this way. This product can be made in three sizes—large, medium, and small—which yield a net unit profit of $12, $10, and $9, respectively. Plants 1, 2, and 3 have the excess manpower and equipment capacity to produce 500, 600, and 300 medium size units per day of this product, respectively, and it takes 1/3 more time to produce the large size and 4/5 of the time to produce the small size. Sales forecasts indicate that 600, 800, and 500 units of the large, medium, and small sizes, respectively, can be sold per day. In order to maintain a uniform workload among plants 1 and 2 and to retain some flexibility, management has decided that the additional production assigned to plants 1 and 2 must use the same percentage of the excess manpower and equipment capacity. Management wishes to know how much of each of the sizes should be produced by each of the plants in order to maximize profit. Formulate the generalized network model for this problem.

5.12 An investor can join a limited partnership which has the options listed in the table below. He can invest up to $1,000,000 at the start but must agree that for every $1 he invests at the beginning of the first year, he will invest $0.5 at the beginning of the third year, in addition to the original money which may be available for re-investment. Construct a netform to maximize the return at the end of 6 years.

Investment Type	Return for $1	Investment Timing Restrictions
Stocks	$1.20 2 years later	Available at beginning
Bonds	$1.40 3 years later	of each of next 6 years
S&L certificates	$1.80 4 years later	Available at start of 2nd year
Real estate	$1.10 1 year later	Available at start of 5th and 6th years
Hold as cash	$1.00 1 year later	No Restrictions

5.13 A new chain of doughnut shops, Dr. J's Dunk'Em Donuts, opened recently in Philadelphia. Dr. J's has one bakery and three retail shops, one of which is located in the same building as the bakery. Dr. J's produces doughnuts in lots of 144. Each lot requires 5 lb of dough. Dr. J's has contracted with suppliers to provide ingredients for at most 150 lb of dough each day. The bakery is capable of making 12 lots of doughnuts in one production run. Because of cleaning and maintenance, at most three production runs can be made in one day. These occur at 6 a.m., 10 a.m., and 3 p.m. (only if required). Doughnuts are estimated to cost 15 cents each (including dough costs) to produce. After a production run, doughnuts are taken immediately to the retail shops. There is no inventory kept at the bakery. There is a handling charge of 75 cents per lot for all doughnuts moved to the shop adjacent to the bakery. Costs of trucking the doughnuts to the remote shops are estimated to be $5.00 per lot (regardless of destination). Truck capacity far exceeds any amounts that might be shipped. Doughnuts are sold by the "bucket" or individually. A bucket contains one dozen and sells for $3.00. Doughnuts sold in quantities less than a bucket sell for 35 cents each. Output of the three possible production runs arrives at the shops at 8 a.m., 12 noon, and 4 p.m., respectively. Dr. J's estimates that loss of freshness results in a 30% spoilage for each 4 h period which passes from the time of production. In other words, only 70% of the doughnuts delivered at 8 a.m. which are not sold between 8 a.m. and 12 noon will be saleable between 12 noon and 4 p.m., and so on. Likewise, there is a 10% immediate spoilage associated with any doughnuts which have to be transported by truck. Estimated demand by store and time periods are as follows:

		Store 1 (at bakery)	Store 2	Store 3
8 a.m.–12 noon	Buckets	50	30	40
	Individually	100	80	70
12 noon–4 p.m.	Buckets	35	25	30
	Individually	80	50	60
4 p.m.–8 p.m.	Buckets	40	40	50
	Individually	90	50	40

All doughnuts which have not been sold after 8 p.m. are given to employees or thrown away. Dr. J's wants to know how many doughnuts to produce in each possible production run and how many to send to each store so as to maximize profits while satisfying anticipated demand only if it is profitable to do so.

a. Create a netform which will model Dr. J's operation.

b. Can your model be converted to a pure network formulation? If no, explain. If yes, how would you do it? (Do not actually attempt a transformation; just explain).

5.14 The Columbian Chemicals Co. produces a product called carbon black for sale to commercial consumers in various industries. Due to the diversity of its use, there are many different grades of carbon black. For example, a strictly refined and extremely pure grade is required by the xerography industry for toners, whereas much less pure grades are required by the automotive and tire industries. In all, Columbian produces 40 different grades of carbon black. Columbian has four production facilities which each produce all 40 grades of carbon black. Due to differences in the refinement process, the number of pounds of carbon black produced from 1 lb of raw material is different for each grade (i.e., production efficiency varies for each grade). Also, since the plants vary in size, age, and staffing, the production efficiencies for a particular grade will be different for each plant. The product (various grades of carbon black) can be stored at the production sites or shipped directly to customers in one of two forms, bag or bulk. Bulk shipments are those for which the product is unpackaged and sent in carload quantities. For bag shipments, the product is actually packed in bags and may be shipped in carload, truckload, or less-than-truckload quantities. Columbian's customers specifically request bag or bulk shipments for each grade they buy. Columbian serves approximately 200 customers on a monthly basis. Customer demands can be categorized by customer, product grade, bag or bulk shipment, and time period. On the average, each of the two hundred customers will require five different product-grade/shipment-type combinations each month. The availability of raw material is known for each plant in each month. Each production facility can potentially serve all of the various demands of each customer, which are also known or can be forecasted. Assume the various costs (production, bagging, inventory, shipping, etc.) and sales prices are available.

a. Create a network model which will give Columbian an optimal (i.e., maximum profit) production schedule and distribution plan for a 12-month planning horizon. The network should represent all problem components, but need not contain every single node and arc.

b. Given what you know about Columbian's situation and your model, estimate as closely as possible the number of nodes and arcs in the entire 12-month model (show your calculations).

5.15 Nutworks Inc. produces both Narcs and Widgets in plants A and B. Production rates are 2 Narcs per hour and 5 Widgets per hour at plant A and 3 Narcs per hour and 4 Widgets per hour at plant B. The capacity of plant A is 60 h regular time plus 30 h overtime and the capacity of plant B is 80 h regular time and 40 h overtime in each period. Production cost is \$100 per hour regular time and \$150 per hour overtime. All production goes into inventory or is shipped to customers. The inventory cost is negligible. Nutworks has received the following orders from customers P and Q for the next two time periods:

Customer	Product	Period 1	Period 2
P	Narcs	100	100
P	Widgets	250	150
Q	Narcs	200	100
Q	Widgets	300	150

The shipping cost per unit (Narcs or Widgets) from plant A to customers P and Q is \$10 and \$5, respectively, and the shipping cost from plant B to customers P and Q is \$5 and \$15, respectively, both for Narcs and Widgets. The selling price per unit for Narcs is \$90. Widgets sell for \$50. Nutworks has guaranteed to fill each order within the planning horizon. If a product has to be backordered, however, Nutworks will give a 10% discount per unit backordered.

a. Draw a netform to determine the most profitable production and distribution plan for Nutworks.

b. Assume 1 Narc requires 3 ft^2 and 1 Widget requires 2 ft^2 of inventory space. Can the network be modified for limited inventory space at plants A and B? Explain!

5.16 The demand for cars at the Big K Rent-A-Car Company is highly seasonal. Consequently, Big K needs varying numbers of trained agents to staff their counters at different airports. The estimated demand for trained agents in month t is given and denoted by $D_t, t = 1, \ldots, T$. If there is a shortage of trained agents in any period, Big K can get (trained) temporary help from an agency at a cost of \$3000 per person per momth. Alternatively, Big K can hire new agents at the beginning of each month. New agents must be trained for 1 month by an experienced (=trained) agent, at a ratio of 10 trainees per teacher. The company never fires its own agents because of insufficient demand but employees without work get paid 70% of regular wages while trainees receive 50%. Agents who work at a counter or train new employees receive \$2000 per month. Approximately 5% of all employees (including those being trained) leave the company at the end of each month. The number of trained agents available at the beginning of the first month is A.

a. Draw a netform to determine a least cost hiring plan, assuming a teacher will always have 10 trainees.

b. Modify your model in part (a) to allow classes of fewer than 10 trainees.

5.17 A hydroelectric power system on a river consists of two dams, their associated reservoirs, and power plants A and B as shown below. Note that water used to produce power at plant A flows on to reservoir B where it can be used to produce power (again) at plant B.

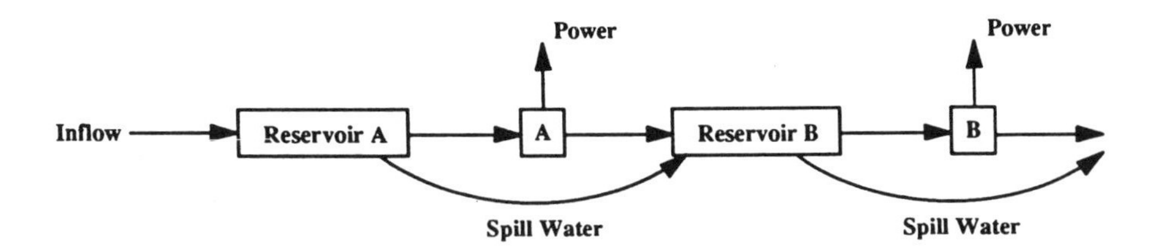

Flow quantities are measured in 10^3 acre feet (KAF) and power is measured in megawatt hours (MWH). The estimated inflows into reservoir A during $t = 1, 2, \ldots, T$ are given and denoted by S_t [KAF]. Inflows and outflows occur at uniform rates during each month, and approximately 5% of the water in each reservoir evaporates per month. If the capacity of a reservoir is exceeded, the excess water runs down the spillway and bypasses the power plant. The storage capacity of reservoirs A and B is 2000 and 1500 KAF, and the water-power conversion rate is 400 and 200 MWH/KAF, respectively. Power can be sold for \$5 per MWH. The capacity of power plants A and B is 60,000 and 25,000 MWH per month.

a. Draw a netform to determine the most profitable water flows.

b. Assume the demand for electric energy in month t, denoted by D_t[MWH], must be met, and shortages have to be filled by importing energy at a cost of $6 per MWH. Can this be modeled as a network? Explain.

5.18 The Big K Trading Company purchases a product at the beginning of periods $t = 1, \ldots, T$, for C_t $/unit. Purchases go into inventory and are subsequently sold at the end of each period for P_t $/unit. The inventory can hold no more than L units, and sales in period t cannot exceed the demand D_t. The variable inventory cost is negligible. At the beginning of the planning horizon Big K has I units in inventory, and their cash position is M dollars. Purchases have to be paid at once, but cash from sales is received with a time delay of one period. In addition, Big K has fixed costs in the amount of F dollars, payable at the beginning of each time period. Up to $B1 financing is available from the bank each period. Borrowed funds must be repaid in the following period with 5% interest. In addition, Big K can obtain a loan of up to $B2, once, at the beginning of the second period. This loan must be repaid after exactly two periods with 4% per period compound interest.

a. Draw the netform which maximizes Big K's cash position at the end of the planning horizon, after all loans have been repaid.

b. Assume that a 15% value-added-tax is due at the end of the planning horizon, payable on the difference between total sales minus total purchases (in $). Can this be modeled? Explain.

5.19 The Double-K-Aluminium Co. operates two bauxite mines in cities A and B. Capacities are 30,000 and 20,000 tons of bauxite per month, respectively. Bauxite mined at A or B is shipped to reduction plants L, M, and N where the bauxite is converted into alumina pellets, an intermediate product. One ton of bauxite yields 0.1 tons of alumina pellets. The capacities of the reduction plants (in 1000 tons of bauxite) and conversion costs (in $ per ton of bauxite) are as follows:

Plant	Conversion Cost ($/ton bauxite)	Capacity (K tons bauxite)
L	8	15
M	7	20
N	9	10

Finally, alumina pellets are transported to smelting facilities in R and S where 1 ton of the final product aluminium is produced from 2 tons of alumina pellets. The capacity of the smelting plants R and S is 2000 and

1000 tons of aluminium, respectively. Management wishes to operate both smelting plants at 80% capacity or above, if possible. A second, but less important, objective is to minimize total conversion plus transportation costs. Transportation costs in $ per ton are as follows:

	L	M	N
A	1	3	4
B	3	2	1
R	5	4	6
S	4	2	3

a. Draw a netform to determine the production schedule which will meet management's objectives.

b. Can this problem be modeled as a pure network? Explain why or why not.

c. Assume that 1 ton of bauxite from mines A or B yields 0.05 or 0.15 tons of alumina pellets, respectively. The capacity of the conversion plants is measured in tons of bauxite, as given above. Can this be modeled as a generalized network? Explain.

***5.20** Plant A has 25 units of resource and plant B has 35 units of resource available to supply three customers who each want at least 10 units. A network model is shown below. Management wants to distribute the resources at equal rates. One way to model this restriction is to add the side constraint

$$(25 - S_A)/25 = (35 - S_B)/35$$

where S_A and S_B represents the leftover (slack) resources at Plants A and B, respectively, as shown on the network.

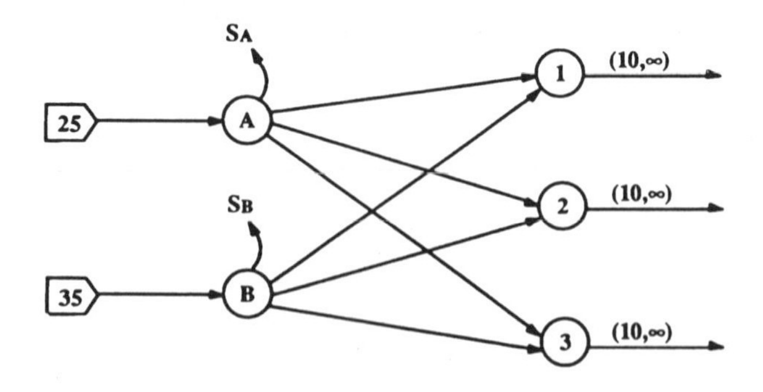

Show that the network together with the side constraint is equivalent to the generalized network in Figure 5.25. (Hint: Write the algebraic statement of the network together with the side constraint. Reduce the side constraint so that all variables will be on the left and the coefficient of S_B will be 1. The side constraint then represents an additional node D which can be eliminated.)

***5.21** Consider the following problem:

$$\text{Minimize} \quad \sum_{t=1}^{T} y_t$$

$$\begin{aligned} \text{Subject to} \quad & y_T + y_1 \geqslant r_1 \\ & y_{t-1} + y_t \geqslant r_t, \quad \text{for } t = 2, 3, \ldots, T \\ & y_t \geqslant 0, \quad \text{for all } t \end{aligned}$$

An example of this model is Exercise 4.17, Transgulf Airlines Personnel Scheduling. In Exercise 4.17, part (c), you showed that this model is not equivalent to a pure network when the number of time periods, T, is an odd number. Now show that when T is odd (use $T = 5$) this model is equivalent to a generalized network. (Hint: Change the constraints to equations by subtracting surplus variables and write the problem in detatched coefficient form. Multiply appropriate rows by -1 to get the algebraic statement into the proper form. Then draw the network.)

5.22 Construct a small generalized network problem having a unique optimal solution that is not integer-valued even though all parameters are integer. Solve using a generalized network optimization or LP software package to verify your expectation.

REFERENCES

Charnes, A. and W. Cooper. 1961, *Management Models and Industrial Applications of Linear Programming*, Vol. I, Wiley, New York, pp. 25–29.

Glover, F. and D. Klingman. 1973, "On the Equivalence of Some Generalized Network Problems to Pure Network Problems," *Mathematical Programming*, Vol. 4, No. 3, pp. 269–278.

Grigoriadis, M. 1980, "Network Optimization Problems and Algorithms: An Annotated Bibliography," DCS-TR-99, Department of Computer Science, Rutgers University, New Brunswick, NJ 08903.

Klingman, D., J. Mote, and N. Phillips. 1988, "A Logistics Planning System at W. R. Grace," *Operations Research*, Vol. 36, No. 6, pp. 811–822.

6

NETFORMS WITH DISCRETE REQUIREMENTS

Although previous chapters have presented many model forms that go beyond the realm of "classical networks," the material of this chapter takes a major step across the dividing line between network and network-related models. We begin by examining the type of discrete conditions embodied in the requirement that certain arcs must have integer-valued flows. By itself, this area greatly extends the horizon of practical models. We will also examine other forms of discreteness, particularly as embodied in "zero-one" and "fixed-charge" problems.

6.1 SIGNIFICANCE OF DISCRETENESS: DEPARTURES FROM CLASSICAL NETWORKS

Historically, the study of certain models with discrete flow requirements in the early 1970s led to the emergence of the general netform perspective and terminology. In particular, the discovery that many problems previously thought to have little or nothing in common with networks could be modeled as networks with discrete flows led to classifying such models as network-related formulations, or netforms. Then, as increasing numbers of management science applications were found to involve network-related constructions (both continuous and discrete), the netform terminology gradually came to be applied to the full range of such problems. In its broadest scope, as treated in this book, the netform model class subsumes a very substantial proportion of the most important problems in the deterministic optimization domain. They are also readily applicable to a variety of problems in nondeterministic optimization.

The question naturally arises: how does the requirement of discreteness open the door to such an array of modeling possibilities? Before the advent of the netform terminology, and the perspective that uncovered network-related structure in problems where it had not been suspected, the field of integer programming (or discrete optimization) was already well established. Theoreticians and practitioners alike had long been aware that when some variables of a problem are required to take integer values, an astonishing diversity of problem features could be given mathematical expression. In fact, given finite bounds on variables, a nonlinear optimization problem can be approximated to any desired degree of accuracy by a linear optimization problem together with the requirement that certain variables be discrete. (Discreteness itself is a form of nonlinearity.) What was not suspected was that coupling discreteness with network models would lead to new perspectives.

We have already encountered the issue of discreteness at an elementary level in the previous chapter, and devoted some attention to its relevance in certain practical settings. In this chapter we show how the integer stipulation can be more subtle than may at first be imagined, and then provide a number of fundamental constructions designed to turn this subtlety to our benefit in modeling new types of conditions.

6.2 THE EFFECT OF THE INTEGER REQUIREMENT IN ROUNDING

Because continuous (nondiscrete) models can be solved very efficiently, the natural inclination is first to solve a discrete problem as though it were continuous, and then attempt to "patch up" the solution by some form of rounding. Rough guidelines for detecting when it may be safe or dangerous to round were provided in Chapter 5, with the caveat that they were not to be presumed infallible. Here we will demonstrate some of the consequences of the integer requirement by showing how far from infallible such guidelines can be.

Consider the situation in which a gardener needs 107 lb of fertilizer for his lawn, and has the option of buying it either in 35 lb bags at $14 each or in 24 lb bags at $12 each. His goal is to buy (at least) the 107 lb he needs at the cheapest cost. The gardener is not allowed to split bags, but must buy a whole bag or none at all. We may model this problem as in Figure 6.1. The option of buying 35 lb bags at $14/lb is depicted by the generalized arc on the upper left, and the alternative option by the arc below it. The #-sign has been used as the symbol to indicate that only integer flows are admissible.

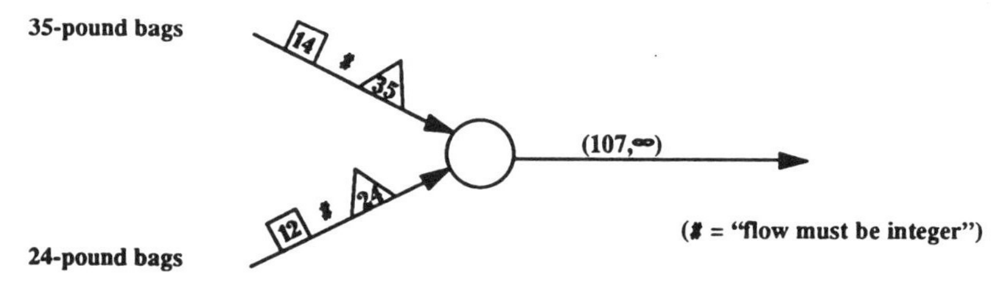

Figure 6.1 Gardener's problem (minimization objective).

We now examine what would happen if this problem were solved as a continuous problem, with the integer requirement temporarily set aside. Since fertilizer in the 35-lb bags costs $14/35, or 40 cents a pound, and fertilizer in the 24-lb bags costs $12/24, or 50 cents a pound, the best policy is evidently to buy 3 and 2/35 of the 35-lb bags. In terms of the diagram, this dictates giving a flow of 3 and 2/35 to the upper generalized arc and a flow of 0 to the lower, exactly satisfying the 107 lower bound requirement on the demand arc. The only feasible way to round this to an integer solution (still satisfying the 107 lower bound requirement) is to round 3 and 2/35 to 4, thus buying 4 of the 35-lb bags of fertilizer at a total cost of $56. However, the optimal integer policy is to send a flow of only 1 on the upper generalized arc and a flow of 3 on the lower (buying 1 bag of 35-lb fertilizer and 3 bags of 24-lb fertilizer) at a total cost of $50. Thus, even in an extremely simple setting such as this, the optimal policy can depart significantly from the policy obtained by rounding.

Now we consider a more realistic example that also introduces some useful modeling ideas. A company is faced with selecting some subset of 5 machines to make 5 different products, consisting of large assembly components. The cost of making a given assembly component on a given machine is shown in Table 6.1. (Blank entries indicate that a product cannot be made on a particular machine.)

The company requires that at least one of each of the components must be made each day. Moreover, if a given machine is selected, then it will make exactly a specified number of components from among those it is capable of making. This information is shown in Table 6.2. Machine capacities prohibit making more components, and union contract prohibits making fewer (because this would generate idle time for machine operators). The issue is to determine which machines to select, and which components to make on selected machines, to minimize daily cost.

A netform model for this problem is shown in Figure 6.2. The bounds of 0 and 1 for each machine indicate that the machine may be unselected (0) or selected (1), and the #-sign, identifying the integer requirement, indicates that a machine cannot be "partly selected." The multipliers on the arcs into the machine nodes specify the number of assembly components the machines are able to produce.

Table 6.1 Cost of Making One Assembly Component

	Assembly Components				
Machines	1	2	3	4	5
1	42	70	93		
2		85	45		
3	58			37	
4	55		55		38
5		60		54	52

Table 6.2

Machines	Total Number of Components Made Per Day (If Machine is Selected)
1	2
2	2
3	2
4	3
5	2

This problem turns out to be largely a pure network, which makes the prospects of rounding more congenial, since pure networks automatically admit integer solutions (for integer data). In fact, the continuous solution to the problem gives integer values of 0 and 1 to most of the arcs. The arcs of the problem that contain nonzero flows in the continuous solution, together with the values of these flows (indicated in half circles) are shown in Figure 6.3.

Analyzing Figure 6.3 for the possibility of rounding the fractional solution values to integers discloses a surprising fact. Not a single rounding possibility will provide a feasible (let alone optimal) solution! Since there are 5 different arcs with flows that can be rounded (to 0 or 1) there are $2^5 = 32$ different possibilities.

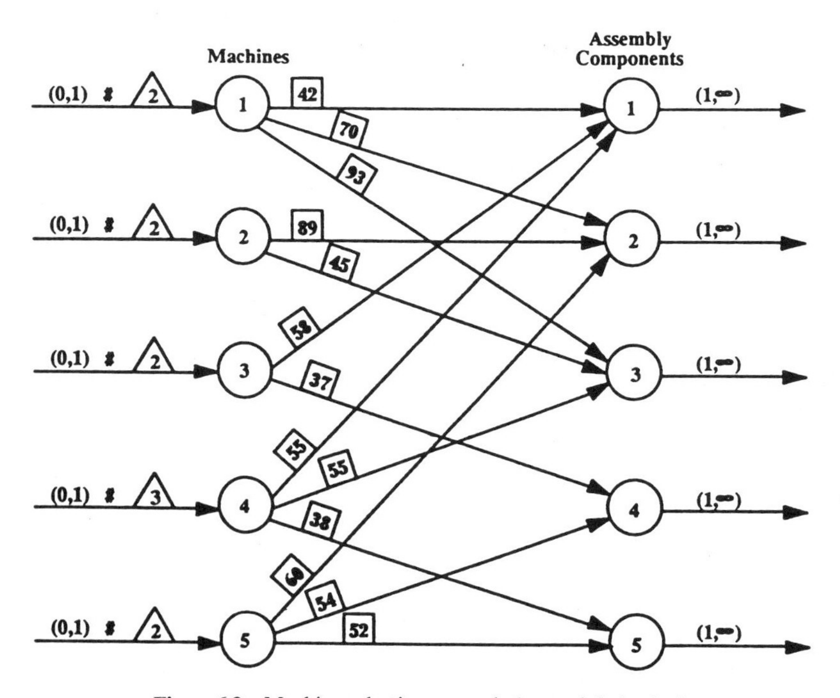

Figure 6.2 Machine selection example (cost minimization).

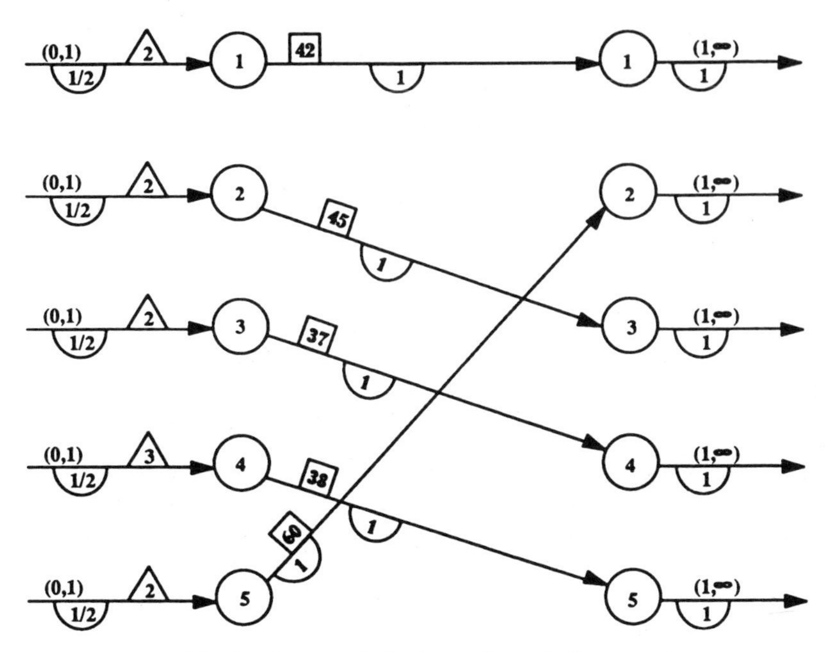

Figure 6.3 Continuous solution to machine selection examples.

A procedure that attempts to obtain an integer solution by trying out each rounding possibility and checking for feasibility would lead to a good deal of wasted effort. (A feasible solution can only be obtained by rounding some flows and then determining wholly new values for others.)

Exercise 6.6 provides a 20 machine, 20 product example with more than 1,000,000,000,000 rounding possibilities, none of which is feasible. For a problem such as this, the wasted effort of trying to round would be extreme!

This discussion on rounding should give you an idea about how difficult it is to solve network problems with discrete requirements. In fact only medium-sized problems (with hundreds or sometimes thousands of nodes and discrete arcs) can be solved in a reasonable amount of time. However, it is still often possible to exploit the underlying network structures by special solution methods that are far more efficient than the methods developed for the algebraic (integer programming) representations of the problems. The important point to keep in mind is that the solution difficulty goes up exponentially with the number of discrete variables (integer arcs).

6.3 INTEGER MODEL TYPES

Problems involving variables with integer values (arcs with integer flows) have a number of standard classifications. Problems whose variables must all be integer-valued are called ***pure integer*** problems; those requiring only some variables to be integer-valued are called ***mixed integer*** problems. When the integer variables are bounded to lie between 0 and 1, as in the case of the immediately preceding example, the problem is called a ***0-1*** ("zero-one") ***pure or mixed integer*** program.

6.4 ZERO-ONE DISCRETE NETWORKS

Problems involving 0-1 variables have a great many applications on their own. Several of these will be indicated in this section, together with appropriate netform representations.

First, consider the situation where a company delivers merchandise to its customers each day by truck (or rail, air, barge, etc.). There are a variety of acceptable delivery routes, each accommodating a specified subset of customers and capable of being traveled by a single carrier in a day. Each route also has an associated cost (such as the length of the route, cost of fuel, etc.). The goal is to select a set of routes that will provide a delivery to each customer and, subject to this, minimize total cost.

To represent such a problem, note first that the customers can be depicted simply as nodes with demand arcs requiring flows of at least 1 (so each customer will be covered by at least 1 route). The way to represent a delivery route will be shown by example. Suppose that a particular route costs $12 and provides deliveries to Customers 1, 2, and 4 (out of Customers 1–6). A way to handle this is provided in Figure 6.4.

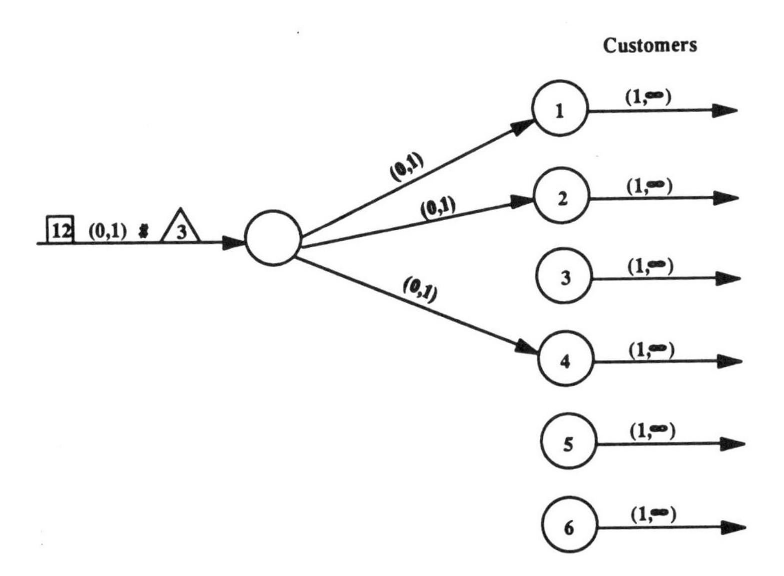

Figure 6.4 Delivery route selection.

The generalized arc with a cost of $12 and multiplier of 3 is the ***decision arc*** of the route, and the three arcs following it are the ***linking arcs***. The multiplier of 3 has been placed on the decision arc because the route covers precisely 3 customers (Customers 1, 2 and 4), identified by the 3 linking arcs. When the decision arc has a 0 flow (corresponding to the decision not to choose the route), the linking arcs likewise have 0 flow. But when the decision arc has a flow of 1 (corresponding to the decision to use the route), the multiplier transmits 3 units of flow to the linking arcs. Because of the upper bounds on each of these arcs, the only possibility is that each will carry exactly 1 unit of flow. In short, the net effect of giving the decision arc a flow of 1 is to send a flow of 1 to each of Customer nodes 1, 2 and 4. This accords with the interpretation that the use of the route corresponds to making a delivery to each of its customers (satisfying their demands). The integer requirement, depicted by the # on the decision arc, conveys the fact that it is not admissible to "partly decide" to use the route.

Question: Construct a netform for a complete delivery problem using the data in Table 6.3.

Figure 6.5 answers this question. This delivery problem is in fact an instance of a classical integer programming problem called the ***covering*** problem. The idea of this problem in general is to cover one set of objects (here customers) by elements from another set (routes). In the classical form of this problem the goal is to minimize the number of elements in the cover instead of minimizing the cost of these elements. (In our example, minimizing the number of elements leads to selecting Routes 1 and 4, while minimizing the cost of the elements leads to selecting Routes 2, 3, and 5.)

Another classical problem, closely related to the covering problem, is the ***partitioning*** problem, which requires that each object be covered exactly once. The delivery example in Figure 6.5 becomes a partitioning problem by changing all demand upper bounds to 1, making lower and upper bounds equal. (In Figure 6.5, the only feasible solution in this case would be to select Routes 1 and 4.)

There is another classical relative of the covering problem, called the ***matching*** problem, in which objects must be covered *at most* once. Our delivery

Table 6.3

Routes	Route Cost	Customers Covered (X) 1	2	3	4	5	6
1	12	X	X		X		
2	7	X		X			
3	4		X		X		
4	13			X		X	X
5	10				X	X	X

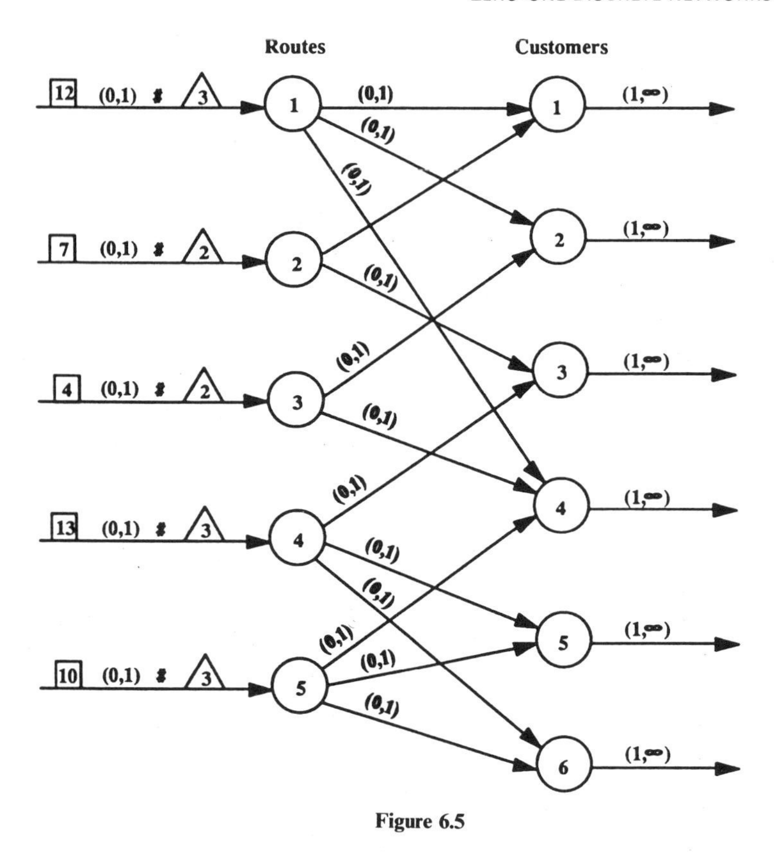

Figure 6.5

example in Figure 6.5 becomes an instance of a matching problem by changing demands to be bounded below by 0 and above by 1. The goal of a matching problem is typically to maximize the number of selected elements, or more generally the total "value" (profit) of these elements. Thus, to represent a matching problem, the delivery example would additionally replace the costs by figures representing, for example, revenues collected from the routes. The special structures of these three types of integer programs has prompted the development of a variety of special methods for solving them.

This points up a key distinction between models and methods. The best type of model, from the standpoint of flexibility and ease of interpretation, may or may not bear a direct relationship to the best method for solving the underlying problem. Thus, a netform representation of a partitioning problem, for example, may give a highly convenient way to depict (and modify) its features, but the best way to solve the problem might be to use a procedure that makes no attempt to exploit (except indirectly) the problem's "network aspect." At the same time, while netform representations do not necessarily imply the value of a "network-based" solution method, it is also true that the perspective afforded by netforms has contributed significantly to the development of effective special purpose algorithms. Many instances have been reported in the literature where such algorithms were successfully applied to discrete flow problems that could not be solved in reasonable lengths of time by alternative methods.

6.5 A SCHEDULING PROBLEM

We now present a simplified version of one of the first practical 0-1 problems to be modeled as a netform. (A specialized network-based method for this problem reduced its solution time from hours to seconds as detailed in the Application at the end of this chapter.)

The setting is that of providing training courses for skilled personnel. Each individual to be trained has a set of possible class schedules to choose from. In turn, each schedule for a particular person has a "profit figure" representing its value for that person. Further, if selected, the schedule will assign that person to a particular collection of classes. Classes themselves have lower and upper limits on the number of people they can accommodate. The goal is to choose exactly one schedule for each person (from that person's set) and to maximize the total value of all schedules selected.

We first show how to model this type of problem for a single person. For concreteness, assume there are 10 classes, each of which must contain between 20 and 30 people. The person considered, however, can only take some subset of the first five classes. There are 3 possible schedules for this person: the first with a value of 8, assigning the person to Classes 1, 2 and 3; the second with a value of 5, assigning the person to Classes 2 and 4; and the third with a value of 6, assigning the person to Classes 3 and 5. A diagram of this situation is shown in Figure 6.6.

The supply arc on the left with lower and upper bounds of 1 indicates that the person must select exactly one of the specified schedules. Each schedule is composed of a decision arc and linking arcs, as in the earlier delivery example. The interpretation in the present setting is similar. For example, a flow of 1 on the "top" decision arc will transmit 3 units of flow to the linking arcs, thereby (due to the upper bounds of 1) donating 1 unit to each—in effect, putting the person into each of Classes 1, 2 and 3. The schedules for people not shown must, of course, provide the additional class assignments that will permit the lower and upper class limits to be satisfied.

A useful (and challenging) extension of the scheduling possibilities shown above is provided by the following question.

Question: Construct an appropriate diagram for a person who must select one of two schedules. The first has a value of 7 and assigns the person to any two of Classes 1, 2 and 3; the second has a value of 5 and assigns the person to at least 1 and at most 3 of classes 3, 4, 5 and 6. (Lower and upper class limits are 20 and 30 for each class.)

Figure 6.7 provides an answer to this question. The decision arc for the first of the two schedules in Figure 6.7 has a multiplier of 2, corresponding exactly to the number of classes that must be selected by this schedule. The decision arc for the second schedule has a multiplier of 3, corresponding to the maximum

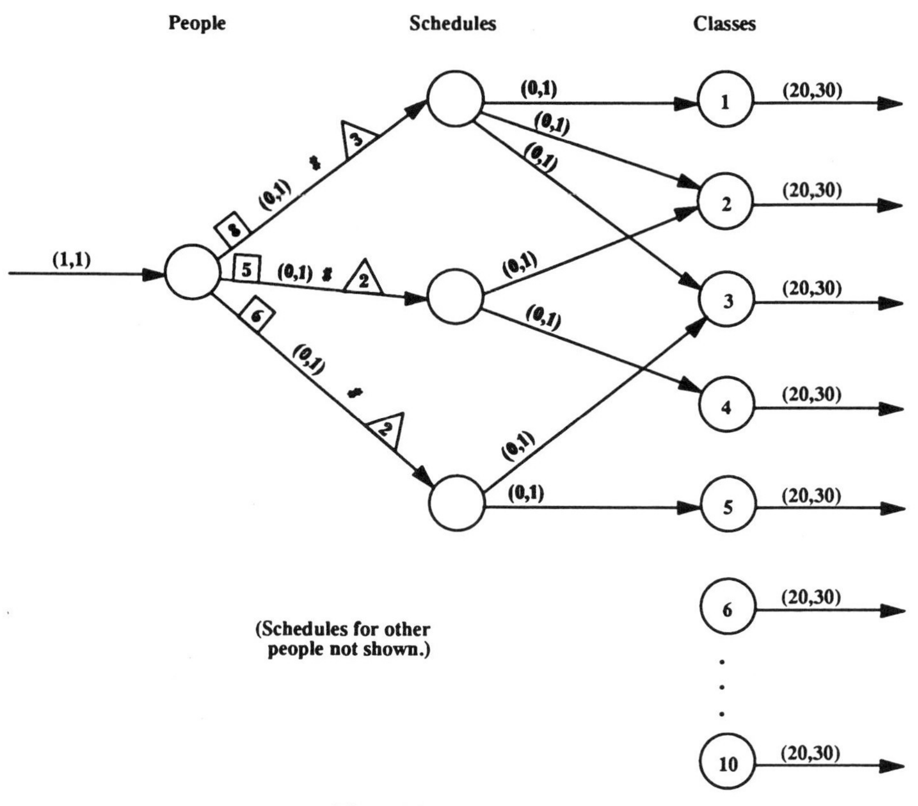

Figure 6.6 Class scheduling.

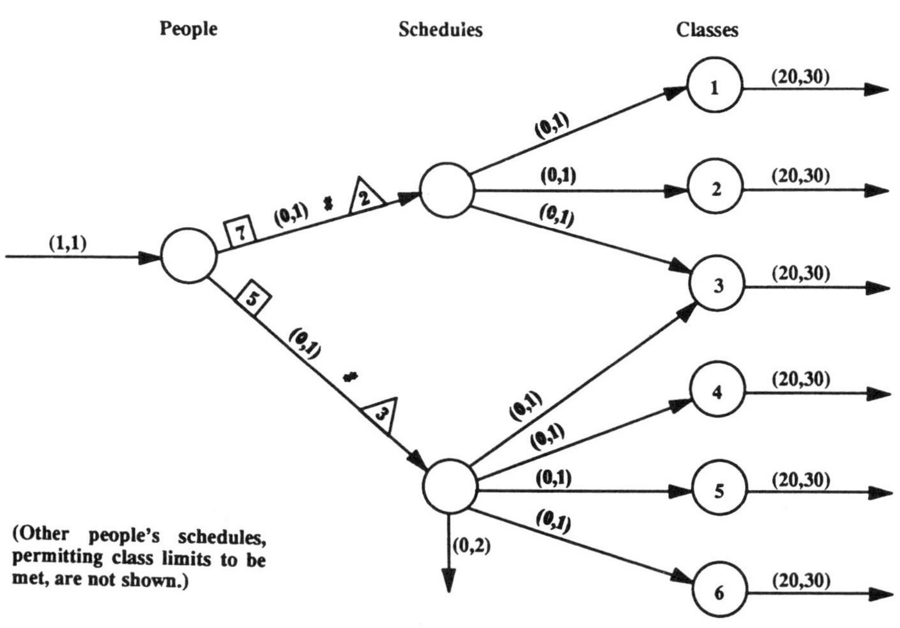

Figure 6.7

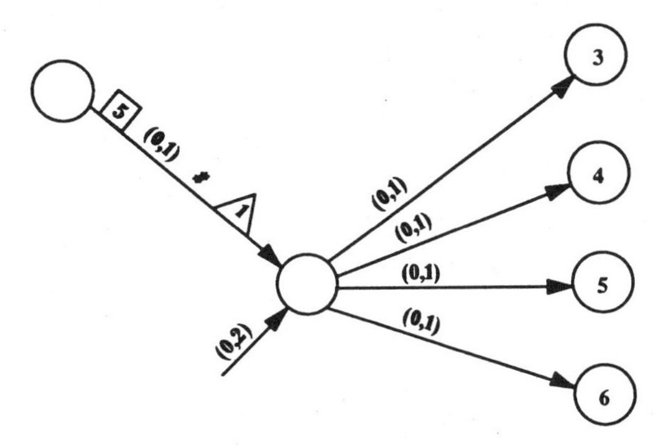

Figure 6.8 Erroneous model.

number of classes this schedule can select. The surplus arc with an upper bound of 2, which leaves the same node as the linking arcs, allows up to 2 of the 3 units reaching this node to escape (when the decision arc has a flow of 1), thereby assuring that between 1 and 3 units of flow will continue on to the linking arcs. (Section 6.7 discusses why this flow will not be distributed fractionally even though the linking arcs have no integer requirements.)

The principle of sending a maximum amount of flow, and allowing the excess to escape (when a discrete condition dictates that flow must lie between two bounds), is also used in other discrete flow constructions to be encountered later. Note in particular that there is no way to model the second schedule in the preceding example if the decision arc transmits the minimum flow of 1 rather than the maximum flow of 3, as attempted in the construction in Figure 6.8.

In Figure 6.8, the two extra units to make up the admissible total of 3 are (presumably) to be brought in by means of the slack arc with an upper bound of 2. The difficulty, however, is that this arc will carry up to 2 units of flow even when the decision arc is "shut off" (has 0 flow). In short, for the decision arc to control all allowable flow, it must provide the maximum amount of this flow.

The supply arc on the left of Figure 6.7, which indicates that the person must select exactly one of the two possible schedules, falls into a generic class called logical conditions. To illustrate other types of logical conditions, consider an investor who may choose whether to invest in three projects, A, B, and C. Suppose the projects have costs of \$1, \$2, and \$3 million, respectively, and each one must be invested in totally or not at all. Figure 6.9 shows how to model various logical conditions.

(a) At most 2 projects can be selected

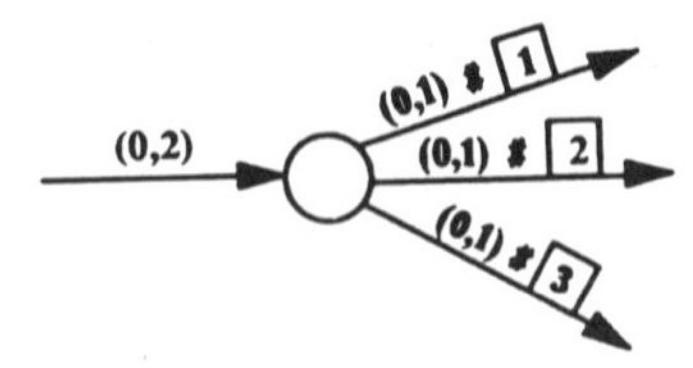

(b) If A is selected, then B must be selected

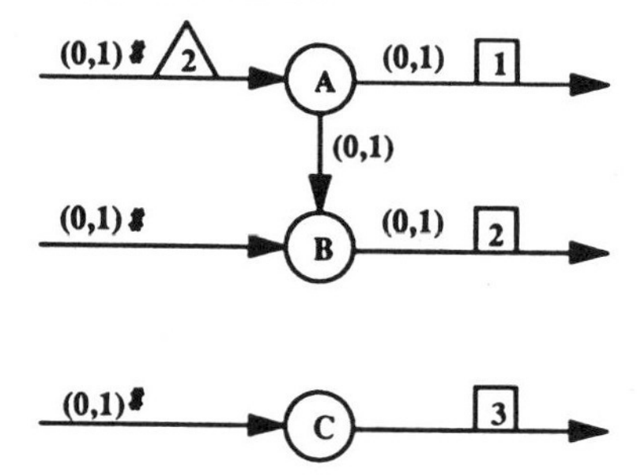

(c) If A and B are selected, then C must be selected

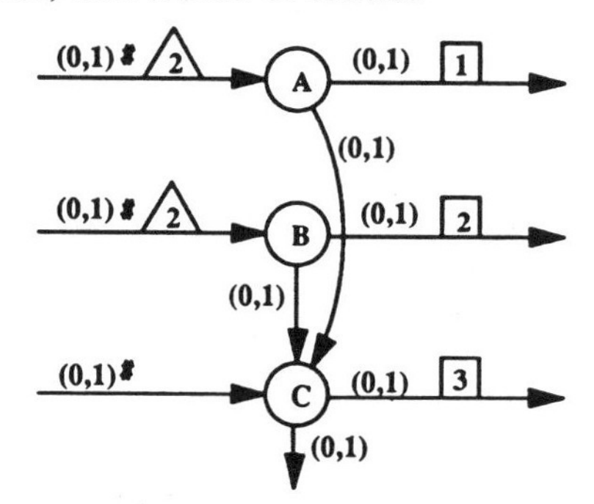

(d) If C is selected, then A and B must be selected

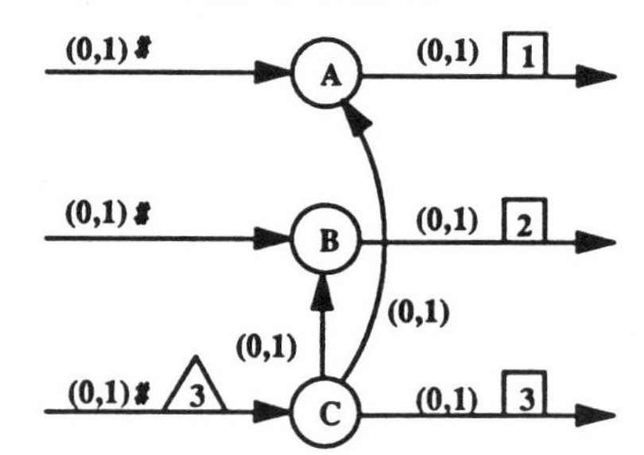

(e) If C is selected, then either A or B must be selected

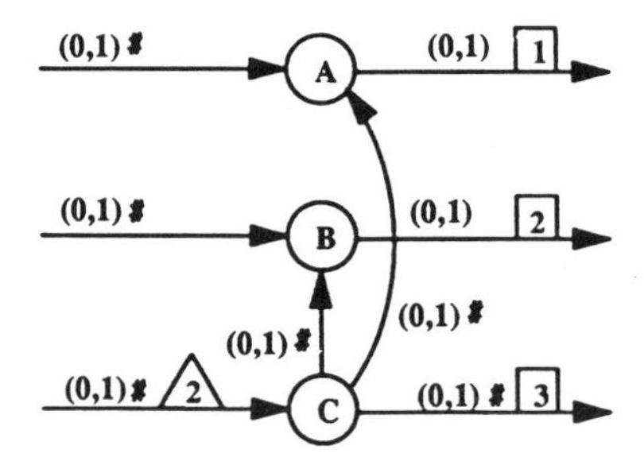

(f) If A and B are selected, then C cannot be selected. (This is equivalent to (a).)

Figure 6.9 Logical conditions.

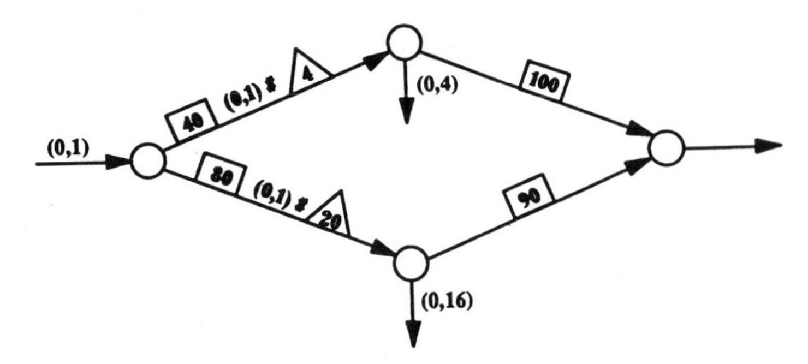

Figure 6.10 Garden supply center.

The modeling concepts illustrated in Figure 6.7 are also useful in modeling concave cost functions. We showed in Chapter 3 that, using pure network models, convex functions could be minimized and concave functions could be maximized. The following example demonstrates how to minimize a concave function using discrete flow decision arcs.

A garden supply center purchases bulk fertilizer by the ton and then bags and resells it to customers. For purchases of 4 tons or less, the cost is $100 per ton with a $40 delivery charge. For purchases over 4 tons, the cost is $90 per ton with an $80 delivery charge. The maximum size of an order is 20 tons. The network component which minimizes purchase costs is shown in Figure 6.10. The supply arc on the left indicates that the garden supply center can select at most one option (4 tons or less, or over 4 tons). The multipliers on the decision arcs represent the maximum number of tons that can be purchased under each option. The surplus arcs allow the excess tons to escape before the per ton cost is charged.

Question: For the Clothing Manufacturer example of Chapter 3, represent the following condition: sew-line 3 handles 100 to 150 garments at a cost of $3 each, above 150 to 175 garments at a cost of $2 for each extra garment, and above 175 to 300 garments at a cost of $1 for each extra garment.

The solution is shown in Figure 6.11. To understand the costs on the decision arcs, you may find it useful to refer to the cost function for this question in Figure 3.22. At the 150 garment price break, the middle production option in Figure 6.11 would change $2 per garment for a total of $300. Thus we need to add a $150 fixed cost to bring the total to $450, the true cost of producing 150 garments. Likewise, at the 175 garment price break, the fixed cost is $500 – 175 ($1) = $325 (where the $500 = 150 ($3) + 25 ($2)). (Graphically, these fixed costs are where the respective linear segments of the cost function would intersect the vertical axis.) In each instance, the garments above the price break will only be charged the discount price. Note that Figure 6.11 reverses the supply-demand orientation of the Clothing Manufacturer network in Figure 3.1. That is, sew-line production capacity is now viewed as supply. To maintain the demand role of the sew-lines, simply add a multiplier of –1 to each of the three

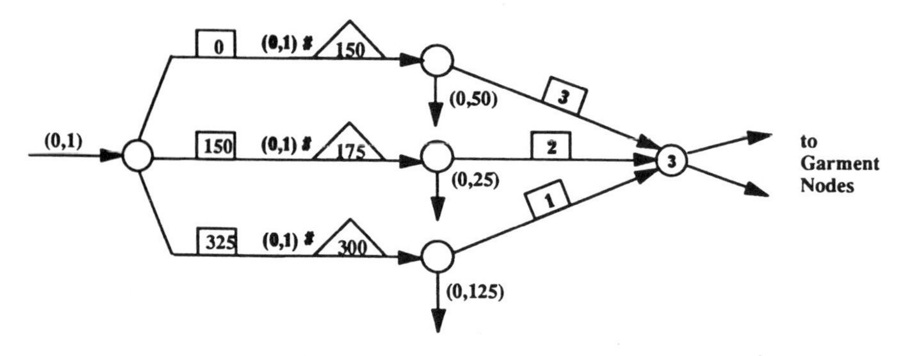

Figure 6.11 Clothing manufacturer.

arcs into node 3 and reverse the two arcs out of node 3 so that they lead from the garment nodes.

6.6 OTHER CONSTRUCTIONS USING NEGATIVE MULTIPLIERS

We now consider another example which incorporates negative multipliers. Each individual of the class scheduling problem in Figure 6.6 has some classes that are required and some that are elective. Assignments of people to their required classes are made in advance, and the scheduling options remaining, for which diagrams such as Figure 6.6 are constructed, represent elective scheduling alternatives. A particular individual has one such elective schedule that will assign him to Classes 1 and 2, and remove him from Class 3, to which he has been pre-assigned. A diagram for this situation follows in Figure 6.12. (Portions of the netform not immediately relevant to the construction are omitted.)

The decision arc has a multiplier of 3, indicating that it will transmit flow to all three of the linking arcs, and not just to the two arcs for Classes 1 and 2. The means of removing the person from Class 3 is provided by the multiplier of -1 on the bottom linking arc.

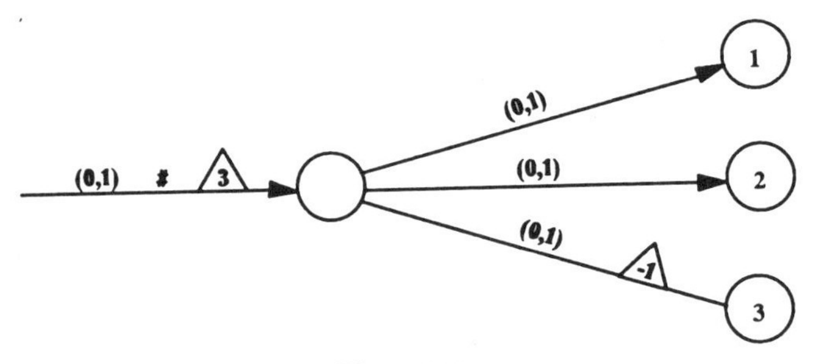

Figure 6.12

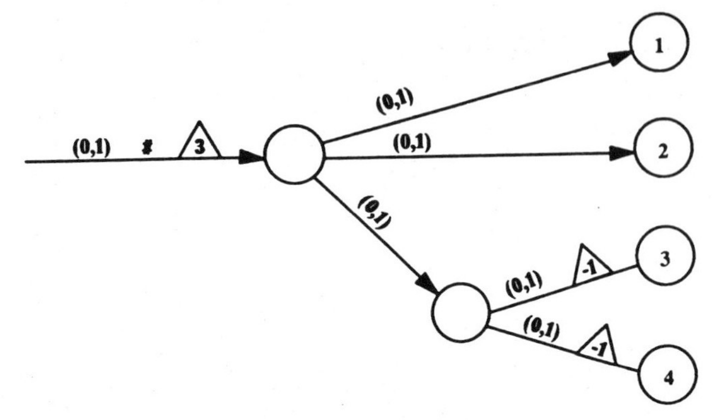

Figure 6.13

This use of negative multiplier constructions in discrete-flow networks is expanded in the following question.

Question: Diagram a schedule that will assign a person to Classes 1 and 2, and remove him from one of the two Classes 3 and 4 (to which he has been preassigned). Diagram a second schedule that will assign him to at least 1, but at most 2, of Classes 1, 2, and 3, and remove him from at least one, but at most 2, of Classes 4, 5 and 6.

The two diagrams for this exercise are shown in Figures 6.13 and 6.14.

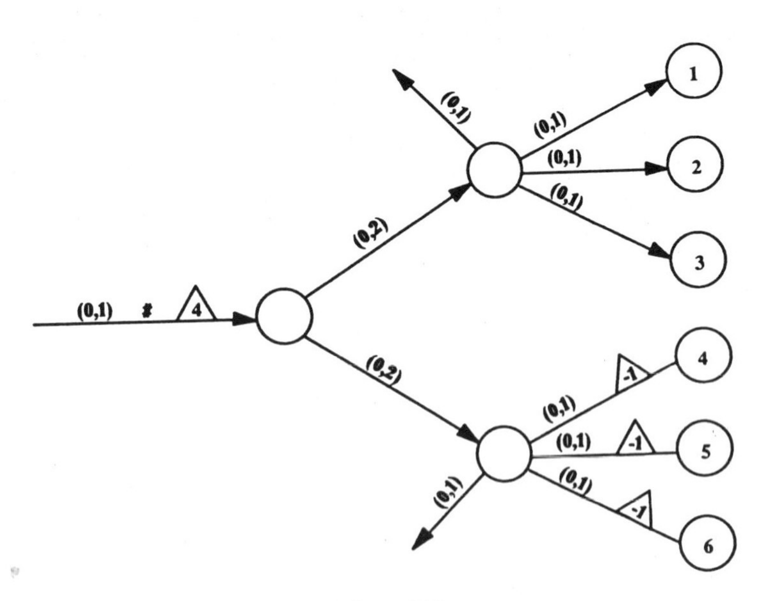

Figure 6.14

In Figure 6.13, the construction assures that at most one of the negative multiplier arcs will carry flow, thus causing a unit of flow to be extracted from either node 3 or node 4. (The 0, 1 bounds on the negative multiplier arcs could be omitted, since they are redundant in view of the 0, 1 bound on the arc that precedes them. See the discussion in the following section.)

In Figure 6.14, the principle of transmitting maximum flow, illustrated earlier, is used once again. The multiplier of 4 on the decision arc assures that the maximum of four classes will be affected (at most 2 will receive flow, and at most 2 will have flow removed). The surplus arcs with 0, 1 bounds allow the escape of up to 1 unit of flow, hence assuring that from 1 to 2 units are carried on the arcs that meet classes. (Thus the arcs with -1 multipliers, for example, will extract flow from 1 or 2 of the nodes for Classes 4, 5 and 6.)

The linking arcs of Figures 6.13 and 6.14 have a more complex configuration than earlier, and may be put in two categories: ***immediate linking arcs***, consisting of those that directly follow and receive flow from the decision arc, and ***secondary linking arcs***, consisting of those not adjacent to the decision arc. This distinction will be important for the development of additional constructions, treated shortly.

6.7 FRACTIONAL FLOW POSSIBILITIES

Figure 6.14 invites closer consideration of the issue of fractional flows. Although the integer requirement assures the flow on the decision arc will be integer, there is nothing to prevent flows on some of the other arcs from being fractional. For example, suppose the decision arc is "turned on" (carries a flow of 1). Then two units of flow will be carried on each of the linking arcs that follow it. But these two units can be distributed fractionally—for example, in the upper portion of Figure 6.14, half a unit might go to each of Classes 1 and 2, and 1 unit to Class 3.

We are protected from this possibility by the following fact: *if all multipliers and bounds (including supplies and demands) are integers, and if all arcs that are not compelled to receive integer flows are pure network arcs, then all flows will be integer* (using normal solution methods). This is a consequence of the fact that, without imposing integer flow requirements, a pure network with integer data always possesses an integer-valued optimal solution (even if it also has optimal solutions that are fractional). Moreover, all standard extreme point solution methods will automatically yield an optimal solution with integer flows.

Yet, in fact, Figure 6.14 does not perfectly satisfy the criterion, because the 3 arcs with negative multipliers are not pure network arcs. Such arcs can lead to situations in which optimal solutions are fractional. Fortunately, however, they will not do so in the present instance. Demonstration of this fact leads to another useful development, resulting from the observation that the construction in Figure 6.14 is also equivalent to Figure 6.15.

To understand the equivalence, we focus on the lower half (since the upper half of the construction is the same as before).

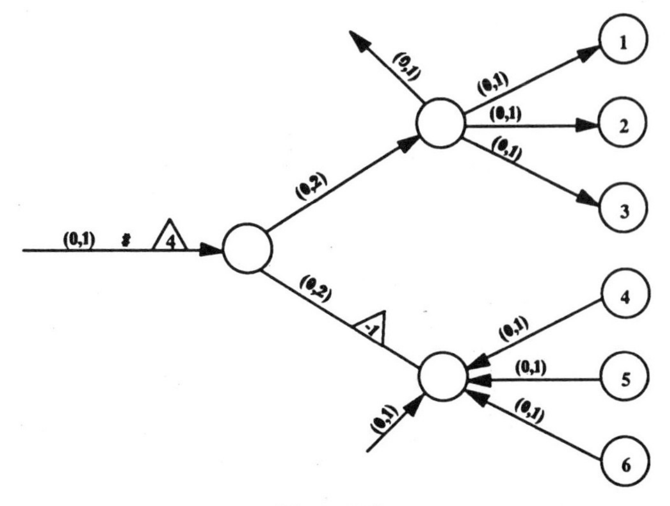

Figure 6.15

This lower portion has given the immediate linking arc a multiplier of -1, and replaced the secondary linking arcs by ordinary (pure) network arcs with the opposite direction. To determine the effect of this construction, note first that when flow on the decision arc is 0, all arcs of the lower half (as well as the upper) must have 0 flow, because there is no outlet for flows on the arcs leaving Classes 4, 5 and 6. On the other hand, when flow on the decision arc is 1, exactly 2 units must be carried on the immediate linking arc with the -1 multiplier, which means that 2 units must be extracted from the node associated with this multiplier, provided by the secondary linking arcs that meet this node. This assures that between 1 and 2 of these units will be pulled from Classes 4, 5 and 6 (since 1 may also come from the slack arc). Thus, the effect of Figure 6.15 is the same as that of Figure 6.14. The difference, however, is that the rule which assures integer solutions now directly applies. Specifically, the only arcs in the diagram that are not pure network arcs are the decision arc and the immediate linking arc with a -1 multiplier. The decision arc is compelled to have integer flow because of the # stipulation. The immediate linking arc with multiplier -1 is compelled to have integer flow because its only possible flow values are 0 and 2, given the integer requirement for the decision arc.

6.8 CANONICAL CONSTRUCTIONS AND HUB DIAGRAMS

The equivalence of Figures 6.14 and 6.15 may lead to the suspicion that other diagrams exist which are equally capable of modeling the same underlying conditions. This is in fact the case. As a result, it is worth questioning whether there may be specific ways to evaluate alternative representations and therefore choose among them. If the goal is to use as few generalized arcs as possible, then a construction of the type represented by Figure 6.15 is appealing. However, there is another type of construction that has especially interesting features, qualifying it for special attention. This construction is illustrated in the following question.

Question: Give a diagram equivalent to Figure 6.15, with the secondary linking arcs unchanged, but with upper bounds of 1 for each of the immediate linking arcs. (Explain why this diagram assures all arcs will receive integer flows.)

Figure 6.16 satisfies the requirements of this question. The features of Figure 6.16 that warrant giving it special attention are: (1) the flow on the decision arc must exactly equal the flow on each immediate linking arc (here, all three flows are simultaneously 0 or 1); (2) the multipliers for the immediate linking arcs identify exactly the amount of flow contributed to each of their associated endpoints (for each unit of flow on the decision arc).

These features can be achieved whenever a flow of 1 on the decision arc will fill the immediate linking arcs (i.e., give them flows equal to their upper bounds). In this case, the technique for guaranteeing these features is simple: give the decision arc a multiplier exactly equal to the number of immediate linking arcs, and give each immediate linking arc bounds of 0, 1 (the same as the bounds of the decision arc). The new multiplier for each immediate linking arc will then be its old multiplier times its old upper bound.

By definition, we will say that any construction possessing features (1) and (2) is a ***canonical construction***. The interest in canonical constructions is severalfold. First, the multipliers on the linking arcs conveniently provide direct information about the effect of flow on the decision arc (as a result of feature (2)). Second, the construction can be diagrammed in a simplified way by means of what we call a

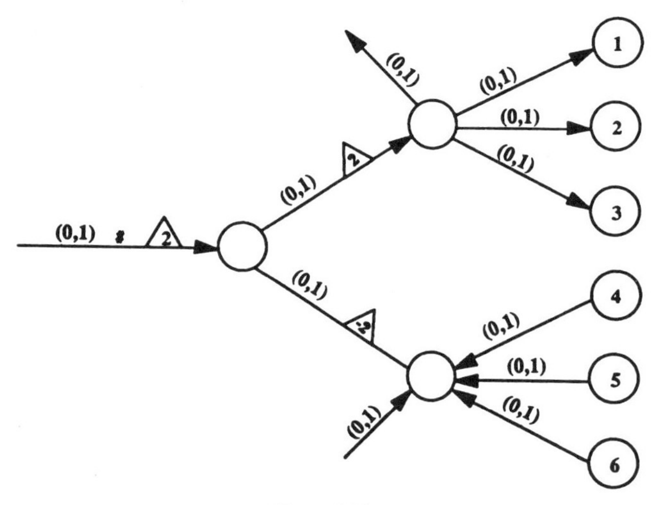

Figure 6.16

hub diagram. As will be shown later, the netform representation provided by the hub diagram is particularly useful in expanded problem settings.

In the hub diagram, the decision arc and its head node are replaced by a box, called the ***hub***. Information about bounds and integer requirements are indicated just outside the box (generally above or to the side) and cost information is given inside (leaving space so the hub can be labeled by a letter or number, just like a node).

The immediate linking arcs of the canonical construction becomes ***spokes*** extending outward from the hub. The reason the hub diagram simplifies the construction is because the multiplier of the decision arc is discarded (it always equals the number of immediate linking arcs, hence spokes), and the bounds on the immediate linking arcs are dropped (they always equal the bounds on the decision arc). Three different canonical constructions and their associated hub diagrams are shown in Figure 6.17.

The hub diagrams can in fact be understood without reference to the

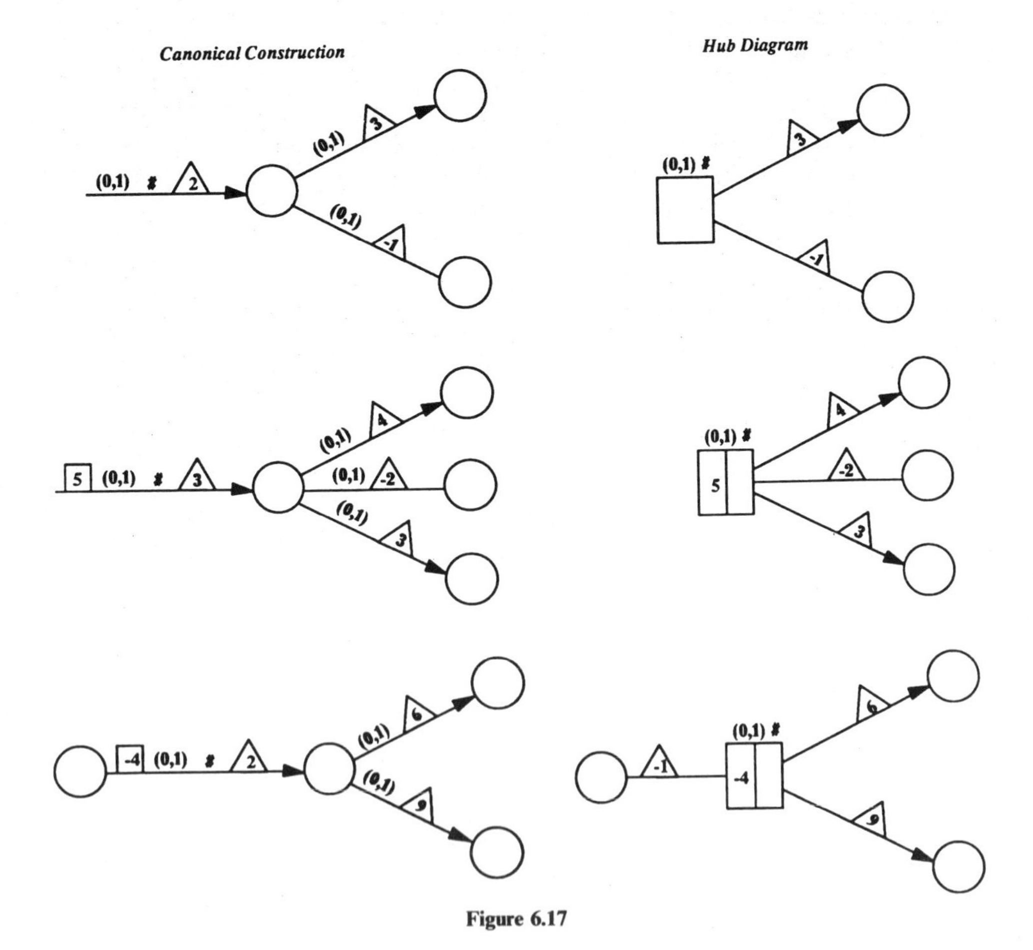

Figure 6.17

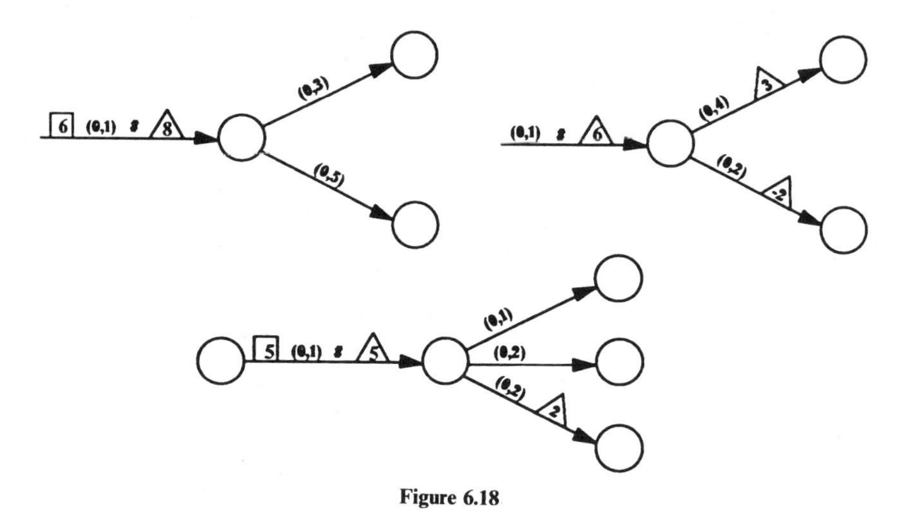

Figure 6.18

canonical constructions by the following conventions: (1) the hub diagram has a single flow value, called the ***hub flow***; (2) this flow must satisfy the bounds and integer requirements attached to the hub, incurring the cost (or profit) enclosed inside; (3) the multipliers on the spokes identify how much flow is contributed to (or, in the case of negative multipliers, extracted from) the nodes associated with these multipliers.

Convention (3) provides the rationale for the hub diagram in the third construction in Figure 6.17. Note in this diagram the hub has a third spoke, in addition to the two that come from the immediate linking arcs. The reason is that the decision arc in this case, unlike the two cases preceding, has a tail node from which it extracts flow (in the amount of 1 unit extracted for each unit of flow on the decision arc). Thus, for the hub to achieve the same effect as the decision arc, it must have a spoke that meets this node with a -1 multiplier.

It may be supposed that the $(0, 1)$ # notation attached to the hub in each of these cases is superfluous, since all decision arcs we have considered are required to have integer 0-1 flows. However, hub diagrams will later be used in settings where other bounds, with and without the integer requirement, apply.

Question: Replace each of the 3 constructions in Figure 6.18 by a canonical construction and its associated hub diagram. In each case, verify that flows of the original diagram directly correspond to flows of the hub diagram.

Figure 6.19 provides the hub diagrams corresponding to each of the diagrams of Figure 6.18. The illustrations in Figures 6.18 and 6.19 help to underscore the chief advantages of the hub diagram—that it expresses all aspects of the construction in terms of a single hub flow, and it does not require computing the products of bounds and multipliers to know at once how nodes are affected by this flow (information available by looking at the spoke multipliers). On the other hand, it should not be forgotten that a hub diagram does not apply if a

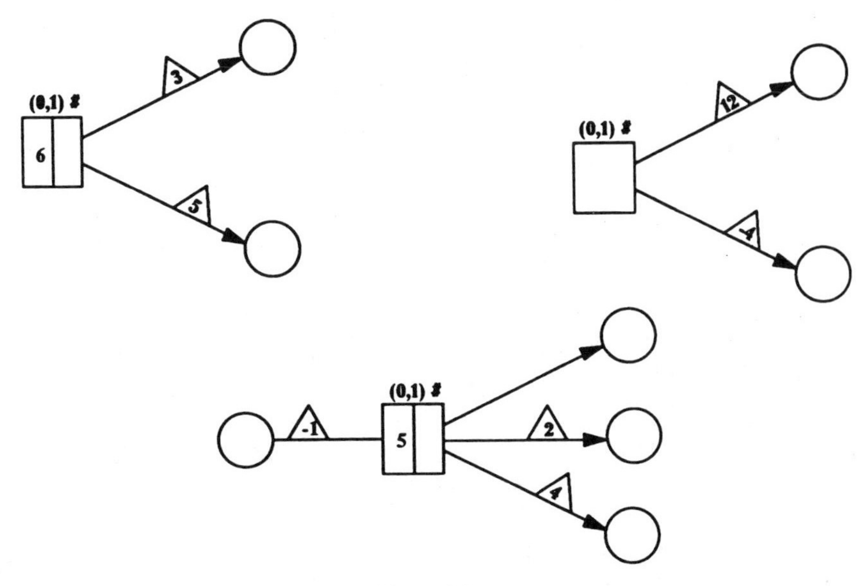

Figure 6.19

decision arc does not transmit enough flow to fill each of its immediate linking arcs. For example, in the earlier class scheduling context, it is a simple matter to diagram a schedule that allows any 2 of Classes 1, 2, and 3 to be taken, using the construction in Figure 6.20. However, there is no corresponding canonical construction or hub diagram, because the bounds on the immediate linking arcs sum to more than the multiplier of the decision arc.

6.9 ZERO-ONE INTEGER PROGRAMMING PROBLEMS AS NETFORMS

The modeling techniques of the preceding sections make it possible to formulate any pure 0-1 integer programming problem as a netform. To illustrate, consider the 0-1 problem below.

$$\begin{array}{lrl}
\text{Minimize} & 3x_1 + 4x_2 + 7x_3 + 5x_4 & \\
\text{Subject to:} & & \\
(1) & 2x_1 \qquad\quad + 4x_3 - 4x_4 & \geq 2 \\
(2) & -1x_1 + 10x_2 - 2x_3 - 6x_4 & \leq 7 \\
(3) & 4x_1 + 1x_2 \qquad\quad + 4x_4 & = 5 \\
(4) & 1x_3 + 1x_4 & \geq 1 \\
 & x_1, x_2, x_3, x_4 = 0 \text{ or } 1 &
\end{array}$$

The right-hand side of the first constraint (≥ 2) may be viewed as a form of

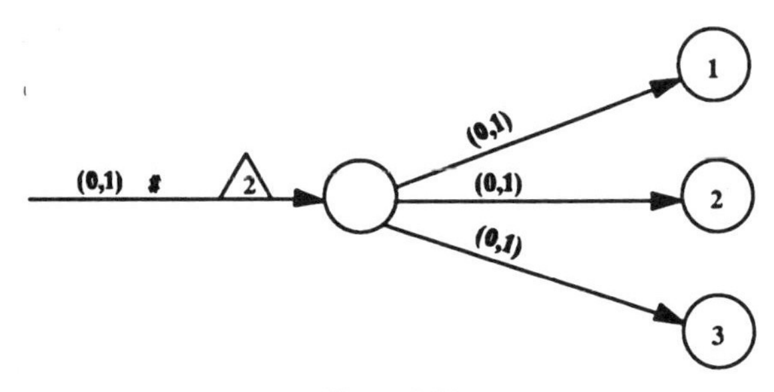

Figure 6.20

demand. If variable x_1 is given a value of 1 rather than 0 then x_1 serves to help meet the demand, and thus plays a supply role. In a similar way, setting $x_1 = 1$ helps meet demands expressed by the other constraints. The "supply/demand" orientation is useful for understanding the intuitive basis of the netform representation.

To construct this representation, create nodes x_1, x_2, x_3, and x_4, one for each 0-1 variable. Also create an origin node 0 and provide arcs called ***variable arcs***, by which node 0 supplies these ***variable nodes***, as shown in Figure 6.21. Flows on these four variable arcs correspond to the values of the four 0-1 variables. The (0, 1) bounds on these arcs together with the # stipulation insure that their flow will be 0 or 1.

Costs on the arcs originating at node 0 correspond to the objective function

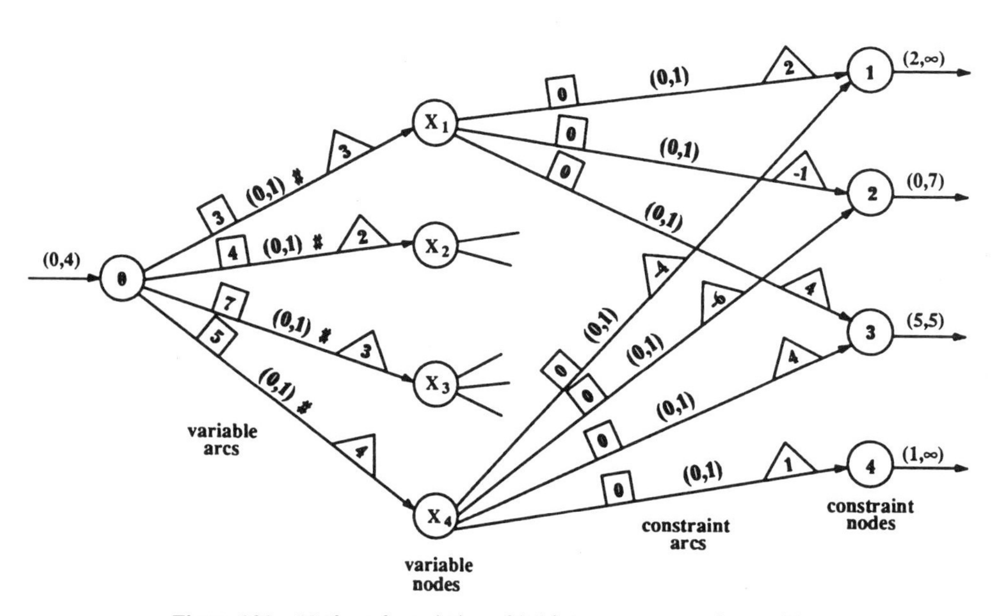

Figure 6.21. Netform formulation of 0-1 integer programming problem.

coefficients of the associated variables. A flow of 1 unit on the arc from node 0 to node x_3 for example, corresponds to setting $x_3 = 1$ in the algebraic representation of the problem and incurs a cost of 7.

Each variable arc has a multiplier whose value is equal to the number of constraints in which that variable appears with a nonzero coefficient in the algebraic statement. Thus, the arc from node 0 to node x_1 has a multiplier of 3, since variable x_1 appears in constraints 1, 2, and 3. Coupling the multiplier assigned to each variable arc with the 0-1 restriction implies that either zero flow is transmitted to a variable node or a flow equal to the number of constraints in which this variable appears in the algebraic statement is transmitted to its variable node.

To complete the netform, create ***constraint nodes*** 1, 2, 3, and 4 for the four constraints. Connect each variable node to the constraint nodes associated with the constraints in which that variable appears with a nonzero coefficient. These arcs are referred to as ***constraint arcs***. (To keep the schematic simple, arcs out of nodes x_2 and x_3 have been left incomplete.) Each constraint arc is given a lower bound of 0, an upper bound of 1, a cost of 0, and a multiplier equal to the actual constraint coefficient of the variable with which this arc is associated. To illustrate, the multipliers on the constraint arcs leaving node x_1 are 2, -1, and 4, which are the coefficients of variable x_1 in the algebraic statement for constraints 1, 2, and 3, respectively.

Finally, each constraint node has a demand equal to the constraint bound (inequalities and equalities involving the right-hand-side coefficients) specified in the algebraic statement.

The equivalence of the algebraic statement and Figure 6.21 will now be made apparent by extending our earlier observations concerning the combined effect of the arc and node attributes. Consider origin node 0 as a supply node, equipped to supply a maximum of four units (the maximum number of 0-1 variables that can receive a value of 1). If flow on the arc from node 0 to node x_1 is 0 (i.e., $x_1 = 0$ in the algebraic statement), then no flow is transmitted to any of the constraint nodes via this variable arc, and variable x_1 contributes nothing to meeting the right-hand-side requirements.

If on the other hand the flow on the arc from node 0 to node x_1 is 1, then the multiplier causes 3 units of flow to enter node x_1. The bounds of (0, 1) on each constraint arc leaving node x_1 assure that each of these arcs receives exactly 1 unit of flow. These flows, amplified by the multipliers on the constraint arcs, reflect variable x_1's contribution to the three constraints in which x_1 appears. In a similar fashion, the contributions of variables x_2, x_3, and x_4 to the four constraints are also reflected correctly in Figure 6.21.

It is important to note from a solution perspective that the netform in Figure 6.21 has the same number of integer variables (those represented by the variable arcs) as the original algebraic formulation. The number of continuous variables added to the problem (the remaining arcs) depends on the particular transformation used. For example, it is also possible to transform 0-1 integer programming problems into "0-u" or "all-or-none" trans-shipment problems where the flows on certain arcs must be equal to either zero or their upper bounds [Glover and Mulvey 1980]. The usefulness of the netform is that it provides a different, perhaps faster, solution approach to the 0-1 integer programming problem. (This solution approach involves "relaxing" the problem by removing the integer requirements. The resulting netform can then be solved using a generalized network solution code, while relaxing the original algebraic statement would require a linear programming solution code. The advantage, then, depends on the speed with which the generalized network can be solved (see, e.g. Glover et al. [1978]).

6.10 CONNECTIONS TO MORE GENERAL DISCRETE PROBLEMS

By duplicating arcs, 0-1 problems can be used to model integer programming problems in which variables have a larger range of values. We illustrate this fact by means of an example that leads to other useful model constructions.

Consider the situation in which it is necessary to decide how many electrical generators to operate in a power plant in a given period. Each generator incurs a startup cost of $700, but once operating, provides 44 megawatt hours of electricity during the period. Suppose there are 13 generators in this category. We may diagram the situation as shown in Figure 6.22.

To convert the integer requirement into a 0-1 integer requirement, it is only required to duplicate the electrical generator arc 13 times (since 13 is the upper bound on the number of generators that can be operated). This yields the diagram in Figure 6.23.

In a like fashion, by duplicating arcs, one can convert other arcs with integer flows into 0-1 arcs. However, in most situations, there is little point in making such a conversion. (Some solution methods handle only 0-1 integer variables, and if there is a compelling argument for using one of these, the preceding type of conversion may be justified. In this case there is a technique illustrated in Exercise 6.11 for getting by with fewer 0-1 variables.)

There are particular problem conditions, however, that leave no choice, but require a general integer-flow arc to be replaced by 0-1 arcs in any event. Other conditions appear to require such a replacement, but in fact allow alternative treatments. One of these is discussed below.

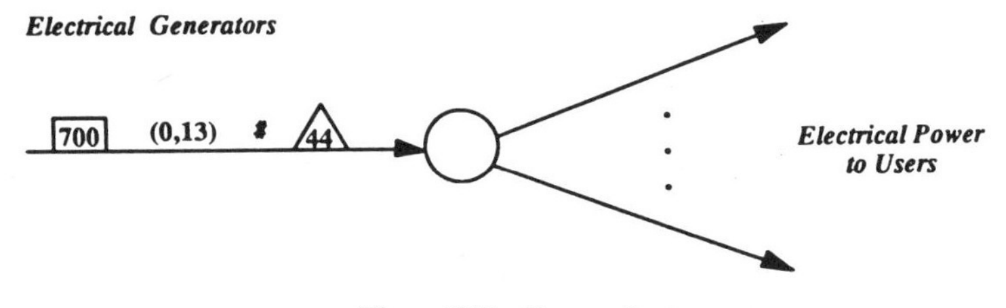

Figure 6.22 Power plant.

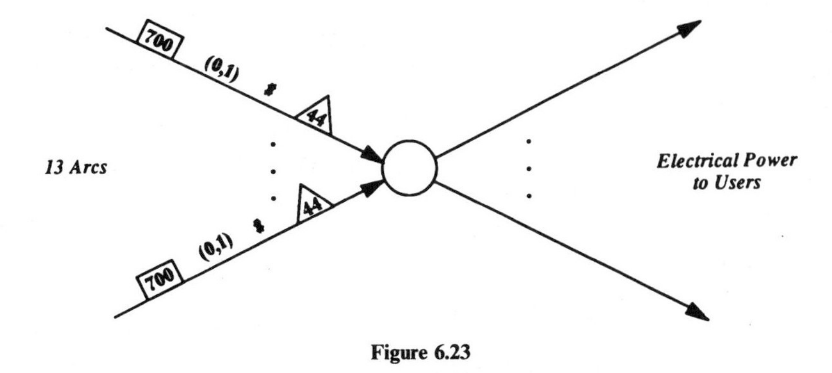

Figure 6.23

In our electrical generator example, suppose that each generator can provide anywhere from 38 to 44 megawatt hours during the period under consideration, depending on user demand. (A generator producing less than 38 megawatt hours during the period cannot be run economically.) Using the principle of sending the maximum flow and allowing the excess to escape, an (unsuccessful) approach to modeling this is Figure 6.24.

This construction may be viewed in the framework of dividing flow into potential and actual components as treated in Chapter 5. Thus, the multiplier of 44 provides the potential electrical power generated, while the amount carried on the arcs to users is the actual amount generated. The remainder, not generated, escapes on the downward arc of the diagram. An upper bound of 78 is placed on this arc because each generator operated may produce as little as 38 megawatt hours instead of 44, and the difference of 6, when multiplied by the total of 13 generators that may be operated, identifies that as much as 78 megawatt hours may need to escape.

However, there is a conspicuous flaw. The 78 megawatt hours should not be allowed to escape unless all generators are in fact operated. If only 1 generator is operated in a given period, the diagram as shown would allow any or all of its output to escape. Similar distortion, though less in magnitude, occurs all the way through operating 12 of the 13 possible generators.

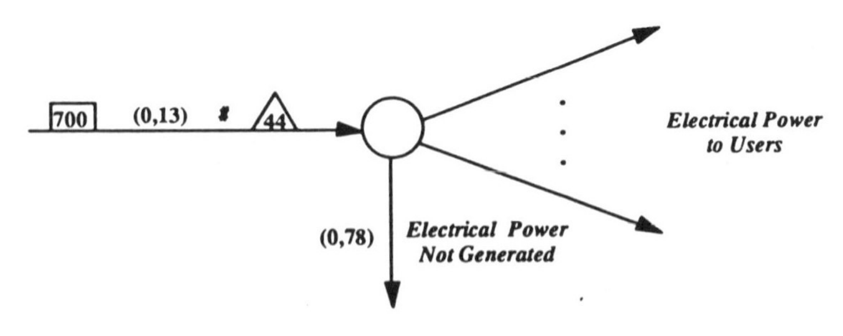

Figure 6.24 Erroneous model.

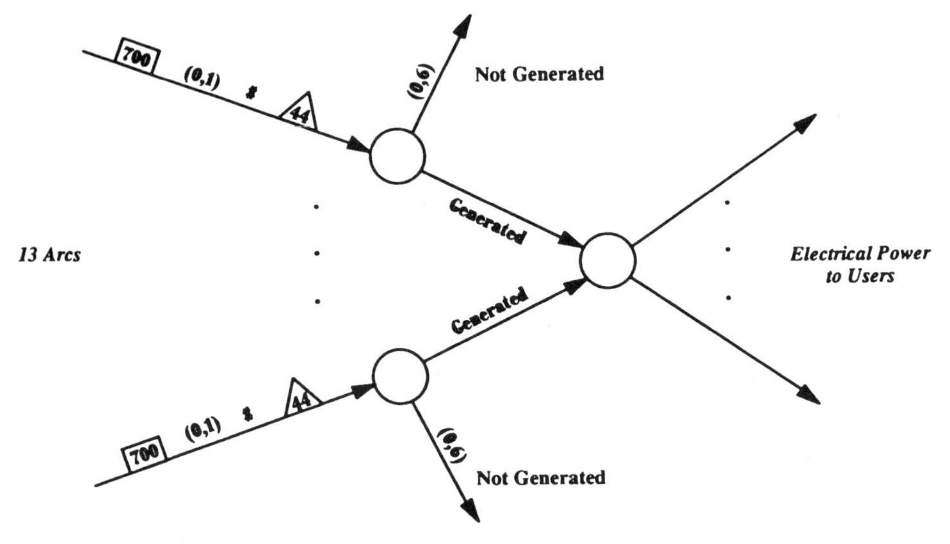

Figure 6.25 Corrected model.

The technique attempted in Figure 6.24 will work, however, if the problem is converted back to one with 0-1 flows, giving each generator its own arc. Then it can be assured that only 6 of the potential 44 megawatt hours will escape for any generator. A diagram for this is Figure 6.25. Although somewhat larger than Figure 6.24, this figure works successfully. As it happens, there is another way to model this situation using a different sort of principle. This alternative approach has the interesting feature that it does not rely on creating a collection of 0-1 arcs. Specifically, the fact that each generator can produce from 38 to 44 megawatt hours is captured by means of the construction in Figure 6.26.

In Figure 6.26, an arc with a multiplier has been added for each of the possibilities from 38 to 44. However, we did not stipulate, for example, that a generator was prohibited from generating 38.5 megawatt hours, yet this alternative is not shown. In spite of appearances, fortunately, the alternative

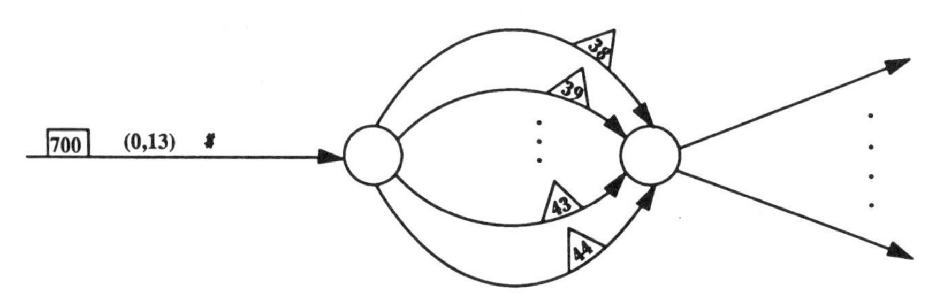

Figure 6.26

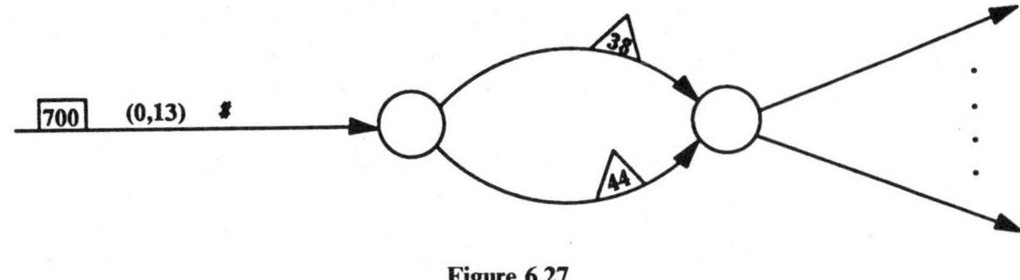

Figure 6.27

is still available, for there is no requirement in the diagram that flows on the arcs with multipliers must have integer values. Thus, for a generator to produce 38.5 megawatt hours, it is only necessary to divide its unit of flow so that 0.5 goes on the 38-multiplier arc and 0.5 goes on the 39-multiplier arc ($0.5 \times 38 + 0.5 \times 39 = 38.5$).

These observations lead to a useful simplification. We can obtain any value from 38 to 44 by dividing a unit of flow between the two extreme arcs. For example, we can get 39.5 by dividing a unit of flow into 11/12 on the 38-multiplier arc and 1/12 on the 44-multiplier arc$(11/12)\times 38+(1/2)\times 44=39.5)$. Consequently, the intermediate arcs, with multipliers from 39 to 43, are not needed. This permits Figure 6.26 to be replaced by Figure 6.27.

6.11 FIXED-CHARGE MODEL

One of the most basic and prevalent forms of netforms with discrete requirements is the fixed-charge network problem whose major offshoots include the extremely important genre known as "location" problems.

A fixed-charge arc is one with the following special property: whenever the arc is "used" (i.e., permitted to transmit flow), a charge is incurred that is independent of the amount of flow across the arc. This property is illustrated in

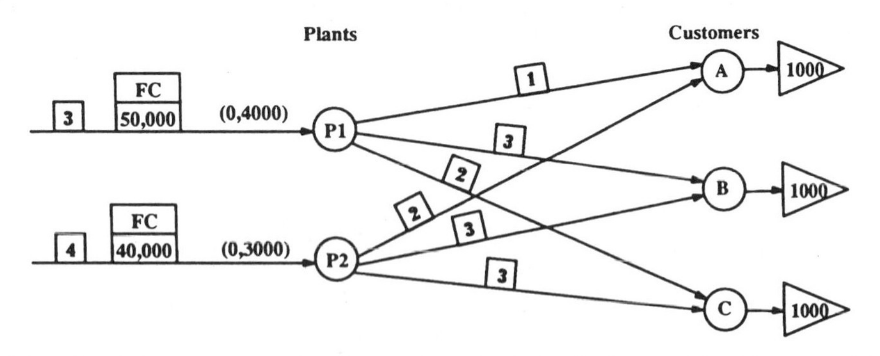

Figure 6.28 Plant location problem. Reprinted by permission. F. Glover and D. Klingman, *AIIE Transactions*, Vol. 9, No. 4, p. 368. Copyright 1977, Institute of Industrial Engineers.

Figure 6.28. The flows on the arcs into nodes P1 and P2 correspond to the number of units of a particular product that are produced at plants P1 and P2, respectively. In turn, the flows on the arcs from P1 and P2 to A, B, and C represent the amounts of the product that are shipped from P1 and P2 to each of the customers A, B, and C, who each demand 1000 units. The plants P1 and P2, however, are not presently in existence, and must be built (or acquired) before the distribution to the customers can occur. (The "location" nomenclature comes from viewing P1 and P2 as possible sites at which to locate plants.)

The figures in the boxes labeled FC on the arcs into nodes P1 and P2 represent the fixed charges of building the two plants (50,000 for P1 and 40,000 for P2). Thus, if any positive flow occurs on the arc into P1 (hence this number of units is produced at P1), then the plant P1 must first be "built" and the full 50,000 must be paid. Plant P1 can produce at most 4000 units. In addition, there is an "ordinary" cost for each unit produced at P1, and this is indicated by the 3 which appears in the unlabeled box on the arc into P1. Likewise, the numbers attached to the arc into P2 indicate that the fixed charge is 40,000, the production limit is 3000 units, and the ordinary unit cost is 4. Transportation costs are shown on the arcs from the plant nodes to the customer nodes.

The fixed-charge distribution problem of Figure 6.28 lends itself to a variety of interpretations other than those involving plants and customers. For example, such a network can be used to model decisions involved in locating warehouses or service centers, purchasing or leasing equipment, and hiring personnel. Another fixed-charge problem of the location variety is selecting the placement of off-shore oil drilling platforms and determining which wells should be drilled from each platform once the platforms are placed. Figure 6.28 provides an instance of this problem by interpreting P1 and P2 to be prospective locations for platforms, and A, B, and C to be drilling sites. In large cities the location of waste disposal collection centers (to which trucks from local collection areas bring refuse to be subsequently transported to main dumps) or fire stations are likewise instances of fixed-charge problems.

A fixed-charge network problem requires problem-solving "machinery" beyond that required to solve an ordinary network problem. However, the cost savings that can result from the solution of these problems generally far outweighs the increased computational effort required to solve them. It is sometimes suggested that one way to deal with such problems is simply to itemize the number of possible ways to locate the facilities, and then solve the distribution problems remaining. However, with as few as 20 prospective sites, this would entail the solution of more than a million distribution problems.

The algebraic statements of the netforms with discrete requirements in earlier sections may be constructed using the rules presented in previous chapters with the addition that the variables representing decision arcs are constrained to be 0 or 1. For fixed-charge models, the procedure is the same except that we must add an extra 0-1 variable and a constraint for each fixed-charge arc. Their purpose is to force the flow on the fixed-charge arcs to be zero if the fixed charge is not paid. To write the algebraic statement of the plant location problem shown in Figure 6.28, let

$$Z_1 \begin{cases} 1, & \text{if plant P1 is built} \\ 0, & \text{otherwise} \end{cases}$$

$$Z_2 = \begin{cases} 1, & \text{if plant P2 is built} \\ 0, & \text{otherwise} \end{cases}$$

If $Z_1 = 1$, then plant P1 can produce up to 4000 units. If $Z_1 = 0$, then plant P1 can produce 0 units. Letting x_{P1} represent the number of units produced at plant P1 (the flow on the arc into node P1), then this restriction becomes $x_{P1} \leqslant 4000\,Z_1$. The variable Z_1 is also used to add the fixed charge for plant P1 to the objective function as shown in the complete algebraic statement below.

Minimize $\quad 3x_{P1} + 4x_{P2} + x_{1A} + 3x_{1B} + 2x_{1C} + 2x_{2A} + 3x_{2B}$
$\quad + 3x_{2C} + 50{,}000Z_1 + 40{,}000Z_2$

Subject to

$$\begin{array}{llllllllllll}
\text{(P1)} & x_{P1} & & -x_{1A} & -x_{1B} & -x_{1C} & & & & = 0 \\
\text{(P2)} & & x_{P2} & & & & -x_{2A} & -x_{2B} & -x_{2C} & = 0 \\
\text{(A)} & & & x_{1A} & & & +x_{2A} & & & = 1000 \\
\text{(B)} & & & & x_{1B} & & & +x_{2B} & & = 1000 \\
\text{(C)} & & & & & x_{1C} & & & +x_{2C} & = 1000
\end{array}$$

$$x_{P1} \leqslant 4000Z_1$$
$$x_{P2} \leqslant 3000Z_2$$
$$x_{P1}, x_{P2}, x_{1A}, x_{1B}, x_{1C}, x_{2A}, x_{2B}, x_{2C} \geqslant 0$$
$$Z_1, Z_2 = 0 \text{ or } 1$$

While the two fixed-charge constraints allow for the possibility that a Z variable could equal 1 and the associated X variable equal zero (we build the plant but do not use it), this solution would be nonoptimal because the objective function seeks to minimize total cost (and would thus set a Z variable to zero if possible).

Fixed-charge arcs, when accompanied by side constraints, provide an alternate way to model concave cost functions. Reconsider the problem about the garden center whose model is shown in Figure 6.10. An alternate model appears in Figure 6.29 with fixed-charge variables noted on the figure. The side constraint, $Z_1 + Z_2 \leqslant 1$, insures that at most one of the ordering options will be used.

Fixed-charge arcs with side constraints may also be used to model logical conditions. While networks with side constraints are more difficult to solve than those without side constraints, this construct allows us to model a wider range of logical conditions. For example, reconsider the project investment example from Figure 6.9. A basic model is shown in Figure 6.30 along with the side contraints to model various logical conditions, including those from Figure 6.9. Note that

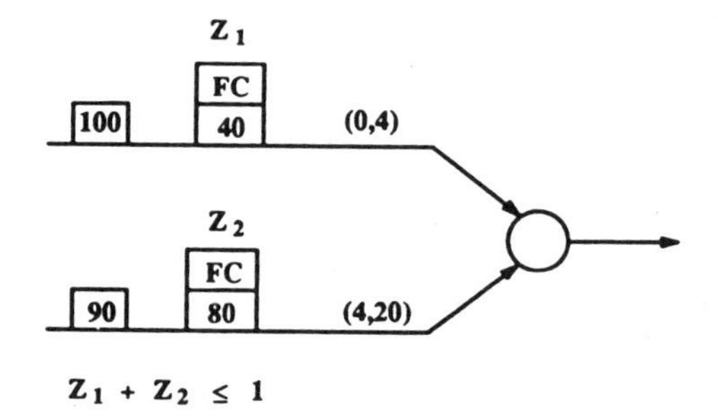

Figure 6.29 Alternate garden supply center model.

since the flow on the arcs in the basic model is either 0 or 1, the costs will automatically be fixed charges.

Another type of fixed-charge model for which there exists efficient solution methods is the locate/allocate model (also called generalized assignment models with fixed charges). This model type is designed to represent facility location and capacity allocation problems for optimal system configurations. The basic model structure is shown in Figure 6.31.

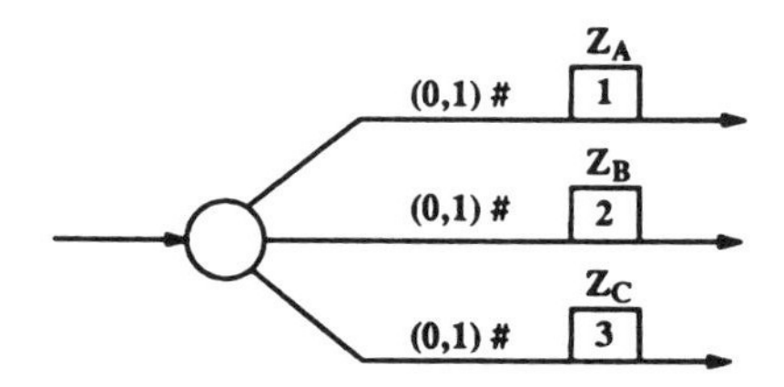

(a) At most 2 projects can be selected
$Z_A + Z_B + Z_C \leqslant 2$

(b) If *A* is selected, then *B* must be selected
$Z_A \leqslant Z_B$

(c) If A and B are selected, then C must be selected
$Z_A + Z_B \leqslant Z_C + 1$

(d) If C is selected, then A and B must be selected
$Z_C \leqslant Z_A$
$Z_C \leqslant Z_B$

(e) If C is selected, then either A or B must be selected
$Z_C \leqslant Z_A + Z_B$

(f) If *A* or *B* is selected, then *C* must be selected
$Z_A \leqslant Z_C$
$Z_B \leqslant Z_C$

(g) If A or B is selected, then C cannot be selected
$Z_A \leqslant 1 - Z_C$
$Z_B \leqslant 1 - Z_C$

Figure 6.30 Logical conditions as side constraints.

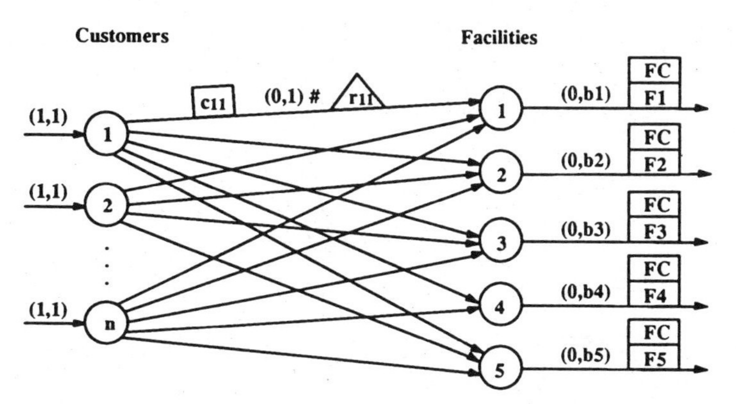

Figure 6.31 Locate/allocate model.

Each customer (represented by a node on the left) must be assigned to a single facility (node on the right) for service. The facilities, however, are not presently in existence and must be built (or acquired) before customer assignments can be made. F_j is the fixed cost of building facility j and is independent of the number of customers assigned to the facility. If facility j is built, its capacity will be b_j. The 0-1 arcs represent the assignment alternatives. The cost of assigning customer i to facility j (transportation and handling costs) is c_{ij}. Customer i requires a specified amount r_{ij} of the facility capacity if assigned to facility j. An optimal solution to this model will provide the selection of facilities and assignments of customers to facilities which minimize total costs (assignment costs plus fixed facility costs).

The mathematical formulation of the model in Figure 6.31 is

$$\text{Minimize} \quad \sum_{i=1}^{n} \sum_{j=1}^{5} c_{ij} X_{ij} + \sum_{j=1}^{5} F_j Z_j$$

$$\text{Subject to} \quad \sum_{j=1}^{5} X_{ij} = 1, \quad \text{for } i = 1, 2, \ldots, n$$

$$\sum_{i=1}^{n} r_{ij} X_{ij} \leqslant b_j Z_j, \quad \text{for } j = 1, 2, \ldots, 5$$

$$X_{ij} = 0 \text{ or } 1, \quad \text{for } i = 1, 2, \ldots, n, j = 1, 2, \ldots, 5$$

$$Z_j = 0 \text{ or } 1, \quad \text{for } j = 1, 2, \ldots, 5$$

$$\text{where } X_{ij} = \begin{cases} 1, & \text{if customer } i \text{ is assigned to facility } j \\ 0, & \text{otherwise} \end{cases}$$

$$\text{where } Z_j = \begin{cases} 1, & \text{if facility } j \text{ is built} \\ 0, & \text{otherwise} \end{cases}$$

Locate/allocate models, like other fixed-charge models, lend themselves to a variety of applications other than customers and facilities. For example, these networks can be used to model decisions involved in assigning checking accounts to auditors (who must be hired), assigning production to machines (which must be set up), and so forth.

APPLICATION: UFT GRADUATES

Undergraduate Flight Training (UFT) graduates in the US Air Force are required upon graduation to take advanced flight training and survival training courses enroute to their first operational assignment. Advanced flight training is offered only in formal schools usually by the Major Air command, the principal aircraft user. UFT graduates must also take from one to four survival courses before being assigned to a crew, for example, basic survival (Washington), water survival (Florida), air weapons delivery (Texas), etc. These courses are only offered at certain times, have enrollment limits, and may have prerequisites. The identification of schedules is further complicated by attendance requirements at combat Crew Training courses, various modes of transportation, the number of dead days in the pipeline, and the opportunity for the UFT graduates to take leave as desired, and so on.

To solve this UFT graduate scheduling problem, the Air Force developed a computer program (called the UFT Pipeline Scheduling Model) which generates from one to five feasible least cost schedules for each graduate. Using these schedules and course enrollment limits, the personnel manager in the Training Pipeline Management division manually assigns each graduate to one of his/her feasible schedules. Clearly, this is a difficult and time-consuming task to do by hand; further, the total cost of these manual assignments may be far from optimal.

The UFT problem may be modeled using an integer generalized network. Figure 6.32 illustrates this model. In this figure, the node Mi represents the ith graduate and has a supply of exactly 1. Each graduate node is connected by arcs to its set of schedule nodes. Each connecting graduate/schedule arc has a multiplier a_{ij} equal to the number of classes in the schedule and a cost c_{ij} equal to the cost of assigning graduate i to his/her jth schedule. The #-sign indicates that flow must be integer-valued. The arcs emanating from a schedule node in Figure 6.32 lead to the individual classes making up the schedule. Each of these arcs has an upper bound of one. Thus, if a particular schedule is "selected," then every class in the schedule is also automatically selected. The objective is to pick a schedule for each graduate that will minimize the cost of the assignments for the overall program, subject to the upper and lower attendance limits for each class, expressed as bounds on the arcs from class nodes to the sink node of Figure 6.32. All arc costs, except for those attached to the graduate/schedules arcs, are thus equal to 0.

Typically the UFT problem involves 120 graduates, 200 classes, and 460 schedules, giving rise to a 0-1 generalized network formulation with 460 0-1 variables, 2200 continuous variables and 780 nodes. The problem was solved

---

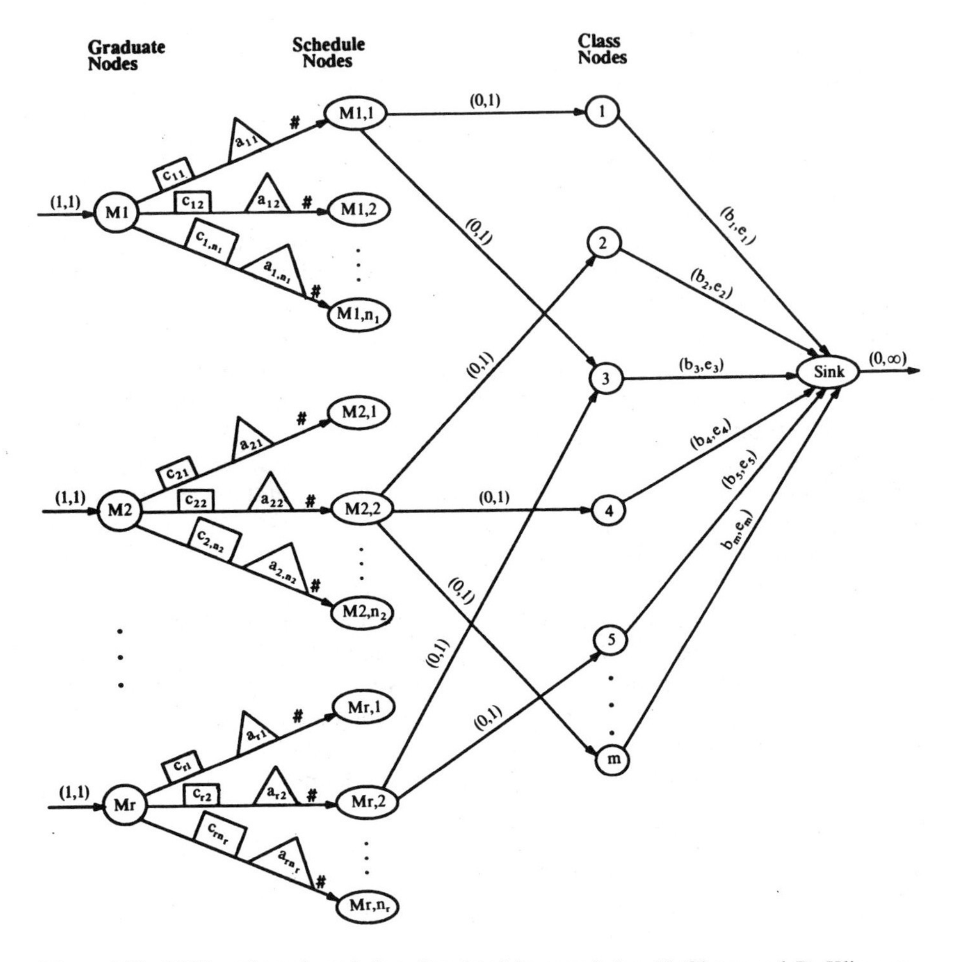

Figure 6.32 UFT netform formulation. Reprinted by permission. F. Glover and D. Klingman, *AIIE Transactions*, Vol. 9, No. 4, p. 373. Copyright 1977, Institute of Industrial Engineers.

using a specialized branch and bound procedure with generalized network subproblems. The optimal solution was often found and verified after only 30 sec and in some cases only required a total solution time of 10 sec on a CDC 6600 computer [Glover and Klingman 1977]. This is exceptionally fast for a zero-one problem of this size on this type of computer.

CASE: COTTON VALLEY

Introduction

The purpose of this study is to systematically lead the student through a complex real-world application of network techniques and sophisticated network solution procedures [Fuller et al. 1976]. It is hoped that by attacking the problem slowly, and in stages, the student will learn how to decompose complicated problems into smaller, more manageable subproblems. These subproblems can then be linked together to form a global model.

The setting for the actual case was the Messilla Valley of New Mexico and the Upper Rio Grande Valley of El Paso County, Texas (Figure 6.33). This area was, and still is, heavily devoted to the farming and gin processing of cotton. However, during roughly a 10-month period, cotton production decreased by 50% while the ginning capacity in the area remained approximately the same. To further compound the problem, innovations in cotton storage eliminated the need for immediate ginning of picked cotton.

Due to these developments, operating all the gins in the area was unnecessary. Because the contributing factors were seemingly overwhelming, decisions as to which gins should be opened for a given cotton crop were made solely by intuitive judgment. These decisions had no sound economic justification and, therefore, resulted in policies that were wasteful of time, labor and money.

It was obvious that the farmers and processors in the area could benefit from certain management science tools. One of the problem's critical components was the transportation of cotton to gins. Due to this, it was decided early on that a model incorporating networks would probably be best. In fact, the actual form the model would take was a major consideration in approaching the problem.

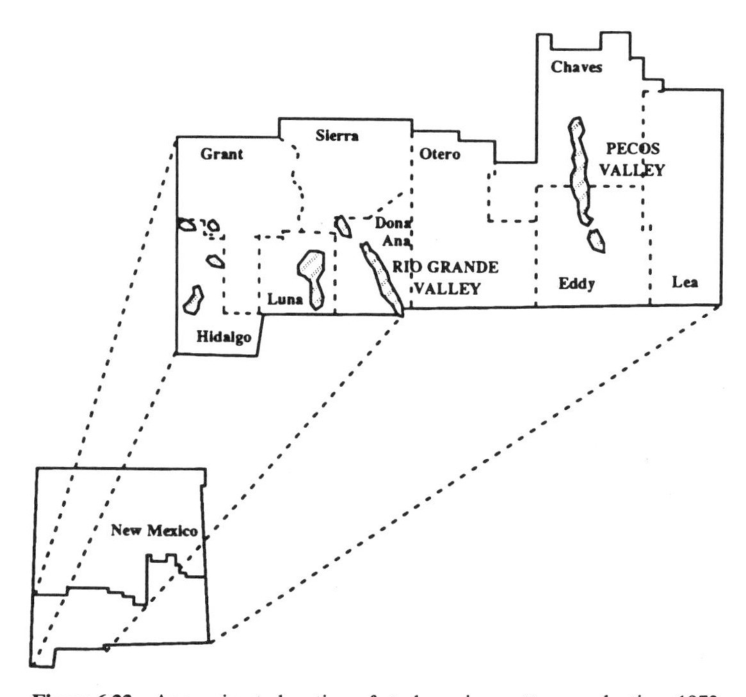

Figure 6.33 Approximate location of study region cotton production, 1973.

The final model was a network that "evolved" through a series of reformulations, revisions, and simplifications.

The next section will present a very elementary version of this cotton ginning problem, including relevant data. It is the student's task to devise a model that accurately portrays the problem's components and relationships. Each subsequent section will add further stipulations and data to the original problem. The student must successively update or revise the model to incorporate the new information and complexity. The final product should be a model capable, with minor modifications, of solving the real-world cotton ginning problem.

Problem 1: The Distribution Problem

The area in question, Cotton Valley, is an irrigated region, approximately 10 miles in length and varying from 0.5–2 miles in width. Situated within this region are three cotton-producing farms. Collectively, these farms yield 80 bales of high-quality Upland cotton and 45 bales of American Pima each cotton season. However, individually the three farms differ in size, production efficiency, and cotton output. The total cotton output for the largest farm, including both Upland and Pima, is 65 bales per season. The smaller farms produce 40 and 20 total bales per season.

Cotton Valley also boasts two large cotton gins. Both gins are owned and operated by the Cotton Valley Co-op in which the three cotton farmers have vested interests. Thus, it makes no difference to the farmers which gin processes their cotton so long as the costs incurred are kept to a minimum.

The larger gin, Turner Gin, received its name from the local politician who donated the land on which it was built to the Co-op. The Turner Gin has the capacity to process 180 bales of cotton per season, regardless of type. The smaller gin is called the Sloe Gin. The name is due to the sloe bushes surrounding the gin. (There never was a Mr. Sloe in the valley). The Sloe Gin can process 150 bales per season, also regardless of type.

It should be noted that either one of the gins could process the valley's entire cotton crop. This was not always the case. At one time, nine cotton farms capable of producing a total of 275 bales per season thrived in the valley. But poor economic conditions, bad weather and a general decline of interest in cotton farming have greatly reduced the valley's cotton production.

The owner of the smallest farm, Mr. Green, must constantly watch his pennies in order to avoid bankruptcy. For this reason, he introduced a motion at a Cotton Valley Co-op stockholder's meeting. The motion required the Co-op to minimize cotton transporting costs between farms and gins. There was a quick and unanimous approval of the motion. "Crazy" Harry, the local crop duster, volunteered to fly over the valley taking pictures with his Instamatic camera. From these photographs, the distance from each farm to each gin was accurately measured. With these distances known, it was a simple matter to determine transportation costs. The following table indicates the costs associated with shipping a single bale of cotton from each farm to each gin.

	Turner Gin	Sloe Gin
(Masterson's) large farm	1	1
(Buck's) medium farm	3	2
(Green's) small farm	1	2

The Co-op has hired you, a nonexpert in management science, to model and solve this problem. Formulate a netform model. Is there more than one optimal solution? If so, why?

Problem 2: Time Periods

When you presented the network problem you modeled and solved, the Co-op farmers were quite impressed. But they realized the information you were given was over-simplified and the resulting model too limited. As farmer Green pointed out, not all the cotton is picked and shipped out at the same time. Instead, cotton is picked continuously through the week, then shipped to the gins on Friday (the farmers are always talking about their Friday "gin" money). Any picked cotton not sent to the gin must be field ricked to protect it from weather. The cost of keeping one bale with ricking is 1 dollar per week. The farmers know from experience that it takes 2 weeks to pick all the cotton planted in the valley. Rip Masterson, the owner of the largest farm, is foreman of the picking crews. Masterson is able to provide you with the following table of cotton to be picked at each farm each week of harvest.

	Week 1	Week 2
Masterson's (large)	25	40
Buck's (medium)	15	25
Green's (small)	10	10

Being a brighter than average student, you deduce that each gin must have weekly capacities too. After some checking around, you discover the Turner Gin will operate for 3 weeks with a processing capacity of 60 bales per week. The Sloe Gin will operate the same 3 weeks with maximum processing capacity of 50 bales per week.

Not being one to back away from challenges, you tell the Co-op farmers that with this additional information you can present them with a model that will minimize total shipping and storage costs. They put it to a vote and give you the go-ahead. Revise your netform model to incorporate this new information.

Problem 3: Holding Costs

On the weekend before you present your new model to the Co-op farmers, you go to Austin to visit some friends. As long as you are there you decide to drop in on your old Management Science professor to discuss the work you are doing for the farmers. You show her your formulation of Problem 2 and ask if she sees any way you can simplify it for your presentation to the farmers. Being a professor and academician, she will not give you a direct answer. However, she points out some factors and poses some interesting questions for your consideration.

- Holding costs are the same for each farm.
- Because the gins are owned cooperatively, all the farmers could store cotton at the gins instead of at their farms.
- How would the model change if cotton was envisioned as being stored at the gins rather than at the farms?
- Does this revised model still represent the real problem where cotton is stored at the farm?

Problem 4: Labor Costs

After your presentation of your latest model, you finally meet up with the owner of the middle-sized farm. His name is Luther Buck Jones, but everyone calls him Luther Buck. His job for the Co-op is management of both gins in Cotton Valley. Since the gins are only open for 3 weeks, Luther Buck mainly sits around drinking beer. Several people have tried to tell Luther Buck that gins and beer do not mix, but he steadfastly ignores them. Actually, when the gins are open, Luther Buck works hard. At any rate, he decided to meet with you to share some information he has concerning the gin operations. He explains that the processing levels you were previously given can only be reached by working overtime. Until 10 years ago that would not have mattered. But now, thanks to government regulations and the appearance of a union, overtime is more expensive than regular shifts. Luther Buck wants to know if your model could be modified to reduce overtime at the gins. Talking late one night over a few beers, Luther Buck fills you in on his crew size, their rate of pay, and how much they can produce on the average. After sleeping on the idea (and sleeping off the beer), you realize you must convert the information into cost per bale for regular and overtime shifts. This is necessary because all other costs are given in dollars per bale. While you have your calculator out, you also figure the average capacity for each gin during the regular and overtime shifts. The regular costs at the Turner Gin are 3 dollars per bale. The same processing at the Sloe Gin costs 2 dollars per bale. Overtime processing at both gins costs an additional dollar per bale. The average weekly capacity for the regular shifts is 40 bales at the Turner Gin and 30 bales at the Sloe Gin. Both gins have an additional capacity of 20 bales per week if overtime is used.

You are careful to explain to Luther Buck that, while you can incorporate his information into your model, you will not necessarily end up with the lowest shipping cost, or the lowest storage cost, or even the lowest labor cost. But you can guarantee the lowest total of all these costs. He says that is precisely what the farmers want. He also mentions that if you write up your work on this project, you might get it published in the Federation of Co-op's magazine, *The Journal of Ginning*. You tell Luther Buck you will certainly give that some consideration after he explains that *The Journal of Ginning* is the most widely read magazine in the valley, next to *Reader's Digest*, of course. Revise your netform model to incorporate labor costs.

PROBLEM 5: OTHER COSTS

Luther Buck was so excited at the prospect of getting the Cotton Valley model written up in *The Journal of Ginning* that he went ahead and wrote to them about your work. Because there are several valleys with problems similar to those in Cotton Valley, *The Journal* felt the article would be worthwhile. Just as you have finished incorporating labor costs into your model, a reporter from *The Journal*, Lois Lane, arrives on the scene. She looks over your model with interest, but points out that you have not included all the relevant costs. Because you are not very familiar with cotton ginning, and because you and Lois would both like to get done and go home, you decided to work together on still another revision of the model.

Over dinner, Lois explains that nonlabor costs associated with operating the gins should be included in your model. When a gin opens for the season, nonlabor costs are fairly low. However, after a certain amount of processing, a higher rate goes into effect. The initial rate at the Turner Gin is 1 dollar per bale. This rate applies to the first 40 bales processed. Beginning with the 41st bale, the rate increases to 2 dollars per bale. Likewise, the Sloe Gin has an initial rate of 2 dollars per bale with a jump to 3 dollars per bale after 70 bales have been ginned. Revise your netform model to incorporate these costs.

PROBLEM 6: FIXED CHARGES

The cotton picking season has arrived. You are growing weary of model revisions and picking cotton lint off your clothes. There is, however, one last detail to be included in your model. The gins have been idle for a year and must be cleaned and readied for another season of production. All the cleaning and oiling done for each gin is a one-time cost that must be paid if the gin is opened, regardless of whether the gin processes one bale or a thousand. These costs are \$100 for the Turner Gin and \$150 for the Sloe Gin. Revise your netform to incorporate fixed costs.

Solve the model. While the fixed costs cannot be accommodated by linear programming or pure networks codes, by specifying which gins are opened or closed (setting gin capacities to zero in the model) you can use LP or pure networks codes to solve your model. If you solve for every possible combination of gins opened and closed, you can then select the plan yielding the minimum total cost. Advanced codes which handle fixed charges will do the comparisons for you using a branch and bound procedure. Simpler codes will solve for a single combination of opened and closed gins, leaving the comparisons and

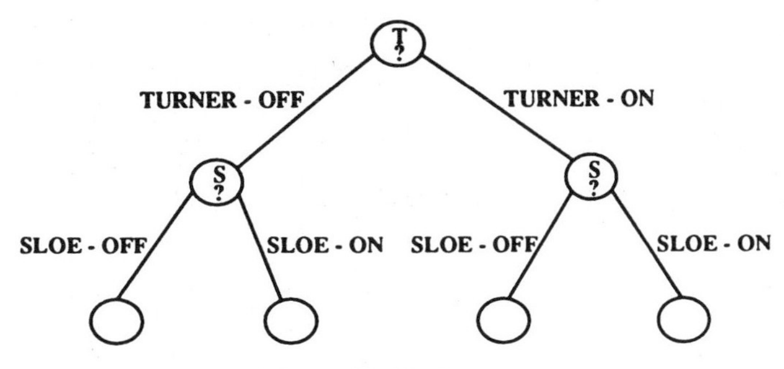

Figure 6.34 Decision tree.

conclusions to you. Because either gin can process all the cotton picked in the valley there are several configurations of gins opened and closed for you to consider. Since you must supply the simpler code with the series of combinations, you make a decision tree to simplify the enumeration process. Figure 6.34 is just such a tree. This tree is not a network and should not be confused with one. Scanning the bottom of the tree you see there are four possible configurations: both off, only the Turner Gin open, only the Sloe Gin open, both gins open. It is possible to determine in advance the number of combinations by taking 2^n where n is the number of gins. The 2 stems from the fact that there are only two possible states for the gins: opened or closed. If there were 10 gins there would be 1024 possibilities, so you can see that enumeration is practical for only a small number of gins (n). It should also be apparent that with both gins off you have an infeasible problem. Only those combinations that allow a processing capacity greater than or equal to cotton production need to be considered.

CASE: ADVANCED CIRCUITRY ELECTRONICS, INC., PART A

Advanced Circuitry Electronics, Inc. (ACE) is a Texas-based firm which manufactures high-technology electronics devices. The Consumer Products Division of ACE has three product lines: (1) hand-held calculators; (2) digital watches; and (3) automobile radios and radio/tapedeck combinations. The company's corporate headquarters and primary production facilities are located in Dallas, Texas. The Dallas plant produces all of the calculators and watches sold by ACE. The company's only other production facility is located in St. Louis. The St. Louis plant produces all of the radio products.

ACE distributes its products to national and regional retail chains through company-owned distribution centers. Currently, there are distribution centers located in Dallas and St. Louis, each of which handles all three consumer product lines. ACE partitions its consumer products market into customer zones (clusters of similar customers in the same geographic region). The customer zones currently being served are Dallas, Houston, St. Louis, Chicago, Atlanta, and Miami. In the past, only a small portion of total sales have come from the southeast area (Atlanta and Miami) and from Chicago because of limited marketing effort in these areas. However, recently ACE has pursued these markets more aggressively since efforts to increase its already large market share in the Southwest were showing little marginal benefit. In addition, the company has acquired new customers in the Los Angeles area for calculators and radios. (ACE does not feel it can compete with an established West Coast manufacturer in the watch product line.)

The staff of the Consumer Products Division has forecasted annual demand by product line and customer zone for the upcoming fiscal year (beginning nine months from now). This forecast is shown in Table 6.4. Demand shown is for cartons. Calculators and watches are packaged with 24 units per carton whereas radios are packaged with one unit per carton. A carton of calculators requires 2 ft^3 of storage space. A carton of watches and a radio each require 1 ft^3 of storage space. Also shown in Table 6.4 is the anticipated distribution center capacity which will be demanded, 80,200 ft^3 for the year.

Currently, the combined capacity of the Dallas and St. Louis distribution centers is 64,000 ft^3 per year of throughput, which is not adequate to handle next

Table 6.4 Forecasted Annual Demand by Product Line and Customer Zone

Customer Zone	Product Lines Required	Annual Demand (cartons/year)	Storage Space Conversion Factor (cubic feet/carton)	Distribution Center Capacity Demanded (cubic feet/year)
Dallas	Calculator	4,000	2.0	8,000
	Watch	1,800	1.0	1,800
	Radio	6,000	1.0	6,000
Houston	Calculator	3,000	2.0	6,000
	Watch	400	1.0	400
Los Angeles	Calculator	5,000	2.0	10,000
	Radio	8,000	1.0	8,000
St. Louis	Calculator	600	2.0	1,200
	Watch	2,400	1.0	2,400
	Radio	3,000	1.0	3,000
Chicago	Calculator	4,000	2.0	8,000
	Watch	3,000	1.0	3,000
Atlanta	Calculator	3,600	2.0	7,200
	Watch	400	1.0	400
Miami	Calculator	2,400	2.0	4,800
	Radio	10,000	1.0	10,000
Total capacity demanded				80,200

year's forecasted demand. Since about 6 months is required to establish a new distribution center, the ACE management realizes that they must soon decide upon a plan for providing additional capacity.

In an executive session with the President of ACE and all of the Division Managers, including Mr. William A. Able, Manager of the Consumer Products Division, the distribution center problem was discussed. The following guidelines were formulated:

1. The Dallas distribution center must remain in operation. If necessary its capacity could be expanded from the current 36,000 ft^3 per year to 66,000 ft^3 per year.
2. The St. Louis distribution center cannot be expanded. If cost effective, this facility could be abandoned as a distribution center since the space could be used for expansion of the manufacturing plant.
3. Any newly established distribution centers must be within 1000 miles of the Dallas corporate headquarters to facilitate supervision by corporate management.
4. All of the demand from a particular customer zone for a particular product line must be provided by only one distribution center. This policy is to reduce administrative coordination efforts and provide better customer service.
5. If a particular customer zone has demand for both calculators and watches, and if this demand for either of these product lines is less than 700 cartons per year, then the two product lines must both be provided to that customer zone from the same distribution center. For example, referring to Table 6.4, the Houston customer zone has demand for both calculators and watches and the demand for watches is only 400 cartons per year. Therefore, Houston's demand for these two product lines must be supplied by the same distribution center. In other words, the calculator and watch product lines are "bundled" together under these circumstances. This policy is to reduce the occurrence of small-size shipments.
6. The final distribution system configuration must not have more than three distribution centers. This policy is also to facilitate supervision by the corporate level staff.

Because of the 1000 mile limit, Los Angeles and Miami are eliminated as possible sites for new distribution centers. Therefore, the members of the executive meeting decided that Mr. Able should investigate the establishment of distribution centers in either Chicago or Atlanta, or both, along with possible expansion in Dallas and possible abandonment in St. Louis. During the discussion, the president stated that he had heard that the city of Denver, Colorado (where the president was born and raised), was offering economic incentives to attract nonpolluting industry. He therefore strongly suggested that Mr. Able also consider the Denver area as a possible distribution center site for servicing the Los Angeles demand.

As the discussion ended, it was agreed that Mr. Able would bring a recommendation for the locations and capacities for distribution centers to next month's meeting.

In the next 2 weeks, the Consumer Products Division staff compiled information about possible locations for distribution centers. They decided to investigate two location strategies. The first, as suggested by the executive session members, considers locations in customer zone cities. The second, as implied by the president, considers locations in smaller towns which are close to more than one customer zone. These locations are expected to have smaller fixed operating costs per cubic foot capacity than the major city locations. The pertinent data for the final list of alternative locations are given in Table 6.5. The handling costs include handling at the facility and inbound freight costs from the manufacturing sites. Note that the proposed Atlanta and Jacksonville facilities will not carry radios, and the proposed Denver and Flagstaff facilities will not carry watches. The transportation costs and mileage between facilities and customer zones are given in Table 6.6. The costs shown are for calculators and radios. Shipping costs for watches are approximately one-half the costs for calculators and radios. If a transportation cost is not given, then shipments from that facility to that customer zone were deemed by the staff to be impractical to consider.

During the course of this investigation, the marketing manager of the Consumer Products Division said that she had been concerned recently about possible deterioration in the level of service due to the increased demand on the

Table 6.5 Data for Distribution Center Location Alternatives

Facility Location	Capacity (cubic feet/year)	Fixed Cost ($/year)	Product Lines Carried/ Handling Cost ($/carton)[a]		
			Calculator	Watch	Radio
Dallas, TX	36,000	150,000	Yes/0.10	Yes/0.10	Yes/2.20
(expanded)	(66,000)	(280,000)	Yes/0.10	Yes/0.10	Yes/2.20
St. Louis, MO	28,000	130,000	Yes/2.10	Yes/1.10	Yes/0.20
Atlanta, GA	26,000	120,000	Yes/2.50	Yes/1.30	No
Chicago, IL	30,000	140,000	Yes/2.90	Yes/1.50	Yes/1.20
Denver, CO	30,000	100,000	Yes/2.40	No	Yes/2.70
Jacksonville, FL	17,000	60,000	Yes/3.00	Yes/1.50	No
Springfield, IL	17,000	54,000	Yes/2.20	Yes/1.10	Yes/0.40
Austin, TX	18,000	60,000	Yes/0.60	Yes/0.30	Yes/2.60
Flagstaff, AZ	17,000	60,000	Yes/3.00	No	Yes/4.00
Memphis, TN	22,000	85,000	Yes/1.40	Yes/0.70	Yes/1.00

[a]Handling cost includes inbound freight charges from the manufacturing locations as well as distribution center handling charges.

Table 6.6 Transportation Costs (Dollars/Carton) for Calculators and Radios[a] and Distance (Miles) Between Distribution Centers and Customer Zones (in Parentheses)

	To						
From	Dallas	Houston	Los Angeles	St. Louis	Chicago	Atlanta	Miami
Dallas	0.30 (20)	3.60 (243)	12.60 (1387)	6.00 (630)	8.40 (917)	7.20 (795)	12.00 (1300)
St. Louis	–	–	16.80 (1845)	0.30 (20)	3.00 (289)	4.80 (541)	10.80 (1196)
Atlanta	–	7.20 (789)	–	–	–	0.30 (20)	6.00 (655)
Chicago	–	–	18.60 (2054)	3.00 (289)	0.30 (30)	6.00 (674)	–
Denver	–	–	9.60 (1059)	7.80 (857)	9.00 (996)	–	–
Jacksonville	–	–	–	–	–	3.00 (306)	3.60 (349)
Springfield	–	–	–	1.00 (100)	1.80 (189)	–	–
Austin	–	1.80 (164)	12.60 (1374)	–	–	8.40 (919)	12.00 (1351)
Flagstaff	–	–	4.20 (484)	12.00 (1361)	14.40 (1604)	–	–
Memphis	–	4.80 (561)	–	3.00 (285)	4.80 (530)	3.60 (371)	9.00 (997)

[a]Transportation costs per carton for watches are approximately one-half of the costs given for calculators and radios.

existing distribution system. Although there are presently no specific corporate guidelines for customer service levels, the marketing manager urged that customer service, as well as cost, be a decision criterion during the distribution center location analysis. She suggested that the distance between distribution centers and customer zones be used as a measure of customer service level and that some average service level for the entire division be the criterion used. Mr. Able agreed.

Mr. Able realized that even with a relatively small problem like this (7 customer zones, 3 product lines, and 10 distribution facility location alternatives), the number of possible distribution system configurations is quite large. Therefore, he decided to call in a management consulting firm with whom he had worked in the past.

Mr. Able was referred to you by the management consulting firm which he contacted for help with his problem. Because you are eager to build a reputation for expertise in this field, you readily agree to work with Mr. Able.

Case Questions

1. Formulate a locate/allocate model to optimize ACE's Consumer Product distribution system. (Hint: Create separate nodes for each product line at each customer zone.)
2. Outline a strategy for determining the best distribution system configuration. What system parameters do you plan to vary between runs, by how much, and in what direction?
3. Solve your model using appropriate computer software and your solution strategy. Analyze the results of the various runs and recommend a "best" distribution system configuration for ACE.
4. Can the demand for calculators at the St. Louis customer zone be satisfied by the proposed Denver distribution center? Why or why not?
5. In the realm of physical distribution systems, what is meant by "single sourcing"? Why is single sourcing desirable?
6. What is meant by "bundling"? Why is bundling desirable?
7. How would the locate/allocate model for Ace change if sales forecasts predicted that customers will buy more when served by a nearby distribution center than when served by a distant distribution center?

CASE: ADVANCED CIRCUITRY ELECTRONICS, INC., PART B

At the next Executive Board Meeting, Mr. Able presented his recommendation for the Dallas–St. Louis–Denver distribution system configuration along with a summary of the analysis procedure. The division managers agreed with Mr. Able's analysis and conclusions. At this point, the president stated that he was impressed with your analysis and would like to use you again to analyze another situation which he had been considering recently.

In addition to the Consumer Products Division, ACE has a Cybernetics Division which designs and manufactures four different lines of computers. Among these is a line of small, interactive computers designed for personal and small business use. These computers are marketed under the name CY-CLOPS (CYbernetics-Consumer Language Oriented Processing System) but are commonly referred to by the staff as ACE's One-Eyed Jacks. Currently, the One-Eyed Jacks are produced on a special order basis only. They are sold through independent sales representatives who service small businesses and specialty computer retailers. The only inventory carried is the demonstrator units used by the sales reps and a few replacement units in Dallas which are kept for customers with emergency service warranty contracts.

ACE's marketing analysts have predicted that sales of the CY-CLOPS line will increase dramatically over the next 5 years. The president stated that he had therefore been considering integration of the CY-CLOPS line into the Consumer Products distribution system on a trial basis. Since this distribution system was now being redesigned and expanded, it seemed like an ideal time to try the new arrangement.

The president instructed the Cybernetics division manager to supply Mr. Able with the relevant cost and demand information about the One-Eyed Jacks. He then asked Mr. Able to develop a distribution system allocation which would include the CY-CLOPS line. Since this was to be a trial arrangement, the addition of CY-CLOPS to the system was to have no impact on the number or location of distribution centers. However, the president said he was interested in seeing if the addition of CY-CLOPS to the system would change the optimal allocation of customers to distribution centers for the other product lines.

The next day Mr. Able received the following information from the Cybernetics division manager:

1. CY-CLOPS is sold by independent reps in Chicago, Washington, DC, Houston, and San Francisco.
2. Each packaged unit requires 50 ft^3 of storage space.
3. ACE pays shipping cost to the sales representatives' location only, not to the ultimate customer.
4. Transportation costs, mileage, handling charges, and forecasted demand for next year are given in the following table.

Transportation Costs in $/Unit (Distance in Miles)

	To				Handling Charge[a] ($/Unit)
From	Chicago	Washington, DC	Houston	San Francisco	
Dallas	206.00 (917)	297.00 (1319)	55.00 (243)	394.00 (1753)	8.00
St. Louis	65.00 (289)	178.00 (793)	175.00 (779)	470.00 (2089)	55.00
Denver	224.00 (996)	364.00 (1616)	229.00 (1019)	278.00 (1235)	67.00
Demand (units)	30	25	40	30	

[a]Includes inbound freight charges from Dallas manufacturing location.

Mr. Able has returned to you with the CY-CLOPS information and asks you if you can include this product line in the original model formulation. You assure him that it can be done quite easily if adequate remuneration is forthcoming. He makes you an offer you cannot refuse.

Case Questions

1. Modify your original model formulation to include the CY-CLOPS line in the Dallas–St. Louis–Denver distribution system.
2. Solve the modified model, analyze the results, and make a recommendation for Mr. Able to present to the Executive Board.

EXERCISES

6.1 A weekend camper, after packing essentials, figures he can carry up to 23 additional pounds. The candidates for extra items to take along have been narrowed down to exactly four possibilities: a battery-operated hair dryer (2 lb), an aluminum lounge chair (9 lb), a favorite bowling ball (11 lb), and a collapsible plastic hot tub (14 lb). On a scale from 1 to 10, the camper ranks how much he would like to have each of these items on the trip, putting the hair dryer at 3, the lounge chair at 5, the bowling ball at 6, and the hot tub at 8. Formulate a netform to maximize the total rating of the extra items taken:

a. Under the conditions stated.

b. Under the added condition that at most two items can be taken.

c. In the situation where there are 3 of each item available to be taken along, a 36 lb total carrying limit for extra items, at most 3 units from the hair dryer and lounge chair group can be taken, and at most 2 units from the bowling ball and hot tub group can be taken.

6.2 The following is an extended version of the machine selection problem in Figure 6.2. Each machine selected makes products in lots. A particular lot on a given machine is devoted to making a specified number of units of a particular product. Information about the number of lots each machine produces (if chosen to be operated), the cost of producing a lot, the number of product units produced in the lot, and the minimum number of units of the product demanded is shown in the following table.

Cost Per Lot ($)/Units of Product Per Lot

No. of Lots Made If Machine Chosen	Machines	Products 1	2	3	4	5
2	1	42/12	70/4	93/6		
2	2		89/7	45/10		
2	3	58/9			37/14	
3	4	55/15		55/9		38/20
2	5		60/9		54/12	52/10
Min demand for product		15	8	8	10	17

For example, the row for Machine 1 shows that if Machine 1 is operated it will produce exactly 2 lots. (To make fewer than the number of lots specified would generate idle time for operators, prohibited by union contract.) Products that can be made on Machine 1 are Products 1, 2, and 3. For instance, each lot of Product 3 run on Machine 1 will cost $93 and produce 6 units of this product. Similarly, the column for

Product 1 shows that Product 1 can be made by lots on Machines 1, 3, and 4. Each lot of Product 1 on Machine 1 costs $42 and makes 12 units of the product. (Thus, the minimum demand of 15 units of Product 1 could be met, for example, by 2 lots of the product on Machine 1 or Machine 3, or 1 lot each on Machines 1 and 3, or 1 lot on Machine 4.) Prepare a netform to minimize total cost, not allowing fractional machines or lots.

6.3 An investor can invest in any of several investment "packages." A package consists of investing in various projects either totally, or not at all. The complete problem allows the investor to select as many as he wants, or can, of packages (a)–(f), following. However, provide a separate netform diagram for each. (All costs are in $10,000 units. Budget constraints and project returns are omitted for simplicity.)

a. The package costs $15, and invests in projects 1, 2 and 3.

b. The package costs $18, and invests in projects 4 and 5 and in one (but not both) of Projects 6 and 7.

c. The package costs $19, and invests in any two of projects 1, 2 and 3 and any one of Projects 4, 5 and 6.

d. The package costs $24, and invests in 4 projects, as long as at least one but not more than 2 of Projects 8, 9, 10 are included.

e. The package costs $21, and invests in at least 2 but not more than 4 of projects 1 through 5.

f. The package costs $27, invests in at least 3 but not more than 5 of projects 4 through 10, provided it invests in at least one of projects 4 and 5 and at most two of projects 7, 8 and 9.

6.4 Prepare a hub diagram, if applicable, for each of the packages (a)–(f) of the preceding problem. Further, provide both a "regular" netform and a hub diagram for packages (b) and (c) combined where at most one of them is allowed to be selected.

6.5 Formulate the following problem as a cost minimization netform, by attaching appropriate bounds, arc costs (or profits), arc multipliers, and integer restrictions to the diagram supplied.
A company has 2 mines, M1 and M2. Each produces 4 bargeloads of ore

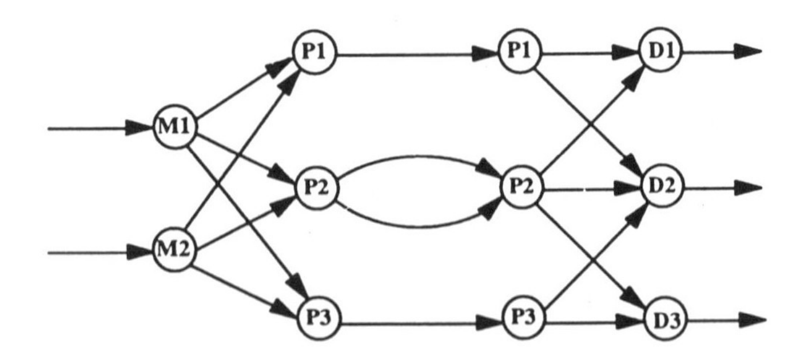

each day. Each mine sends its bargeloads to any of three processing plants, P1, P2, P3, where ore is processed and sent along to 3 demand centers, D1, D2, D3. Processing plant P1 can accommodate at most 4 bargeloads a day. It can transform each bargeload into 9 truckloads of finished material, and ship these truckloads in any desired fashion to demand centers D1 and D2 (so that either center receives any number of truckloads). Processing plant P2 is less efficient, but can handle 6 bargeloads of ore per day. Each of the first 4 bargeloads can be transformed into 8 truckloads of finished material, but processing efficiency drops after that, and each remaining bargeload yields only 7 truckloads. Plant P2 can send its truckloads to any of the 3 demand centers. Plant P3 is still less efficient, and can only produce 6 truckloads for each bargeload. This plant can send its truckloads (in any fashion) to D2 and D3. Processing costs per bargeload are 30 at plant P1, 26 at plant P2 and 23 at plant P3. Tables providing additional relevant information appear below. For efficiency, the company will not allow fractional bargeloads or truckloads.

Cost Per Bargeload Shipped From Mines to Processing Plants

	P1	P2	P3
M1	30	20	10
M2	15	25	15

Cost Per Truckload Shipped From Plants to Demand Centers

	D1	D2	D3
P1	17	19	×
P2	14	14	18
P3	×	12	12

Daily Requirements and Price Paid for a Truckload at Each Demand Center

	Minimum Truckloads Required	Price Paid/ Truckload
D1	13	80
D2	16	70
D3	14	75

6.6 The following diagram shows the continuous solution to the problem of Exercise 6.2 (i.e., the solution with the integer requirements relaxed.)

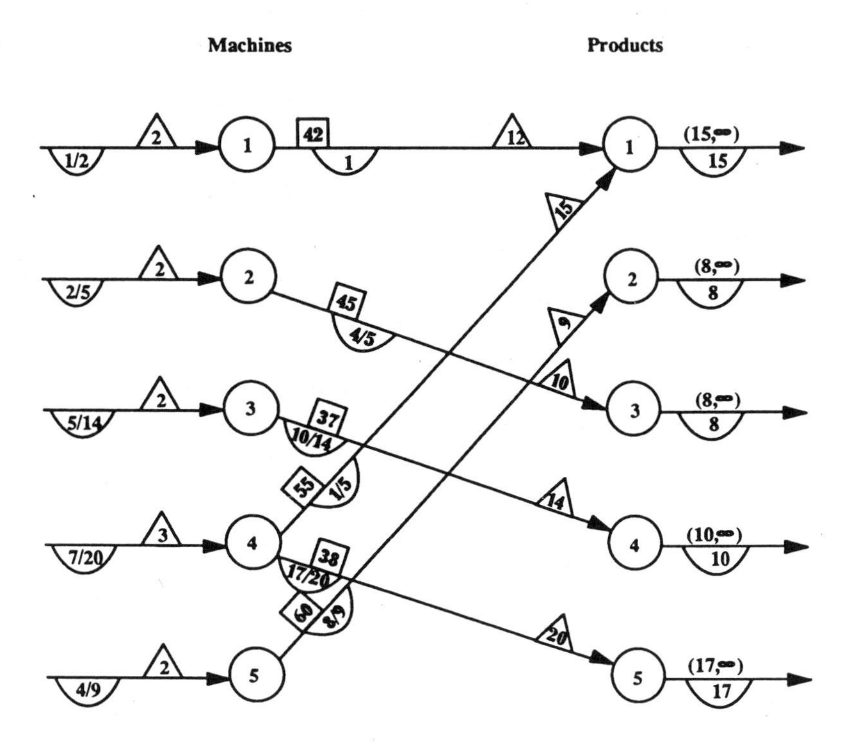

Analyze this diagram to determine how many possible ways the continuous solution can be rounded (so that the fractional-valued flows will be integers). Can you establish that none of these rounded solutions is feasible? Note that if the extra freedom is allowed to round some of the flows, and then determine wholly new values for others, a feasible solution can be obtained. However, show that even this strategy will not work if at most 16 units of Product 2 can be produced. Finally, suppose a problem consisting of 20 machines and 20 products is created, which consists simply of four copies of the problem of Exercise 6.2, plus any number of additional arcs linking the copies together, where the additional arcs cost just enough that their optimal continuous flow values are 0. (This does not imply that their optimal integer flow values are 0.) Explain why this problem will have 2^{40} (more than 1,000,000,000,000) rounding possibilities, none feasible.

6.7 For a special project that requires delivery in 1 month, a company can hire two types of skilled labor (Skill Category 1 and Skill Category 2) to produce two products, A and B. A person of either skill can work on producing either product, although individuals of Skill 1 are more efficient at producing Product A, and individuals of Skill 2 are more efficient at producing Product B. The following table shows the number of units of each product that can be produced in a day by a person of a given skill type working only on that product (e.g., if a person of Skill 2 works half a day on each product, he will produce 5 units of Product A and 8 units of Product B).

Units of Product That Can Be Produced By Person of Particular Skill Type in 1 Day, Working Only on a Single Product

Skill Category	Product A	Product B
1	15	8
2	10	16

To determine the number of people of each skill to hire, the company has the following information:

Skill Category	Maximum No. of People Available to Hire	Hiring Cost Per Person ($)	Days Available During Month at Regular Daily Wage (By Union Contract)	Regular Daily Wage ($)
1	7	80	20	40
2	4	90	23	45

Each person cannot work more than the days available specified for his skill category. Assuming that the company requires 2500 units of Product A and 2600 units of Product B to be produced at the end of the month, and assuming a given employee can split his time in any fashion between working on the two product types, construct an integer generalized network model to determine how many people of each skill should be hired to minimize cost.

6.8 Indicate how the model for Exercise 6.7 would change under the assumption that each employee hired must be assigned to work on a single product for the full duration of employment.

6.9 Show how the model of Exercise 6.7 can be modified to handle each of the following situations (introduced cumulatively):

a. The union contract requires that a person of Skill 1 be paid an additional $3 for each day spent working on Product A, and that a person of Skill 2 be paid an additional $4 for each day spent on Product A.

b. The 2500 units of Product A and 2600 units of Product B that the company requires to be produced are the result of a contract calling for delivery of these amounts, for which the company is paid an attractive lump sum. The purchasers indicate that they are willing to accept up to 300 additional units of each product at a price of $4 per unit for Product A and $3.50 per unit for Product B.

c. The purchasers will allow delivery of up to 200 units fewer than the previously required amount of each Product A and B, but will reduce the amount they pay to the company by $6.50 for each unit short of Product A and by $5 for each unit short of Product B.

d. In addition to the maximum number of days available for each individual previously specified, the company must agree to hire each person for at least 15 days for Skill 1 and 17 days for Skill 2.

e. It is possible to assign an employee up to 5 extra days of work a month beyond the "maximum," provided the employee is paid an extra duty wage of $50/day for Skill 1 and $57/day for Skill 2 for each day worked beyond the stipulated limit.

***6.10** Expand the final diagram of Exercise 6.9 to include the following:

a. The purchasers agree to conditions (b) and (c) of Exercise 6.9, and beyond this will accept up to 100 additional units of each product (over the agreed 300 units) at a price of $3.75/unit for Product A and $3.25/unit for Product B. Similarly, they will allow a further delivery shortage, of up to 100 additional units for each product (beyond the agreed 200 units) at a reduction in their payments of $7 for each additional unit short of Product A and of $5.50 for each additional unit short of Product B.

b. The figures of condition (a) are instead as follows:

	Product A	Product B
Unit price for up to 100 units beyond the first 300 extra ($)	5	4
Unit price reduction for up to 100 units beyond the first 200 short ($)	6	3

6.11 The supply arc

can be replaced by 15 arcs:

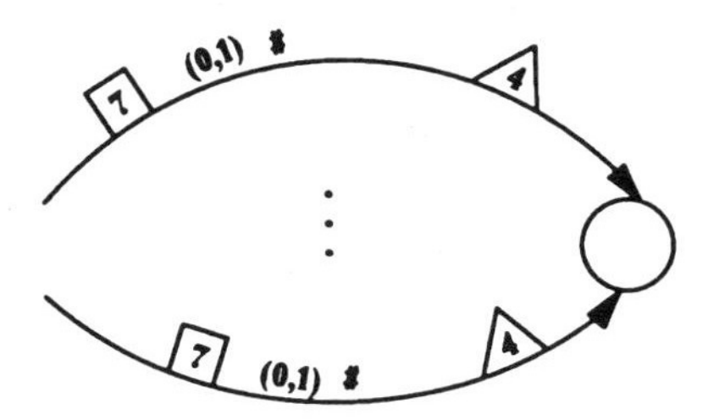

The original arc can also be replaced by 4 arcs:

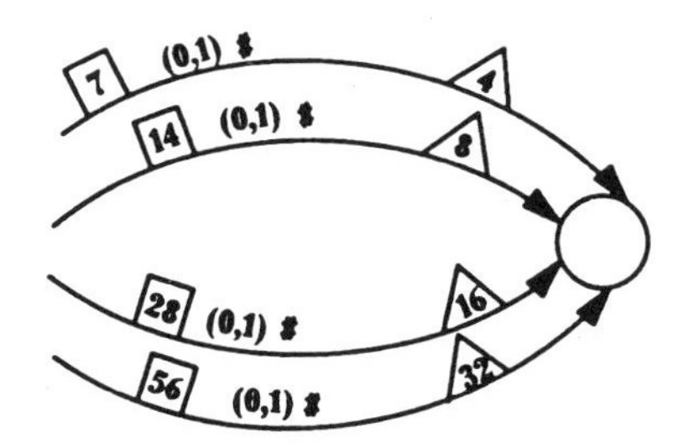

To illustrate this, identify how to assign integer flows to this latter diagram to correspond to each of the following:

a. A flow of 2 on the original arc

b. A flow of 3 on the original arc

c. A flow of 6 on the original arc

d. A flow of 9 on the original arc

e. A flow of 15 on the original arc

f. Why is the last diagram only an alternate model for supply arcs? (Hint: Suppose the original arc was between two nodes A and B, having no supply or demand. Would each diagram require the same flow into node A in order to have 60 units of flow arrive at node B?)

6.12 How could you modify the last diagram of Exercise 6.11 (by adding an additional arc and node) if the original arc has an upper bound of 13? In addition, show an appropriate corresponding diagram equivalent to the following:

6.13 A bus company has three different types of buses (standard, limousine, and shuttle) that it seeks to assign to four routes. The following table gives the capacity limits (in passengers), numbers of buses available, numbers of trips that can be made on different routes, and the daily passenger demand for each route.

Bus Type	Passenger Capacity	No. of Buses Available	Maximum Number of Daily Trips Per Bus on Route			
			1	2	3	4
Standard	40	6	2	3	1	2
Limo	12	11	3	4	2	3
Shuttle	9	13	5	5	2	4
Demand in passengers			180	70	90	85

Operating costs on different routes and the opportunity cost (lost profit) for not carrying a passenger are as follows:

Bus Type	Dollar Operating Cost Per Trip on Route			
	1	2	3	4
Standard	220	200	300	240
Limo	180	160	200	200
Shuttle	160	120	180	160
Opportunity cost per customer	12	9	14	10

Provide an integer netform to determine how many buses to assign to each route to minimize total cost.

6.14 For each of the following facility configuration conditions, expand the locate/allocate model in Figure 6.31 to include the condition. These conditions are equivalent to side constraints expressed in terms of the Z_j variables. The task is to create these constraints in terms of the basic locate/allocate model structure. (Hint: A facility can be closed by assigning it to a dummy customer which uses its entire capacity. Similarly, a facility can be opened by assigning it to a dummy customer which uses only a very small part of its capacity.)

a. At least one of the first three facilities must be chosen to be open (equivalent to $Z1 + Z2 + Z3 \geqslant 1$).

b. At least two of the first four facilities must be chosen to be open (equivalent to $Z1 + Z2 + Z3 + Z4 \geqslant 2$).

c. At most three of the five facilities may be chosen to be open (equivalent to $Z1 + Z2 + Z3 + Z4 + Z5 \leqslant 3$).

d. Facilities 4 and 5 are either both chosen to be open or both closed (equivalent to $Z4 = Z5$).

6.15 Below is an actual letter describing a problem situation and a proposed model. Create a netform model which represents the problem situation. You may make reasonable assumptions about anything which is not specified. Try to remain open minded (i.e., do not be overly influenced by the proposed model).

Professor Darwin Klingman
Department of Computer Science
University of Texas
Austin, Texas 78712

September 25, 1981

Dear Professor Klingman,

I have just seen your paper appearing in the June issue of *Operations Research.* The problem you solve in your paper is very like a problem recently posed to me, and I am wondering whether or not your methods could be adapted to solve my problem.

The problem is as follows: A steel company produces n different products. Each product is produced from an ingot which can be any one of several different sizes depending on the product. There are m different ingot sizes, and the cost of producing the ith product from an ingot of the jth size is c_{ij}. If the products are produced in the most economical way, the number of different ingot sizes used will be large. The company prefers to restrict the number of sizes used to $k(k < m)$. What k sizes should they choose in order to minimize the total cost of producing all n products?

The problem can be formulated as a mixed, integer linear program, but the number of 0,1 variables would be m, and m can be as large as 150. It can also be formulated as a trans-shipment problem with n sources (products), m trans-shipment nodes (ingot sizes), and one sink. A reduction in the number of trans-shipment nodes used in the solution can be forced by imposing the same fixed charge, say C, on all arcs connecting the trans-shipment nodes to the sink. By varying C, it might be possible to obtain a solution using exactly k of the trans-shipment nodes.

If you have any suggestions for solving this problem, I would very much appreciate hearing from you.

Sincerely,

Associate Professor of
Industrial Engineering
Lehigh University
Bethlehem, Pennsylvania 18015

6.16 A small cartage firm has 6 trucks with a 3-ton carrying capacity and 4 trucks with a 5-ton carrying capacity. A special job arises that requires immediate shipment of 22 tons of material from a customer's factory to one of his dealers in a nearby city. The cost of sending a truck to make a shipment is $35 for each 3-ton capacity truck and $55 for each 5-ton capacity truck.

a. Devise a netform to identify the least cost way for the cartage firm to handle this job.

b. Suppose that if a 5-ton capacity truck carries less than its full capacity, there will be a cost savings (due to reduced gas consumption) for the trip of $7 for each ton below the 5-ton capacity. Prepare a netform to minimize shipping costs.

c. The cartage firm will be paid $13 for each ton delivered up to 19 tons and $10.50 for each additional ton up to 22 tons, as long as the firm agrees to deliver a minimum of 15 tons. Construct a netform to maximize the firm's income, including the conditions of part (b).

d. The cartage firm has two customers both requiring immediate shipments. The first customer requires 14 tons to be shipped, and the second requires 17 tons to be shipped, where the costs to the cartage firm are as follows:

	Cost ($) Per Trip for	
Truck Capacity (tons)	Customer 1	Customer 2
3	35	37
5	55	58

Construct a netform to minimize shipping costs.

e. The two customers of part (d) have altered their shipping requirements to 36 tons for Customer 1 and 35 tons for Customer 2. After studying the situation carefully, the cartage firm determines there is enough time for each truck to make up to 3 trips for Customer 1 and up to 2 trips for Customer 2, provided a given truck is assigned to only one of the customers (due to the geographical separation of the shipment origins and destinations for the two customers). Further, the firm estimates that under these circumstances, it will cost $9 for each truck assigned to Customer 1 and $13 for each truck assigned to Customer 2, in addition to the trip costs identified in part (d). Construct a netform to minimize the firm's costs.

6.17 On January 1 a distributorship for a manufacturer of heavy machinery estimates its monthly sales for the first quarter to be as follows:

	January	February	March
Expected sales (in units of product)	144	132	158

The current inventory is 65 units, and inventory purchases can be made on the 1st or 15th of each month, requiring an approximate one-half month lag time before delivery (at which time they are paid for). Sales are expected to cluster around the middle and the end of each month, equally divided between them. Delivery on up to 30 units of sales in any half-month period may be postponed to the following half-month period at a cost of $10/unit, and delivery on up to 10 additional units may be postponed for a full month at a cost of $15/unit. The cost for holding a unit in inventory is set at $4/unit for every unit carried between successive half-month periods. Company policy requires an inventory of 80 units (with no backlogged deliveries) to be available for the start of the second quarter. Each inventory purchase is limited to a maximum of 81 units. The distributor wants to minimize total cost.

a. Create a netform model which includes a set-up cost of $35 for placing an inventory purchase order by introducing fixed costs on appropriately created purchase arcs.

b. Disregarding the fixed cost of placing an order, how would the netform be altered if purchases had to be in multiples of 20 units (using discrete flow arcs)?

6.18 Provide an illustrative part of the UFT model that will handle each of the following considerations:

a. A given schedule (of those available to a student) allows the student to take any 3 of 5 specified classes.

b. A schedule requires taking from 2 to 4 (any 2 to 4) of the 5 specified classes.

c. A schedule requires taking 2 particular classes and any 2 of 3 remaining.

d. A schedule requires taking 2 particular classes but between 1 or 2 of 3 remaining.

e. A class may close down, but otherwise requires between 20 and 30 people to attend. (Hint: Use an all-or-none arc, whose flow is compelled to equal either 0 or the arc's upper bound. This type of bound is indicated on the arc as (0 or *u*).)

6.19 The Big-K-Investment Co. has $10 which can be invested in projects A, B, C, D, or any combination of these. The cash flow C_{it} associated with project *i* at the beginning of year *t* is given below, where $C_{it} < 0$ denotes a cash outflow and $C_{it} > 0$ denotes a cash inflow.

	Beginning of Year		
Project	1	2	3
A	−8	3	8
B	−2	0	3
C	0	−9	11
D	−1	−4	7

Each project can be selected once or can be rejected. Fractional projects are not allowed, that is, a project has to be totally invested in or not at all. Returns can be reinvested. Cash not used can be invested in a bank account yielding 5% interest per year. Up to $3 additional financing is available each year from the bank which must be repaid the following year with 15% interest.

a. Draw a netform to maximize cash on hand at the end of the planning horizon. (Hint: Model the cash flows as multipliers which transform an investment decision into cash.)

b. Assume that fractional projects are allowed, resulting in the same fractions of the corresponding cash flows. Can this be modeled as a generalized network by deleting the integer requirements (#) in your model for (a)? Explain.

6.20 The Small-K-Manufacturing Co. has signed a contract to deliver the following quantities of special widgets at the beginning of each week:

Week	1	2	3
Quantity	0	38	18

Small-K has 2 machines, A and B, capable of producing widgets. Widgets are produced in lots. The minimum lot size for Machine A is 10 and the maximum is 15. Similarly, for Machine B the lot size must be between 20 and 30 units. Each machine can produce at most one lot per week. There is a set-up cost of $18 for A and $40 for B each time a lot is to be produced, and the variable production cost is $5 per widget for A and $3 per widget for B. All widgets produced go into inventory first and are shipped to the customer the following week. The inventory holding cost is $1 per week. Initial inventory is zero and any leftover widgets after the 3rd week must be recycled at a cost of $2 per widget.

a. Draw a netform to minimize the cost of fulfilling the contract.

b. What changes must be made to your netform from part (a) to represent: if Machine B is used in period 1 then it must also be used in period 2?

c. What changes must be made to your netform from part (a) to represent: if Machine A is used in period 1 and in period 2 then Machine B must be used in period 2?

6.21 The Super-K-Trading Co. purchases a product in each time period $t = 1, 2, \ldots, T$, stores it in inventory, and sells it in subsequent time periods for a price of $\$p$ per unit. The inventory holding cost is negligible, the maximum inventory is L. The total purchasing cost per unit depends on the quantity purchased. If 10 or fewer units are purchased, they cost \$5 each. Units purchased in excess of 10 are available for a discounted cost of \$2 each. At most 25 units are available for purchase each time period. Draw two netforms to maximize profit for period t using:

a. Fixed-charge arcs of the form

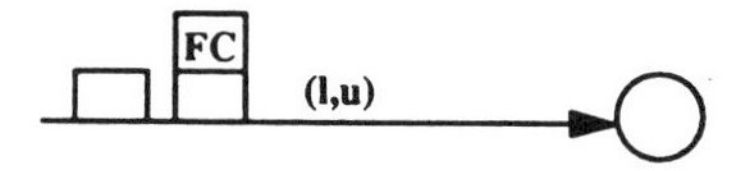

plus, if necessary, logical side constraints.

b. Decision arcs of the form

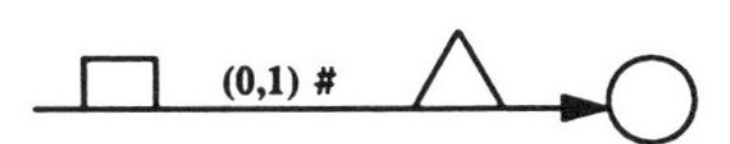

(Hint: Let the multiplier supply the maximum purchase quantity.)

6.22 Briefly outline two application areas for locate/allocate models which do not involve physical distribution systems. Specify the entities which are analogous to the following physical distribution entities (be creative):

Customer
Facility
Transportation cost
Handling cost
Capacity available at facilities
Capacity demanded by customers

REFERENCES

Fuller, S., D. Klingman, and P. Randolph. 1976, "A Cotton Ginning Problem," *Operations Research*, Vol. 24, No. 4, pp. 700–717.

Glover, F., J. Hultz, D. Klingman, and J. Stutz. 1978, "Generalized Networks: A Fundamental Computer-Based Planning Tool," *Management Science*, Vol. 24, No. 12, pp. 1209–1220.

Glover, F. and D. Klingman. 1977, "Network Application in Industry and Government," *AIIE Transactions*, Vol. 9, No. 4, pp. 363–376.

Glover, F. and J. Mulvey. 1980, "Equivalence of the 0-1 Integer Programming Problem to Discrete Generalized and Pure Networks," *Operations Research*, Vol. 28, No. 3, Part 2, pp. 829–836.

APPENDIX A

LINEAR PROGRAMMING

A.1 REVIEW

Suppose we have a problem which is to optimize an objective subject to a set of constraints. When both the objective and constraints are described using mathematical symbols, functions, and relationships, the problem is called a ***mathematical program.*** Further, if these relationships are linear, the program is a ***linear program*** or ***linear programming problem*** (LP problem).

The critical linearity assumptions of an LP problem are:

1. *Additivity*
   i. Profit measure and resource usage must be additive
   ii. No by-products
2. *Proportionality*
   i. Profit measure and resource usage increase proportionally
   ii. No increasing or decreasing returns to scale
   iii. No set-up costs
3. *Divisibility*
   i. Fractional units allowed
   ii. No integer requirements

The general form of an LP problem with n decision variables $x_j, j = 1, 2, \ldots, n$, and m constraints is

$$\text{Maximize} \quad z = c_1x_1 + c_2x_2 + \cdots + c_nx_n$$

$$\text{Subject to} \quad \left.\begin{array}{l} a_{11}x_1 + a_{12}x_2 + \cdots + a_{1n}x_n \leqslant b_1 \\ a_{21}x_1 + a_{22}x_2 + \cdots + a_{2n}x_n = b_2 \\ \vdots \qquad\qquad\qquad\qquad\qquad\quad \vdots \\ a_{m1}x_1 + a_{m2}x_2 + \cdots + a_{mn}x_n \geqslant b_m \end{array}\right\} m \text{ constraints}$$

$$x_1, x_2, \ldots, x_n \geqslant 0$$

Or in matrix notation:

$$\begin{aligned} \text{Maximize} \quad & Z = C^T X \\ \text{Subject to} \quad & AX \begin{matrix} \leqslant \\ = \\ \geqslant \end{matrix} B \\ & X \geqslant 0 \end{aligned}$$

In ***standard form*** an LP problem is

$$\begin{aligned} \text{Maximize} \quad & Z = C^T X \\ \text{Subject to} \quad & AX = B \\ & X \geqslant 0 \end{aligned}$$

To convert a general LP problem to standard form, add a slack variable to every $\leqslant$ constraint and subtract a surplus variable from every $\geqslant$ constraint. Restrict the slack and surplus variables to be nonnegative.

For a standard form LP problem with n variables (including the decision variables, slack variables, and surplus variables) and m equations, where n is greater than m, a ***basic solution*** may be obtained by setting $n - m$ of the variables equal to zero and solving the m constraint equations for the remaining variables. The variables set equal to zero are called ***nonbasic variables***, and the others are called ***basic variables***. A set of basic variables is referred to as a ***basis***. A ***basic feasible solution*** is a basic solution that also satisfies the nonnegativity conditions. The optimal solution to an LP problem, if one exists, occurs at a basic feasible solution. Thus, at an optimal solution, at most m of the decision variables will have nonzero value. (If the problem has decision variables with upper or lower bounds, then these variable bounds do not need to be included in the m constraint equations for most solution procedures. In this case the bounded variables may be nonbasic at their upper or lower bounds instead of at zero. This increases the potential number of nonzero decision variables to m plus the number of bounded variables. It is still true, however, that at most m of the unbounded decision variables may be nonzero in an optimal solution.)

Example 1: A company has the opportunity of producing two products. The first product yields a net profit of \$3 per unit while the second product yields \$4 per unit. To produce one unit of the first product requires 6 h of processing on machine 1 and 4 h of processing on machine 2. To produce one unit of the second product requires 4 h of processing on the first machine and 8 h on the second machine. There are at most 12 h of processing time available on machine 1. Machine 2 must be used at least 16 h. The company wants to maximize its net profit when producing these products.

Let x_1 be the number of units of the first product produced and x_2 be the number of units of the second product produced. This can be formulated as the following linear programming problem:

$$\begin{aligned} \text{Maximize} \quad & 3x_1 + 4x_2 \\ \text{Subject to} \quad & 6x_1 + 4x_2 \leqslant 12 \\ & 4x_1 + 8x_2 \geqslant 16 \\ & x_1, x_2 \geqslant 0 \end{aligned}$$

An equivalent statement of this problem in detached coefficient form is

$$\begin{array}{lcl} & \begin{matrix} x_1 & x_2 \end{matrix} & \\ \text{Maximize} & [3 \quad 4] & \Leftarrow \text{Objective function} \\ \text{Subject to} & \begin{bmatrix} 6 & 4 \\ 4 & 8 \end{bmatrix} \begin{matrix} \leqslant \\ \geqslant \end{matrix} \begin{bmatrix} 12 \\ 16 \end{bmatrix} & \Leftarrow \text{Right-hand side (RHS)} \\ & \Uparrow & \\ & \text{Coefficient matrix } (A) & \end{array}$$

To transform this problem to standard form let s_1 be slack time on machine 1 and s_2 be surplus time on machine 2.

$$\begin{aligned} \text{Maximize} \quad & 3x_1 + 4x_2 \\ \text{Subject to} \quad & 6x_1 + 4x_2 + s_1 = 12 \\ & 4x_1 + 8x_2 - s_2 = 16 \\ & x_1, x_2, s_1, s_2 \geqslant 0 \end{aligned}$$

Example 2: Daily requirements of 70 g of protein, 1 g calcium, 12 mg iron, and 3000 calories are needed for a balanced diet. The following foods are available for consumption with the cost and nutrients per 100 g as shown.

	Protein (g)	Calories	Calcium (g)	Iron (mg)	Cost ($)
Brown bread	12.0	246	0.1	3.2	.5
Cheese	24.9	423	0.2	0.3	2
Butter	0.1	793	0.03	0.0	1
Baked beans	6.0	93	0.05	2.3	.25
Spinach	3.0	26	0.1	2.0	.25

The objective is to find a balanced diet with minimum cost. Let x_1 = number of 100 g units of brown bread employed, x_2 = number of 100 g units of cheese employed, x_3 = number of 100 g units of butter employed, x_4 = number of 100 g units of baked beans employed, x_5 = number of 100 g units of spinach employed.

The linear programming problem is then

$$\begin{aligned}
\text{Minimize} \quad & 0.5x_1 + 2x_2 + 1x_3 + 0.25x_4 + 0.25x_5 \\
\text{Subject to} \quad & 12x_1 + 24.9x_2 + 0.1x_3 + 6.0x_4 + 3.0x_5 \geqslant 70 \\
& 246x_1 + 423x_2 + 793x_3 + 93x_4 + 26x_5 \geqslant 3000 \\
& 0.1x_1 + 0.2x_2 + 0.03x_3 + 0.05x_4 + 0.1x_5 \geqslant 1 \\
& 3.2x_1 + 0.3x_2 + 0.0x_3 + 2.3x_4 + 2.0x_5 \geqslant 12 \\
& x_1, x_2, x_3, x_4, x_5 \geqslant 0
\end{aligned}$$

Example 3: Consider a mixing problem within the context of a nut factory. Suppose that the factory wishes to sell peanuts, cashews, almonds, walnuts, hazel nuts, and two types of mixed nuts. The problem is to determine the mix of the seven types of nuts that will maximize profit during the Christmas season. The availability and costs of the five constituents are:

Constituent	Abbreviation	Maximum Quantity Available (lb)	Cost ($/lb)	Selling Price ($/lb)
Peanuts	P	20,000	.7	2
Cashews	C	8,000	2.5	5
Almonds	A	6,000	2	4
Walnuts	W	7,000	1.5	3
Hazel nuts	H	9,000	1.7	3

The desired mixture specifications of mixed nut Types A and B are indicated below:

Type	Specification	Selling Price ($/lb)
A	50% peanuts 40% cashews 10% almonds	3.50
B	Not more than 40% peanuts Not less than 25% cashews Not less than 15% hazel nuts	4.50

Assume that all other cash flows are fixed so that the profit to be maximized is total sales income minus the total cost of the constituents. Set up a linear programming model for determining the amount and blend of each nut type.

Before attempting to write down a linear programming model, careful consideration should always be given to the proper definition of the decision variables. Notice that the constituents are both inputs and outputs in this problem; however, the inputs are not the only outputs. Usually, if this is the case, the easiest way to set up the problem is to define a decision variable for each possible input-output combination. Thus, we define the variable x_{ij} to be the total amount of input i allocated to output j. For example, x_{pA} represents the total amount of peanuts (p) allocated to mixed nut Type A. It should be noted that the only possible decision variables are

$$x_{pp},\ x_{pA},\ x_{pB},\ x_{cc},\ x_{cA},\ x_{cB},\ x_{aa},\ x_{aA},\ x_{ww},\ x_{hh},\ x_{hB}$$

The total profit is

$$\begin{aligned}(2-0.7)x_{pp} &- 0.7x_{pA} - 0.7x_{pB} + (5-2.5)x_{cc} - 2.5x_{cA} - 2.5x_{cB}\\ &+ (4-2)x_{aa} - 2x_{aA} + (3-1.5)x_{ww} + (3-1.7)x_{hh}\\ &- 1.7x_{hB} + 3.5(x_{pA} + x_{cA} + x_{aA}) + 4.5(x_{pB} + x_{cB} + x_{hB})\end{aligned}$$

Hence, the problem is to maximize this objective, subject to the restrictions imposed by the availability of inputs, the mixture requirements, and by the restrictions that $x_{ij} \geqslant 0$. The availability restrictions clearly are

$$\begin{aligned} x_{pp} + x_{pA} + x_{pB} &\leqslant 20{,}000\\ x_{cc} + x_{cA} + x_{cB} &\leqslant 8{,}000\\ x_{aa} + x_{aA} &\leqslant 6{,}000\\ x_{ww} &\leqslant 7{,}000\\ x_{hh} + x_{hB} &\leqslant 9{,}000\end{aligned}$$

The mixture restrictions for Type A are as follows: Since $x_{pA}/(x_{pA} + x_{cA} + x_{aA})$ is the proportion of peanuts in Type A mixture we have the restriction

$$x_{pA}/(x_{pA} + x_{cA} + x_{aA}) = 0.5$$

Or, rewriting,

$$x_{pA} = 0.5(x_{pA} + x_{cA} + x_{aA})$$

or

$$0.5x_{pA} - 0.5x_{cA} - 0.5x_{aA} = 0$$

Similarly, the other constraints are

$$x_{cA} = 0.4(x_{pA} + x_{cA} + x_{aA})$$
$$x_{aA} = 0.1(x_{pA} + x_{cA} + x_{aA})$$

The mixture restrictions for Type B are as follows:

$$x_{pB} \leqslant 0.4(x_{pB} + x_{cB} + x_{hB})$$
$$x_{cB} \geqslant 0.25(x_{pB} + x_{cB} + x_{hB})$$
$$x_{hB} \geqslant 0.15(x_{pB} + x_{cB} + x_{hB})$$

Notice that the variable x_{ww} is restricted as follows:

$$0 \leqslant x_{ww} \leqslant 7000$$

This type of linear programming problem is called a ***bounded variable (capacitated) linear programming problem*** since some of the variables (e.g., x_{ww}) are explicitly restricted to lie between two numbers (bounds).

Example 4: A manufacturer produces particle board panels which consist of a low-quality filler sheet sandwiched between two higher quality veneer sheets, each made from different combinations of sawdust and plastic resin. The manufacturer has available 400 lb of sawdust and 500 lb of plastic resin. Three departments produce the low- and high-quality sheets with each department using a slightly different manufacturing method. The following table gives the raw material requirements per production run and the resulting output.

	Input per Run (lb)		Output per Run (sheets)	
Department	Sawdust	Plastic Resin	Low Quality	High Quality
1	9	5	4	8
2	6	8	9	6
3	3	7	7	3

The objective is to determine the number of production runs for each department which will maximize the total number of particle board panels.

Let x_1, x_2, and x_3 be the number of production runs for departments 1, 2, and 3, respectively. The total number of low-quality sheets produced by the three departments is $4x_1 + 9x_2 + 7x_3$. Similarly, the total number of high-quality sheets is $8x_1 + 6x_2 + 3x_3$. The corresponding constraints on the raw materials are given by $9x_1 + 6x_2 + 3x_3 \leqslant 400$ for sawdust and $5x_1 + 8x_2 + 7x_3 \leqslant 500$ for plastic resin.

Since the objective is to maximize the total number of particle board panels and since each panel requires one low-quality sheet and two high-quality sheets, it follows that the maximum number of panels cannot exceed the smaller value of

$$4x_1 + 9x_2 + 7x_3 \quad \text{and} \quad \frac{8x_1 + 6x_2 + 3x_3}{2}$$

Thus the problem becomes

$$\begin{aligned} \text{Maximize} \quad & \min\left\{4x_1 + 9x_2 + 7x_3, \frac{8x_1 + 6x_2 + 3x_3}{2}\right\} \\ \text{Subject to} \quad & 9x_1 + 6x_2 + 3x_3 \leqslant 400 \\ & 5x_1 + 8x_2 + 7x_3 \leqslant 500 \\ & x_1 \geqslant 0, x_2 \geqslant 0, x_3 \geqslant 0 \end{aligned}$$

The above formulation violates the linear programming properties since the objective function is nonlinear. A trick can be used, however, which will reduce the above model to the acceptable linear programming format.
Let

$$z = \min\left\{4x_1 + 9x_2 + 7x_3, \frac{8x_1 + 6x_2 + 3x_3}{2}\right\}$$

This means that

$$4x_1 + 9x_2 + 7x_3 \geqslant z \quad \text{and} \quad \frac{8x_1 + 6x_2 + 3x_3}{2} \geqslant z$$

These two inequalities are equivalent to

$$\begin{aligned} 4x_1 + 9x_2 + 7x_3 - z \geqslant 0 \\ 8x_1 + 6x_2 + 3x_3 - 2z \geqslant 0 \end{aligned}$$

It immediately follows that the above problem reduces to the following linear programming problem:

$$\begin{aligned} \text{Maximize} \quad & z \\ \text{Subject to} \quad & 4x_1 + 9x_2 + 7x_3 - z \geqslant 0 \\ & 8x_1 + 6x_2 + 3x_3 - 2z \geqslant 0 \\ & 9x_1 + 6x_2 + 3x_3 \leqslant 400 \\ & 5x_1 + 8x_2 + 7x_3 \leqslant 500 \\ & x_1 \geqslant 0, x_2 \geqslant 0, x_3 \geqslant 0, z \geqslant 0 \end{aligned}$$

A.2 THE DUAL

Imagine that you own a company producing two products and that your operations can be modeled in the following linear program:

$$\begin{aligned} \text{Maximize} \quad & 2x_1 + 3x_2 \\ \text{Subject to} \quad & 1x_1 + 2x_2 \leqslant 3 \\ & 3x_1 + 1x_2 \leqslant 4 \\ & x_1, x_2 \geqslant 0 \end{aligned}$$

The right-hand side, $B = \begin{bmatrix} 3 \\ 4 \end{bmatrix}$, represents the resources that make up the company (e.g., units of time, labor, capital, etc.), and the variables represent the number of units of each product produced, with revenue shown in the objective function.

The optimal production policy for your company is to produce one unit each of x_1 and x_2 to yield an objective value of 5. Under this policy, you totally use up the resources available to the company.

Consider what would happen if you were to sell the company (i.e., the resources that comprise the company). What is the minimum value that you would place on your total resources?

The answer is 5 units, since this is what you would get if you were to keep the company.

A second and even more important question is: How much is each unit of each resource worth to you? For example, your company has two resources, resource 1 and resource 2. For what price would you sell one unit of resource 1? And, vice versa, what would you be willing to pay for an additional unit of resource 1?

These questions can be answered by using another linear program called the ***dual***. We begin by creating a dual variable for each of the resources. Let w_1 be the value of resource 1 and w_2 be the value of resource 2. The question concerning the total value of our resources is answered by the dual objective function:

$$\text{Minimize} \quad 3w_1 + 4w_2$$

Note this is a minimization since it is always easy to over-value resources.

The constraints of the dual are constructed by determining how the company is currently using its resources. For example, it is using one unit of resource 1 and three units of resource 2 to produce each unit of product 1. In return, it receives a revenue of 2. Therefore, the first constraint is

$$1w_1 + 3w_2 \geqslant 2$$

implying that the value assigned to the resources must net at least what the company could get if it used those resources.

Likewise, the second dual constraint is

$$2w_1 + 1w_2 \geqslant 3$$

This constraint stems from how the company uses the resources in producing each unit of product 2.

Finally, we stipulate that the dual variables, w_1 and w_2, are restricted to have nonnegative values. The original linear program, called the ***primal***, and its associated dual are shown below.

Primal

$$\begin{aligned} \text{Maximize} \quad & 2x_1 + 3x_2 \\ \text{Subject to} \quad & 1x_1 + 2x_2 \leqslant 3 \\ & 3x_1 + 1x_2 \leqslant 4 \\ & x_1, x_2 \geqslant 0 \\ \text{Optimal solution:} \quad & x_1 = 1, x_2 = 1 \\ \text{Objective function:} \quad & 5 \end{aligned}$$

Dual

$$\begin{aligned} \text{Minimize} \quad & 3w_1 + 4w_2 \\ \text{Subject to} \quad & 1w_1 + 3w_2 \geqslant 2 \\ & 2w_1 + 1w_2 \geqslant 3 \\ & w_1, w_2 \geqslant 0 \\ \text{Optimal solution:} \quad & w_1 = 7/5, w_2 = 1/5 \\ \text{Objective function:} \quad & 5 \end{aligned}$$

If the primal is in ***standard form:***

Primal

$$\begin{aligned} \text{Maximize} \quad & C^T X \\ \text{Subject to} \quad & AX = B \\ & X \geqslant 0 \end{aligned}$$

Then the dual is

Dual

$$\begin{aligned} \text{Minimize} \quad & W^T B \\ \text{Subject to} \quad & W^T A \geqslant C^T \\ & W \text{ unrestricted} \end{aligned}$$

However, if the primal is not in standard form it is not necessary to convert it. The following primal ↔ dual transformation table can be used. (Note that the dual of the dual is the primal.)

Maximize	Minimize
Variables	*Constraints*
Nonnegative ($\geqslant 0$)	Greater than or equal to ($\geqslant$)
Nonpositive ($\leqslant 0$)	Less than or equal to ($\leqslant$)
Unrestricted	Equality ($=$)
Constraints	*Variables*
Greater than or equal to ($\geqslant$)	Nonpositive ($\leqslant 0$)
Less than or equal to ($\leqslant$)	Nonnegative ($\geqslant 0$)
Equality ($=$)	Unrestricted

Example 5

Primal

$$
\begin{aligned}
\text{Maximize} \quad & 1x_1 + 2x_2 + 3x_3 \\
\text{Subject to} \quad & 4x_1 + 1x_2 + 5x_3 \geqslant 6 \\
& 3x_1 + 2x_2 + 6x_3 \leqslant 7 \\
& 2x_1 + 3x_2 + 7x_3 = 8 \\
& x_1 \geqslant 0,\ x_2 \leqslant 0,\ x_3 \text{ unrestricted}
\end{aligned}
$$

Dual

$$
\begin{aligned}
\text{Minimize} \quad & 6w_1 + 7w_2 + 8w_3 \\
\text{Subject to} \quad & 4w_1 + 3w_2 + 2w_3 \geqslant 1 \\
& 1w_1 + 2w_2 + 3w_3 \leqslant 2 \\
& 5w_1 + 6w_2 + 7w_3 = 3 \\
& w_1 \leqslant 0,\ w_2 \geqslant 0,\ w_3 \text{ unrestricted}
\end{aligned}
$$

A.3 MODEL CHARACTERISTICS

The following model characteristics affect the ability to solve an LP problem:

1. Constraints
   i. Capacities (bounded variables $x_j \leqslant u_j$) are easy

ii. Number of constraints; more constraints increases difficulty
iii. "soft" constraints (inequalities) are easier than "hard" ones (equalities)

2. Magnitude of parameters; most codes provide accuracy up to 10^6 for:
   i. Coefficient matrix
   ii. Right-hand side
   iii. Objective function
3. Density of coefficient matrix; more dense (percentage of nonzero parameters) increases difficulty.
4. Size of problems; large problems require use of:
   i. Matrix generator to generate coefficient matrix
   ii. Report writer to report solution in a readable form
   iii. Database management systems to collect data
   iv. Modeling languages to generate the model

APPENDIX B

DECISION SUPPORT SYSTEMS FOR NETWORK MODELS

B.1 INTRODUCTION

Decision support systems (DSS) are computer-based systems used to support the needs of managers for data and analysis. They are flexible and adaptable to accommodate changes in the environment and in the decision-making approach of the user. They possess features that make them easy to use by all levels of management across all functional areas.

Most DSS contain four basic components: the control subsystem, the data subsystem, the model subsystem, and the report subsystem. The control subsystem handles the user interface with the DSS and also manages the interactions among the three other subsystems. The functions of the data subsystem are data definition, query, and manipulation. The model subsystem handles model generation and solution. The report subsystem provides solution information for the user in the form of printed reports, graphics, etc.

Recently, artificial intelligence and expert systems techniques have been incorporated into the DSS. The resulting system, called an intelligent decision support system (IDSS), uses the knowledge of analysts, decision makers, and experts accumulated from experience in the field and with the system to further enhance the decision-making support provided by the system. Thus a knowledge base component forms a fifth subsystem of an IDSS. The next section presents a framework for an intelligent decision support system.

We then present an organization strategy for initializing and implementing an IDSS. This strategy is based on experience with several large companies. We conclude with a brief discussion of an optimization-based IDSS implemented at Citgo Petroleum Corporation to address the short-term planning and operational issues associated with the supply, distribution, and marketing of refined

---

Reprinted by permission. D. Klingman and N. Phillips, Proceedings from International Conference on Expert Systems and the Leading Edge in Production Planning and Control (May 10–13) pp. 79–110, 1987.

petroleum products. The estimated dollar benefits of the system in its first year of use include a $116 million reduction in product inventories resulting in a $14 million reduction in interest costs. In addition, the system is credited with saving over $2.5 million as a result of improved decision making.

B.2 INTELLIGENT DECISION SUPPORT SYSTEM

Figure B.1 depicts the basic elements of an IDSS and outlines each of their functions. As mentioned earlier, there are five components: the data subsystem, the model subsystem, the report subsystem, the knowledge base subsystem, and the control subsystem. The functions of each of these subsystems are discussed below.

B.2.1 Data Subsystem

The primary purpose of the data subsystem is to organize and maintain all of the data required by the IDSS so that they may be effectively used in the decision-making process. The capabilities of the data subsystem include: (1) data definition to specify and organize the data items; (2) data query to retrieve data items from the database; and (3) data manipulation to update data items or change data items for alternative scenario analysis. These tasks can be accomplished efficiently and easily through the use of an appropriate database management system (DBMS). Such a system provides for the maintenance, control, and sharing of the database, simplifies the interface between the IDSS

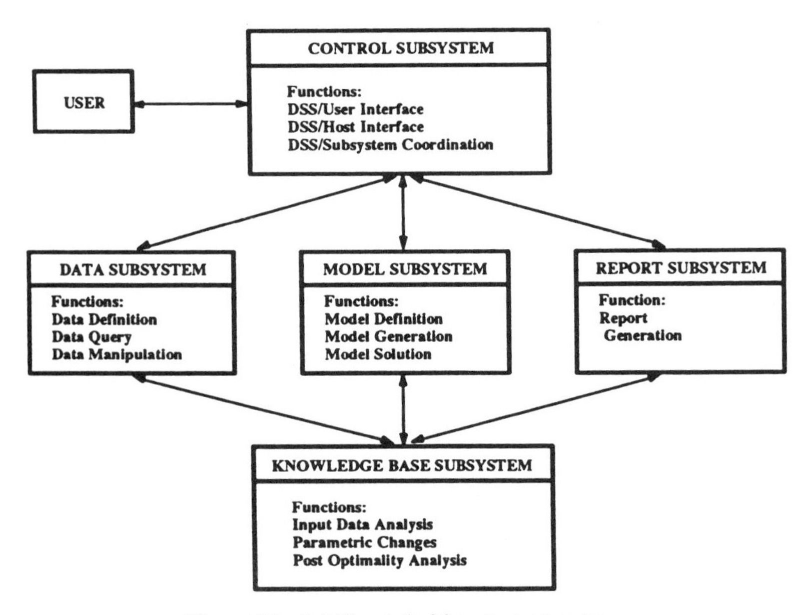

Figure B.1 Intelligent decision support system.

and the database, and reduces data redundancy by facilitating conversion of existing data into the form required by the IDSS. A DBMS also simplifies consistency checking and updating or expanding the database. Finally a corporate-wide DBMS facilitates operational decision making by enabling all users across all functional areas to have access to common data in addition to the support provided by the IDSS.

For example, in an integrated distribution planning problem, because supply and distribution decisions are inextricably tied to forecasted demand, the data subsystem requires inputs from all functional areas. Production may provide such inputs as production capacity, process conversion rates, material supplies and supplier locations, and production/supply costs. Marketing may contribute forecasts of sales and ranking of the relative importance of market areas, market segments or products. Distribution may provide such information as warehouse location and capacity, handling and storage costs, and freight rates. Some data, such as freight rates or forecasted supply costs, may be purchased from an outside vendor. Data which depicts management policies must be obtained directly from managers. These data include such items as disallowed shipping patterns, minimum and maximum inventory levels, backordering limitations, and customer service priorities.

The consequence of using data from throughout the company and including management dictated policies among the data is significant in several ways. First, the operating plans produced by the IDSS will conform to these policies and parameters. Second, because policies are input to the system as data rather than encoded in the analytical process per se, it is easy for management to specify alternative "what if" policies and gauge their impacts. Third, because the data cover the breadth of activities that must be coordinated, the system provides a quantitative basis for assessing the influence of decisions in one functional area on other areas. This offers a perspective from which to evaluate the combined corporate effects of decisions in each functional area. Fourth, as noted earlier, because input data are often drawn from distinct separate systems, discrepancies are often noted and reconciled. This improvement in data consistency in turn leads to greater consistency of effort throughout the firm.

B.2.2 Model Subsystem

The model subsystem enables the IDSS to consider decision activities as an integrated whole rather than separately. It does this by quantifying the combined impact over the planning horizon of resource restrictions, costs, management policies, market conditions, etc., on alternative decisions. The specific functions of the model subsystem are: (1) model definition and generation—using the data to generate the mathematical equations which specify the appropriate model; and (2) model solution—determining the recommended decisions for the problem at hand. Network optimization models form an important component of the model subsystem because, as discussed in Chapter 1, network solution algorithms are very efficient, and network models facilitate communication between technical and nontechnical staff due to their pictorial nature.

Over time, as the problem definition changes, the model should evolve accordingly. This is facilitated if, as much as possible, both the logic underlying the model and description of data items available to the model are expressed as inputs rather than coded in computer program logic. In this way changes in program inputs bring about the necessary evolution of the model without requiring changes in computer software. We refer to the generator for such a model as being "table-driven." An additional advantage of a table-driven model generator is that it allows the system to be used to support decisions associated with the sizing and configuration of the network (i.e., strategic or long-range planning) by making it easy to change, for example, the number or location of warehouses, plants or supply points, and customer demand points included in the analysis.

B.2.3 Report Subsystem

The report subsystem is a critical link in the communication process between the model subsystem and the users of the IDSS. This subsystem accesses the basic data and model solutions, which become a part of the database. From these data, it generates the set of reports specified by the user. Report media options include hard (paper) copy and soft (screen or disk) copy. The report writer, like the model generator, should be table-driven to facilitate changes in the evolving system and strategic "what if" analysis. The design of the report subsystem should also be flexible enough to allow new standard reports to be "plugged in" to the system easily. This feature is necessary since managers typically discover additional information requirements as they become more sophisticated users.

In addition to this flexibility, the reports must be custom-tailored to suit the needs and desires of the manager. The best analytical tools and techniques are worthless if their results are not communicated properly. Information content, level of detail, formatting, labeling, graphics, even type style and color greatly influence the usability of reports. This objective has priority over all other considerations including hardware, software, technical, aesthetic, efficiency, and generality.

For example, the types of reports which should be available in an integrated distribution planning system include:

1. *Cost Summaries:* Annual, monthly, and/or weekly costs of the recommended plan in total and for individual activities (replenishment shipments, inventory, customer shipments, etc.).
2. *Volume Summaries:* Inventory levels by time period and shipment levels by mode of transportation and time period.
3. *Facility Reports:* Cost and volume for throughput and inventory for each plant, distribution center, and/or warehouse by time period.

4. *Customer Service:* On-time deliveries by source and mode of transportation, backorders, and lost sales for each customer zone by time period.
5. *Variance Reports:* Volume and cost differences between the recommended plan and some performance standard, or last year's actual operation, or an alternative plan, etc.
6. *Master Production Schedule:* Cost and volume by time period and facility.

B.2.4 Knowledge Base Subsystem

The knowledge base subsystem, which employs artificial intelligence and expert systems techniques, enhances management support by incorporating into the system the knowledge of analysts, decision makers, and experts in the field. Its primary functions are: (1) input data analysis—to analyze, verify, and modify the data which supports the system; (2) parametric changes—to automate tactical "what if" analysis by resetting certain parameters of the model or to set parameters based on logical conditions; and (3) post optimality analysis—to extract key information from the model solution and various reports in order to facilitate evaluation of the model recommendation.

Input data analysis is extremely important for two main reasons. First, projections and forecasting techniques are used to generate much of the input data, and thus complete accuracy is a never-to-be attained dream. By maintaining both historical data and forecasts in the database, management may use input data analysis to track forecasting accuracy. Second, input data is collected (often through an automated process) from different functional areas. Thus intensive error checking, validation, and cross checking should be performed to avoid the GIGO (garbage in garbage out) phenomenon and prevent system "crashes" due to improperly formatted data. Output from this analysis should include the erroneous data value, location in the file, and relevant information from the record which might help management locate the source of the error.

Parametric changes may be used to adjust for imperfect data by determining how sensitive the optimal solution is to changes in these data values. The knowledge base subsystem may also be used to determine the initial values of certain parameters which are dependent on "if-then" conditions. For example, maximum inventory levels may be increased if total demand for the planning horizon exceeds a certain value. The knowledge base contains the rules and facts used to extract key information from the model and data regarding these parameters.

Analysis of the model solution is the final area that can be aided considerably by the knowledge base subsystem. Artificial intelligence provides helpful support when the model output is very lengthy. It can be used to create exception reports, extracting key points and flagging unusual model recommendations, thus allowing top management to reap the maximum benefits from the system without having to invest a large amount of personal time in output analysis. It also plays an important role in helping management resolve infeasibilities.

To perform the above functions, the knowledge representation scheme known as production systems, which consist of a database of "if-then" rules called productions, is frequently used. Individual productions in the rule base can be added, deleted, or changed independently and are easy to specify. The knowledge base subsystem, then, is composed of these production rules along with the input data files, the model solution, and the output reports.

B.2.5 Control Subsystem

The control subsystem links the data, model, report, and knowledge base subsystems together and provides the primary user interface for system operation. Through this interface, the user issues a set of instructions to be carried out by the IDSS in order to perform a particular analysis. For example, a manager may wish to solve a "base case" (basic model with basic data), then solve an alternative scenario which asks "what if we allow shipments directly from plants to preferred customers?", and then compare the results. This analysis requires performance of several functions in each of the subsystems: data access, model generation and solution, reporting, and knowledge base interface. The control subsystem will invoke each of the other subsystems as required to carry out the request in addition to satisfying the job step and file manipulation requirements of the host computer systems. All of these technicalities should be transparent to the user. Rather than being required to master the intricate operating procedures of his computer system, the manager should only be required to understand his problem situation and analysis strategy. The control subsystem removes the burden of technology from the manager, which frees him to perform his managerial function.

The design of the user interface is extremely important. It should consist of a set of high-level, English-language statements which are custom-tailored to the manager's conceptualization of the problem situation. These statements are used to issue the instructions to perform particular analyses. It is essential that this instruction set include the capability of generating alternative scenarios for "what if" type analyses. In addition, the structure and syntax of the language must be as natural as possible to enhance the usability of the system.

We should re-emphasize here that flexibility is a most important attribute for decision support systems. All problem situations are dynamic. In addition, since IDSS technology is fairly new, most users cannot be aware of its full potential at the outset. We have found that managers' understandings, perception, and needs change and grow as they gain more experience using the technology. The system must be flexible enough to grow and change along with them if decision support is going to achieve its full potential. Addition of new models, new data, new policies, new reports, and new analysis capabilities must be a routine process which does not entail major system overhauls.

B.3 SALIENT REQUIREMENTS FOR SUCCESS

Several Fortune 500 companies in various industries (including Agrico [Glover et al. 1979], Citgo Petroleum [Klingman et al. 1986, 1987], Cahill May Roberts Pharmaceutical [Harrison 1978], Hunt Wesson Foods [Geoffrion and Graves 1974], W. R. Grace [Klingman et al. 1988], Sloane [Information Processing 1980], General Mills, General Motors [Blumenfeld et al. 1987], Phillips, Exxon, Amoco, International Paper [Bender et al. 1981], Ciba-Geigy, Kelly-Springfield [King and Love 1980], IBM [Cohen et al. 1990], and Dunlop) have designed, built, and implemented DSS for integrated distribution planning. Based on their reported experiences and our first-hand experiences in working with many of them, we have identified the following requirements as being absolutely necessary to the success of a decision support system:

1. Adequate data availability
2. Management involvement as well as approval
3. High-quality reports
4. Evolutionary approach to implementation

Each of these items have been discussed, or at least alluded to, in the previous section along with the other IDSS requirements. However, we feel that these four are absolutely essential and therefore warrant additional attention in order that the reader will be fully aware of their importance.

B.3.1 Adequate Data Availability

The availability of correct and current data is critical to achieving the ultimate goal of a decision support system, that of providing useful, credible information which managers can trust as a basis for their decisions. If managers cannot trust the data, they will not (and should not) trust the analysis results. Correct and current data are usually difficult to obtain. The establishment of formal procedures for data collection and maintenance must be given a high priority from the outset.

We strongly recommend assigning data collection and maintenance responsibilities across the functional areas with specific time deadlines and management sign-off functions. This process must be clearly managed. We have found data maintenance to be a major problem, and this procedure has proved to be highly successful. Without validated data, planning meetings often result in discussions about why the current reports cannot be used for planning purposes.

Organizations which already have a well-established management information system (MIS) for operations are a quantum leap ahead of those who do not in successfully coping with the data problem. Once the investment has been made in an MIS, the next logical step is to add an intelligent analyzer in order to gain the multiplicative advantage of the IDSS technology. In this case, most of the data required to support decision making can be obtained directly from the corporate database. The organizations which do not have an existing MIS must be willing to endure a significant effort to develop organizational policies and structures which will insure good data maintenance. However, experience indicates that the results easily justify the effort.

B.3.2 Management Approval and Involvement

The decision to use an IDSS approach must come from the upper levels of corporate management for two reasons. First, a significant quantity of company resources must be committed with no expectation of real return for up to 12 months. The costs of system development may range from $500,000 to $2,000,000. In addition, increased staff for data collections, system maintenance, and possibly actual analysis may be necessary.

Second, the ultimate usefulness of an IDSS depends upon the joint cooperation and efforts of the users (bottom-level management) in the various corporate areas which the model concerns. However, different areas will always have some inherently conflicting individual goals. Therefore, top management involvement is necessary. With this involvement, IDSS can support a truly holistic approach to managing and planning operations.

To facilitate user acceptance and cooperation, many companies have formed management planning committees composed of representatives from the different areas involved. The functions of the committee include (1) primary responsibility for promoting and executing two-way communication between areas, (2) establishing data maintenance policies and monitoring their compliance, and (3) making final recommendations.

Planning committees have found it helpful to initially specify a single user of the IDSS. This user then performs all analysis desired by the planning committee and functional areas and disseminates all reports. The advantages of this approach are numerous. Functional areas are able to use the results of the system much sooner. It allows managers to gain confidence in the results of the system and often creates desire on their part to learn how to use the system.

As managers become accustomed to using IDSS reports, they should be encouraged to use the system by themselves. As they conduct their own independent scenario analysis they garner valuable insights which are difficult, if not impossible, to obtain in any other manner.

B.3.3 High-Quality Reports

The requirement for high-quality reports was discussed in detail in the previous sections, but it deserves reinforcement. System reports must be in the exact format and style which the user prefers. This may require creating different versions of the same report information for different managers or management groups. In addition, it is essential that each report contain the proper amount of detail for its intended function. Forcing a manager into the time-consuming and frustrating task of using unfamiliar or inappropriate report structures is certain to alienate him from the system and bias him against accepting its results. The term "user friendly" has been widely applied as a desirable operational characteristic of computer systems. This term is also appropriate to describe this critical characteristic of the output from decision support systems. User friendly reports relieve the decision maker from clerical tasks and provide him more time to think, analyze, decide, and manage.

B.4 CONSIDERATIONS FOR DESIGN AND IMPLEMENTATION

Most companies have found it best to begin the process of creating an IDSS by setting up a task force. It should be composed of representatives from the functional areas involved, the computer services departments, and an outside staff who have prior experience in building IDSS.

Through meetings with other members of the corporation, the task force must prepare a detailed list of the specific information requirements that the system needs to provide. Simultaneously the data to be used by the system to meet these information requirements should be specified and its availability should be verified. Next the set of reports which are to be used by management should be designed. This is crucial for the system to successfully meet its intended information requirements. After a series of modifications, which usually take from one to three months, the data and report specifications of the system should be firmly established.

Subsequently, a design or blueprint of the proposed system should be developed with supporting documents on its capabilities, limitations, and anticipated computer and staff requirements. Once the design is finalized and accepted, which normally requires two to three months, the construction phase of the system begins.

The most important implementation strategy which we can recommend is plan to build a prototype. A prototype system is a "first cut" toward the final system. It will be used, misused, expanded, contracted, molded, redesigned, and possibly discarded. IDSS technology is so new to decision makers that they cannot possibly pre-define all of their decision support requirements. IDSS builders can help, but the decision maker is the authority on his problem situation and ultimately the system must conform to his needs.

The prototype system typically limits data requirements by, for example, restricting the planning horizon to 12 months or less and restricting the number of products handled by the system to a major product or product group. The advantages of building a prototype are that it can be created relatively quickly (in comparison to building the total system); it gives the decision maker a real system to interact with and react to; it allows data validation; and it allows evaluation of ongoing computer resource requirements. Typically the lead time required from project build approval to making meaningful evaluations with a prototype is from 6–9 months. (A significant portion of this time is spent in the data collection effort.)

There are several factors which motivate the need for this evolutionary approach to implementation of an IDSS. Typically users and builders of such systems are not able to totally predict all possible requirements. Only by actual experience with the system is the full range of its necessary and feasible capabilities uncovered. In addition, individual users as well as organizations require a transition period of learning new processes and restructuring old ones. During this period, changes in the system are required to accommodate the changes in its uses. The evolutionary approach is best supported by the development of a prototype system and by making a concerted effort to build as much flexibility as possible into the full system. System flexibility can best be accomplished by building a data- or table-driven system. As mentioned earlier, using a table-driven model generator and report writer facilitates changes in the evolving system without requiring changes in the computer software.

B.5 APPLICATION: CITGO PETROLEUM CORPORATION

Citgo Petroleum Corporation, with 1985 sales of approximately $3.8 billion, is the nation's largest independent downstream marketer of petroleum products. Citgo has recently realized multi-million dollar savings through decision support provided by an optimization-based IDSS for supply, distribution, and marketing planning. This system, called the SDM system, integrates an optimization-based network model, state-of-the-art database management system, fourth-generation modeling language, and flexible report generation capabilities with a knowledge base component.

The model subsystem contains a separate multi-period minimum cost flow network model for each of Citgo's four major products (regular, unleaded, and super unleaded gasoline, and fuel oil). These models, generated using a fourth-generation table-driven modeling language called GENASYS, integrate Citgo's key economic and physical supply, distribution, and marketing characteristics over a short-term (11-week) planning horizon. Product sources include refining at Citgo's refinery in Lake Charles, Louisiana, spot purchases on each of the five major spot markets, and over 400 product exchange contracts with other industry refiners. Demand for product occurs as wholesale and retail demand at over 350 Citgo and exchange partner product terminals, spot sales at the spot markets, and product exchanges and trades. All costs such as purchases, refining, transportation, time value of money, and marketing overhead are reflected in the model. The time horizon incorporates both manufacturing and distribution time lags. The article by Klingman et al. [1987] contains a derivation of the entire model in network format.

The SDM system serves as both a financial and operational planning tool. Top management uses the system to make many operational decisions such as where to sell product, what price to charge, where to buy or trade product, how much product to hold in inventory, and how much product to ship by each mode of transportation (pipeline, tanker, and barge). All information is provided by location, by line of business, and by week. The model results are also used in the preparation of monthly and quarterly financial statements and in capital decisions such as product terminal acquisitions and inventory investments.

The development of the data subsystem was an important prerequisite for the successful operation of the SDM system. Prior to the implementation of the system, each department at Citgo collected, organized, and stored its own operational data. Due to the lack of corporate price and volume forecasts and limited access to other departmental decisions, different departments often made operational decisions which were non-optimal or even conflicting. In order to overcome these problems and to adequately support the SDM system, a corporate-wide, on-line database, called the Product Acquisition and Supply System (PASS) was developed. PASS was implemented using ADABAS, a state-of-the-art database management system with dictionary, query, and fourth-generation language features.

PASS, fed by information gathering and forecasting systems, contains up-to-date operational information on items such as sales, inventory, trades, and exchanges for all refined products for the past (historical data), present (scheduled activity), and future (forecasts). Data from PASS also feed another system called TRACS (Tracking, Reporting, and Aggregating Citgo Sales) which analyzes data to check forecast accuracy and manager performance.

The output reports of the SDM system, also written in GENASYS, are aimed toward four distinct user groups: the product managers, the pricing manager, the product traders, and the budget manager. Since each user is in a different aspect of the operation, the input data and the output reports are presented in the units and formats they normally use, such as thousands of barrels for traders, thousands of gallons for marketers, thousands of dollars for the budget manager, etc. The level of detail in the reports is also varied based on the purpose of the report in the decision-making activity. These report formats evolved gradually as the users became more knowledgeable about the system. Klingman et al. [1986] discuss in more detail the uses of the SDM system and the implementation challenges and success factors.

The knowledge representation scheme called production systems was chosen to implement the intelligent aspect of the SDM system, and SAS (Statistical Analysis System) was chosen as the implementation vehicle. The first function of the knowledge base subsystem is to perform intensive error checking on the input data. For example, the input data file containing information on product terminals is examined for missing values, invalid distribution network connections, initial inventory values below safety stock, etc. Relevant product terminal information, such as product terminal type and location in the file, is displayed in the error message along with the erroneous value so that management can locate the source of the error at a glance as well as know why the error occurred.

Second, the process of resetting certain parameters of the model was automated using expert systems. For example, the formulation of the model for exchange agreements was both complicated and problematic. Deterministic linear formulations cannot effectively capture real-world situations such as "if the contract balance is positive (Citgo owes), then Citgo should not be allowed to draw more product from the contract partner." This highlighted the need to add rules to the knowledge base which would prescreen the problem input to correctly handle such 0-1 conditions. In this example, production rules are used to compute the upper limits (bounds) for payback variables (arcs), which represent bulk product shipped from Citgo to the exchange partners or from the partners to Citgo.

Analysis of reports is the final area that has been aided considerably by the knowledge base subsystem. For example, exception reporting plays a key role in analyzing the voluminous reports on product exchanges. Rules are built into the system that will identify exchanges that have no wholesale sales, occurrences of imbalances exceeding specified amounts, and binding limits on the payback variables. The exchanges where these problems occur are displayed along with the condition tested, conclusion reached, and the solution recommended.

The control subsystem, written in the IBM EXEC language, prompts the user for necessary information and issues instructions to the operating system so that an appropriate sequence of operations is initiated to solve the problem at hand. The results are fed into the knowledge base which supplies changed information required to rerun the model if desired. Using a medium-sized computer, an IBM 4381 mod 2, each product model consisting of approximately 3000 equations (nodes) and 15,000 variables (arcs) can be solved with an efficient network optimization code in approximately 30 sec. About 2 min are required for data checking and model generation and about 7 min for report processing.

Implementation of the SDM system followed the general procedure outlined in the previous section. After the Southland Corporation purchased Citgo in 1984, Mr. John P. Thompson, CEO of Southland, appointed a task force headed by an outside consultant, Darwin Klingman. This task force studied ways to increase the profitability of Citgo and recommended development of the SDM system. A prototype system was constructed by outside consultants with help from internal Citgo managers who provided insight into the operational and organizational issues at Citgo and unique industry-specific concepts which should be included in the model. Managers from supply, distribution, and marketing were responsible to collect data for the system and to sign for the accuracy of this data. An internal manager was appointed to resolve format, validity, and consistency problems in the data. This person provided invaluable support in getting the data "clean" enough to run in the system. Several organizational changes were made to support the objectives of the SDM system. A position of senior vice-president of operations coordination was created and staffed to evaluate and coordinate model-recommended decisions which spanned area boundaries. The supply and distribution departments were combined to centralize the required responsibility and authority to deal with inventory control and terminal outage goals. During prototype development, numerous meetings were held with representatives from each functional area to validate the model and decide on report formats. The final system evolved gradually as managers became familiar with its capabilities.

The estimated dollar benefits of the SDM system in its first year of use (1985) include a $116 million reduction in product inventories (based on historical inventory-to-sales ratios), resulting in a $14 million decrease in interest costs. In addition, Citgo avoided at least a $58 million devaluation of its product inventories during the first half of 1986 as a result of this inventory reduction. The SDM system is also credited with saving over $2.5 million in 1985 as a result of improved decision making.

REFERENCES

Bender, P. S., W. D. Northup, and J. F. Shapiro. 1981, "Practical Modeling for Resource Management," *Harvard Business Review*, Vol. 59, No. 2, pp. 163–173.

Blumenfeld, D., L. Burns, C. Daganzo, M. Frick, and R. Hall. 1987, "Reducing Logistics Costs at General Motors," *Interfaces*, Vol. 17, No. 1, pp. 26–47.

Cohen, M., P. Kanesam, P. Kleindorfer, H. Lee, and A. Tekerian. 1990, "Optimizer: IBM's Multi-Echelon Inventory System for Managing Service Logistics," *Interfaces*, Vol. 20, No. 1, pp. 65–82.

Geoffrion, A. H., and G. W. Graves. 1974, "Multicommodity Distribution System Design by Benders Decomposition," *Management Science*, Vol. 20, No. 5, pp. 822–844.

Glover, F., G. Jones, D. Karney, D. Klingman, and J. Mote. 1979, "An Integrated Production, Distribution and Inventory Planning System," *Interfaces*, Vol. 9, No. 5, pp. 21–35.

Harrison, H. 1978, "A Planning System for Facilities and Resources in Distribution Networks," *Interfaces*, Vol. 9, No. 2, Part 2, pp. 6–22.

1980, "Information Processing—'What If' Help for Management," *Business Week*, pp. 73–74.

King, R. H. and R. R. Love, Jr. 1980, "Coordinating Decisions for Increased Profits," *Interfaces*, Vol. 10, No. 6, pp. 4–19.

Klingman, D., J. Mote, and N. Phillips. 1988, "A Logistics Planning System at W. R. Grace," *Operations Research*, Vol. 36, No. 6, pp. 811–822.

Klingman, D., N. Phillips, D. Steiger, R. Wirth, and W. Young. 1986, "The Challenges and Success Factors in Implementing an Integrated Petroleum Products Planning System for Citgo," *Interfaces*, Vol. 16, No. 3, pp. 1–19.

Klingman, D., N. Phillips, D. Steiger, R. Wirth, R. Padman, and R. Krishnan. 1987, "An Optimization Based Integrated Short-Term Refined Petroleum Product Planning System," *Management Science*, Vol. 33, No. 7, pp. 813–830.

APPENDIX C

SELECTED READINGS

Adolphson, D., M. Baird, and K. Lawrence. 1991, "A New Approach to Optimal Selection of Services in Health Care Organizations," *Socio-Economic Planning Science*, Vol. 25, No. 1, pp. 35–48.

Allen, J., R. Helgason, and J. Kennington. 1987, "The Frequency Assignment Problem: A Solution via Nonlinear Programming," *Naval Research Logistics*, Vol. 34, pp. 133–139.

Altinkemar, K. and B. Gavish. 1990, "Heuristics for Equal Weight Delivery Problems with Constant Error Guarantees," *Transportation Science*, Vol. 24, No. 4, pp. 294–297.

Amini, M. and R. Barr. 1990, "Applications of Network Reoptimization," *Proceedings of the 21st Annual Conference of the Southwest Decision Sciences Institute*, Atlanta, pp. 77–79.

Aronson, J., T. Morton, and G. Thompson. 1985, "A Forward Simplex Method for Staircase Linear Programs," *Management Science*, Vol. 31, No. 6, pp. 664–679.

Balakrishnan, A., T. Magnanti, and R. Wong. 1989, "Models for Planning Capacity Expansion in Local Access Telecommunication Networks," Working Paper No. 3048-89-MS, Sloan School of Management, Massachusetts Institute of Technology, Cambridge, MA.

Balas, E. 1970, "Project Scheduling with Resource Constraints," in E. Beale (ed.). *Applications of Mathematical Programming Techniques*, The English Universities Press, pp. 187–200.

Balas, E. 1989, "The Prize Collecting Traveling Salesman Problem," *Networks*, Vol. 19, pp. 621–636.

Balas, E., V. Chvatal, and J. Nesetril. 1987, "On the Maximum-Weight Clique Problem," *Mathematics of Operations Research*, Vol. 12, pp. 522–536.

Balas, E. and P. Ivanescu. 1964, "On the Generalized Transportation Problem," *Management Science*, Vol. 11, No. 1, pp. 188–202.

Balinsky, M. 1981, "On a Selection Problem," *Management Science*, Vol. 17, No. 3, pp. 230–231.

Ball, M., B. Golden, and R. Vohra. 1989, "Finding the Most Vital Arcs in a Network," *Operations Research Letters*, Vol. 8, pp. 73–76.

Bard, J. and T. Feo. 1989, "The Cutting Path and Tool Selection Problem in Computer Aided Process Planning," *Journal of Manufacturing Systems*, Vol. 8, No. 1, pp. 17–26.

Barlow, J. and F. Glover. 1987, "A Multicriteria Stratification Framework for Uncertainty and Risk Analysis," *International Journal of Policy and Information*, Vol. 3, No. 12, pp. 327–339.

Barr, R. and S. Turner. 1990, "Quality Issues and Evidence in Statistical File Merging," in G. Liepins and Uppuluri (eds.). *Theory and Pragmatics of Data Quality Control*, Marcel Dekker, New York, pp. 245–313.

Barr, R. and S. Turner. 1981, "Microdata File Merging Through Large-Scale Network Technology," *Mathematical Programming Study 15*, pp. 1–22.

Barras, J., S. Alec, C. Pasche, A. Gesmond, and D. deWerre. 1987, "Network Simplex Method Applied to AC Load-Flow Calculation," *IEEE Transactions on Power Systems*, Vol. 2, pp. 197–203.

Bazarra, M. and J. Jarvis. 1978, *Linear Programming and Network Flows*, Wiley, New York.

Belford, P. and H. Ratiff. 1972, "A Network-Flow Model for Racially Balancing Schools," *Operations Research*, Vol. 20, No. 3.

Bell, W., L. Dalberto, M. Fisher, A. Greenfield, R. Jaikumar, P. Keida, R. Mack, and P. Prutzman. 1983, "Improving the Distribution of Industrial Gases with an Online Computerized Routing and Scheduling Optimizer," *Interfaces*, Vol. 13, No. 6, pp. 4–23.

Bellmore M., H. Greengerg, and J. J. Jarvis. 1970, "Multicommodity Disconnecting Sets," *Operations Research*, Vol. 16, No. 6.

Bodin, L., B. Golden, and W. Stewart. 1979, *Report on the Subscriber Bus Routing Problem*, U.S. Department of Transportation, Urban Mass Transit Authority, Washington, D.C.

Boros, E., P. Hammer, and R. Shamir. 1990, "Balancing Rooted Data Flow Graphs," RUTCOR Research Report RRR 28-90, Rutgers University, New Brunswick, NJ.

Boros, E., P. Hammer, and X. Sun. 1991, "Network Flows and Minimization of Quadratic Pseudo-Boolean Functions," RUTCOR Research Report RRR 17–91, Rutgers University, New Brunswick, NJ.

Bowers, M. and J. P. Jarvis. 1992, "A Hierarchical Production Planning and Scheduling Model," *Decision Sciences*.

Brooks, R. 1984, "The Tenrac Gas Pipeline Competition Model," in B. Lev, F. Murphy, J. Bloom, and A. Gloit (eds.). *Analytic Techniques for Energy Planning*, Elsevier North-Holland, New York.

Brooks, R. 1981, "Using the GASNET3 Model to Forecast Natural Gas Distribution and Allocation During Periods of Shortage," *Math Programming Study 15*, North Holland, New York.

Brown, G., A. Geoffrion, and G. Bradley. 1981, "Production and Sales Planning with Limited Shared Tooling at the Key Operation," *Management Science*, Vol. 27, No. 3, pp. 247–259.

Brown, G., R. McBride, and R. Wood. 1985, "Extracting Embedded Generalized Networks from Linear Programming Problems," *Mathematical Programming*, Vol. 32, No. 1, pp. 11–31.

Chachra, V., P. Ghare, and J. Moore. 1979, *Applications of Graph Theory Algorithms*, Elsevier North-Holland, New York.

Chahal, N. and D. deWerre. 1989, "An Interactive System for Constructing Timetables on a PC," *European Journal of Operational Research*, Vol. 40, pp. 32–37.

Chinneck, J. 1989, "Viability Analysis: A Formulation Aid for All Classes of Network Models," Technical Report SCE-89-20, Department of Systems and Computer Engineering, Carleton University, Ottawa, Ontario, Canada K1S 5B6.

Chinneck, J. 1990, "Processing Network Models of Energy/Environment Systems," Technical Report SCE-90-10, Department of Systems and Computer Engineering, Carleton University, Ottawa, Ontario, Canada, K1S 5B6.

Chinneck, J. 1990, "Formulating Processing Network Models: Viability Theory," *Naval Research Logistics*, Vol. 37, pp. 245–261.

Chinneck, J. and M. Chandrashekar. 1984, "Models of Large-Scale Industrial Energy Systems—II, Optimization and Synthesis," *Energy*, Vol. 9, No. 8, pp. 679–692.

Chvatal, V. 1983, *Linear Programming*, W. H. Freeman, New York.

Cochand, M., D. deWerra, and R. Slowinski. 1989, "Preemptive Scheduling with Staircase and Piecewise Linear Resource Availability," *Zeitschrift fur Operations Research*, Vol. 33, pp. 297–313.

Collins, M., L. Cooper, R. Helgason, J. Kennington, and L. LeBlanc. 1978, "Solving the Pipe Network Analysis Problem Using Optimization Techniques," *Management Science*, Vol. 24, No. 7, pp. 747–760.

Collins, R., C. Meinhardt, D. Lemon, M. Gillette, and S. Gass. 1988, "Army Manpower Long Range Planning System," *Operations Research*, Vol. 36, No. 1.

Cornuejols, G., M. Fisher, and G. Nemhauser. 1977, "Location of Bank Accounts to Optimize Float: An Analytic Study of Exact and Approximate Algorithms," *Management Science*, Vol. 23, No. 8, pp. 789–810.

Cornuejols, G., G. Nemhauser, and L. Wolsey. 1980, "A Canonical Representation of Simple Plant Location Problems and Its Applications," *SIAM Journal of Algebraic and Discrete Methods*, Vol. 1, pp. 261–272.

Crum, D., D. Klingman, and L. Tavis. 1979, "Implementation of Large-Scale Financial Planning Models: Solution Efficient Transformations," *Journal of Financial and Quantitative Analysis*, Vol. 14, No. 1, pp. 137–152.

Dantzig, G. 1990, "The Diet Problem," *Interfaces*, Vol. 20, No. 4, pp. 43–47.

Dembo, R., A. Chiarri, J. Martin, and L. Paradinas. 1990, "Managing Hidroelectrica Espanolas Hydroelectric Power System," *Interfaces*, Vol. 20, No. 1, pp. 115–135.

Dembo, R., J. Mulvey, and S. Zenios. 1989, "Large-Scale Nonlinear Network Models and Their Application," *Operations Research*, Vol. 37, No. 3, pp. 353–372.

Dembo, R. and U. Tulowitski. 1988, "Computing Equilibria on Multicommodity Networks: An Application of Truncated Quadratic Programming Algorithms," *Networks*, Vol. 18, pp. 273–284.

Derigs, U. 1988, *Programming in Networks and Graphs*, Lecture Notes in Economics and Mathematical Systems, Vol. 300, Springer-Verlag, New York.

Deutsch, S., J. J. Jarvis, and R. Parker. 1979, "A Network Flow Model for Criminal Displacement and Deterrence," *Evaluation Quarterly*, Vol. 3, No. 2.

Dorsey, R., T. Hodgson, and H. Ratliff. 1975, "A Network Approach to a Multi-Facility, Multi-Product Production Scheduling Problem Without Backordering," *Management Science*, Vol. 21, No. 7.

Dutta, A., G. Koehler, and A. Whinston. 1982, "On Optimal Allocation in a Distributed Processing Environment," *Management Science*, Vol. 28, No. 8, pp. 839–853.

Elmaghraby, S. and J. Molder (eds.). 1978, *Handbook in Operations Research*, Van Nostrand-Reinhold, New York.

Evans, J. 1978, "A Single Commodity Transformation for Certain Multicommodity Networks," *Operations Research*, Vol. 26, No. 4, pp. 673–680.

Faaland, B. and T. Schmitt. 1987, "Scheduling Tasks with Due Dates in a Fabrication/Assembly Process," *Operations Research*, Vol. 35, No. 3, pp. 378–388.

Farina, R. and F. Glover. 1983, "The Application of Generalized Networks to Choice of Raw Materials for Fuels and Petrochemicals," *Energy Models and Studies*, Special Issue of the Institute of Management Sciences, B. Lev (ed.), North-Holland, New York, pp. 513–524.

Feo, T. and J. Bard. 1989, "Flight Scheduling and Maintenance Base Planning," *Management Science*, Vol. 35, No. 12, pp. 1415–1420.

Fisher, M., A. Greenfield, R. Jaikumar, and J. Lester. 1982, "A Computerized Vehicle Routing Application," *Interfaces*, Vol. 12, No. 4, pp. 42–52.

Fisher, M., R. Jaikumar, and L. Van Wassenhove. 1986, "A Multiplier Adjustment for the Generalized Assignment Problem," *Management Science*, Vol. 32, No. 9, pp. 1095–1103.

Fong, C. 1980, "Planning for Industrial Estate Development in a Developing Economy," *Management Science*, Vol. 26, No. 10, pp. 1061–1067.

Frey, S. and G. Nemhauser. 1972, "Temporal Expansion of a Transportation Network," *Transportation Science*, Vol. 6, pp. 306–323.

Gass, S. 1984, "Documenting a Computer-Based Model," *Interfaces*, Vol. 14, No. 3, pp. 84–93.

Gass, S. 1987, "Managing the Modeling Process: A Personal Perspective," *European Journal of Operational Research*, Vol. 31, No. 1.

Gass, S. 1991, "Military Manpower Planning Models," *Computers and Operations Research*, Vol. 18, No. 1, pp. 65–73.

Gautier, A. and F. Granot. 1992, "A Parametric Analysis of a Nonlinear Cash-Flow Asset Allocation Management Model," *Operations Research Letters*, Vol. 11, No. 5.

Gautier, A. and F. Granot. 1991, "Forest Managment: A Multicommodity Flow Formulation and Sensitivity Analysis," Commerce and Business Administration Research Paper, The University of British Columbia, Vancouver, British Columbia, Canada.

Gautier, A. and F. Granot. 1990, "A Parametric Analysis of a Constrained Nonlinear Inventory-Production Model," Commerce and Business Administration Research Paper, The University of Bristish Columbia, Vancouver, British Columbia, Canada.

Gavish, B. 1981, "A Decision Support System for Managing the Transportation Needs of a Large Corporation," *AIIE Transactions*, Vol. 13, No. 1, pp. 61–85.

Gavish, B. and K. Altinkemar. 1990, "Backbone Network Design Tools with Economic Tradeoffs," *ORSA Journal on Computing*, Vol. 2, No. 3, pp. 236–245.

Gavish, B. and H. Pirkul. 1986, "Computer and Database Location in Distributed Computing Systems," *IEEE Transactions on Computers*, Vol. 35, No. 7, pp. 583–590.

Gavish, B., P. Trudeau, M. Dror, M. Gendreau, and L. Mason. 1989," Circuit Network Design Under Reliability Constraints," *IEEE JSAC*, Vol. 7, No. 8, pp. 1181–1187.

Geoffrion, A. 1976, "Better Distribution Planning with Computer Models," *Harvard Business Review*, Vol. 54, No. 4, pp. 92–99.

Geoffrion, A., G. Graves, and S. Lee. 1982, "A Management Support System for Distribution Planning," *INFOR*, Vol. 20, No. 4, pp. 287–314.

Glover, F. 1981, "Creating Network Structure in LP's," in Greenberg and Maybee (eds.). *Computer-Assisted Analysis and Model Simplification*, Academic Press, pp. 361–368.

Glover, F., J. Hultz, and D. Klingman. 1978, "Improved Computer-Based Planning Techniques, Part I," *Interfaces*, Vol. 8, No. 4, pp. 16–25.

Glover, F., J. Hultz, and D. Klingman. 1979, "Improved Computer-Based Planning Techniques, Part II," *Interfaces*, Vol. 9, No. 4, pp. 12–20.

Glover, F., and K. Jones. 1988, "A Stochastic Generalized Network Model and Large Scale Mean-Variance Algorithm for Portfolio Selection," *Journal of Information and Optimization Sciences*, Vol. 9, No. 3, pp. 299–316.

Glover, F., D. Klingman, and N. Phillips. 1990, "Netform Modeling and Applications," *Interfaces*, Vol. 20, No. 4, pp. 7–27.

Glover, F. and D. Sommer. 1975, "Pitfalls of Rounding in Discrete Management Decision Problems," *Decision Sciences*, Vol. 6, No. 2, pp. 211–220.

Golden, B. and T. Magnanti. 1978, "Transportation Planning: Network Models and Their Implementation," in A. Hax (ed.). *Studies in Operations Management*, North-Holland, New York, pp. 465–518.

Golden, B., Q. Wang, and L. Liu. 1988, "A Multifaceted Heuristic for the Orienteering Problem," *Naval Research Logistics*, Vol. 35, pp. 359–366.

Goldman, A. and G. Nemhauser. 1967, "A Transport Improvement Problem Transformable to a Best Path Problem," *Transportation Science*, Vol. 1, pp. 295–307.

Greenberg, H. and F. Glover. 1987, "Netforms Provide Powerful Tools for Enhancing the Operations of Expert Systems," *Proceedings of the 1987 Rocky Mountain Conference on Artificial Intelligence*, pp. 259–268.

Greenberg, H. and J. Kalan. 1976, "Representations of Networks," *National Computer Conference Proceedings*, pp. 939–943.

Greenberg, H. and F. Murphy. 1985, "Computing Market Equilibria with Price Regulations Using Mathematical Programming," *Operations Research*, Vol. 33, pp. 935–954.

Gregg, S., J. Mulvey, and J. Wolpert. 1988, "A Stochastic Planning System for Siting and Closing Public Service Facilities," *Environment and Planning A*, Vol. 20, pp. 83–98.

Hadj-Alouane, A. and K. Murty. 1991, "A Constrained Assignment Problem," *Arabian Journal of Science and Engineering*, Vol. 16, No. 2B, pp. 233–238.

Hammer, P. 1965, "Some Network Flow Problems Solved with Pseudo-Boolean Programming," *Operations Research*, Vol. 13, No. 3, pp. 388–399.

Harary, F. 1969, *Graph Theory*, Addison-Wesley, Reading, MA.

Hu, T. 1970, *Integer Programming and Network Flows*, Addison-Wesley, Reading, MA.

Jarvis, J. J., R. Rardin, V. Unger, R. Moore, and C. Schimpeler. 1978, "Optimal Design of Regional Wastewater Systems: A Fixed-Charge Network Flow Model," *Operations Research*, Vol. 26, No. 4, pp. 538–550.

Jarvis, J. P., A. Perticone, and D. Shier. 1989, "Evaluation and Design of Voice Telecommunications Networks," in E. Sharda, B. Golden, E. Wasil, O. Balci, and W. Stewart (eds.). *Impacts of Recent Computer Advances on Operations Research*, North-Holland, New York.

Jarvis, J. P. and D. Shier. 1990, "NETSOLVE: Interactive Software for Network Optimization," *Operations Research Letters*, Vol. 9, No. 4, pp. 275–282.

Kelly, J., B. Golden, and A. Assad. 1991, "Cell Supression: Disclosure Protection for Sensitive Tabular Data," Working Paper MS/S 91-014, College of Business and Management, University of Maryland, College Park, MD 20783.

Kincaid, R., D. Nicol, D. Richards, and D. Shier. 1990, "A Multistage Linear Array Assignment Problem," *Operations Research*, Vol. 38, pp. 993–1005.

Kolen A. and J. Lenstra. 1990, "Combinatorics in Operations Research," Technical Report No. 925, School of Operations Research and Industrial

Engineering, College of Engineering, Cornell University, Ithaca, NY 14853-7501.

Lawler, E. 1976, *Combinatorial Optimization: Networks and Matroids*, Holt, Rinehart and Winston, New York.

Leung, J., T. Magnanti, and V. Singhal. 1988, "Routing in Point to Point Delivery Systems," Working Paper OR 174–88, Operations Research Center, Massachusetts Institute of Technology, Cambridge, MA.

Love, R. and R. Vemuganti. 1978, "The Single Plant with Mold Allocation Problem with Capacity and Changeover Restrictions," *Operations Research*, Vol. 26, No. 1, pp. 159–165.

Ma, P., F. Murphy, and E. Stohr. 1989, "Representing Knowledge about Linear Programming Formulation," *Annals of Operations Research*, Vol. 21, pp. 149–172.

Magnanti, T. and R. Wong. 1990, "Decomposition Methods for Facility Location Problems," in R. Francis and P. Mirchandani (eds.). *Discret Location Theory*, Wiley, New York, pp. 209–262.

Magnanti, T. and R. Wong. 1984, "Network Design and Transportation Planning: Models and Algorithms," *Transportation Science*, Vol. 18, pp. 1–56.

Mazzola, J. and A. Neebe. 1986, "Resource-Constrained Assignment Scheduling," *Operations Research*, Vol. 34, No. 4, pp. 560–572.

Mellalieu, P. and K. Hall. 1983, "An Interactive Planning Model for the New Zealand Dairy Industry," *Journal of the Operational Research Society*, Vol. 34, No. 6, pp. 521–532.

Merchant, D. and G. Nemhauser. 1978, "A Model and an Algorithm for the Dynamic Traffic Assignment Problem," *Transportation Science*, Vol. 12, pp. 183–199.

Minieka, E. 1978, *Optimization Algorithms for Networks and Graphs*, Marcel Dekker, New York.

Montreuil, B. and H. Ratliff. 1989, "Utilizing Cut Trees as Design Skeletons for Facility Layout," *IIE Transactions*, Vol. 21, No. 2.

Morton, T. and A. Srinivasan. 1989, "LOGJAM—A Logistics Planning System," Paper Presented at the Rochester Logistics Conference.

Mulvey, J. and H. Vladimirou. 1990, "Stochastic Network Programming for Financial Planning Problems," Technical Report SOR-89-7, Department of Civil Engineering and Operations Research, Princeton University, Princeton, NJ 08544.

Murphy, F., E. Stohr, and A. Asthana. 1991, "Representation Schemes for Linear Programming Models," Working Paper, School of Business, Temple University, Philadelphia, PA 19122.

Murphy, F. and Z. Wang. 1991, "A Network Reformulation of an Electric Utility Expansion Planning Model," Working Paper, School of Business and Management, Temple University, Philadelphia, PA 19122.

Narasimhan, S., H. Pirkul, and P. De. 1988, "Route Selection in Backbone Data Communication Networks," *Computer Networks and ISDN Systems 15*, North-Holland, NY, pp. 121–133.

Nulty, W. and H. Ratliff. 1991, "Interactive Optimization Methodology for Fleet Scheduling," *Naval Research Logistics*, No. 38.

Pirkul, H., J. Current, and V. Nagarajan. 1991, "The Hierarchical Network Design Problem: A New Formulation and Solution Procedures," *Transportation Science*, Vol. 25, No. 3, pp. 175–180.

Pirkul, H. and D. Schilling. 1988, "The Siting of Emergency Service Facilities with Workload Capacities and Backup Service," *Management Science*, Vol. 34, No. 7, pp. 896–900.

Rardin, R. 1982, "Tight Relaxations of Fixed Charge Network Flow Problems," Industrial Systems and Engineering Report Series No. J-82-3, School of Industrial and Systems Engineering, Georgia Institute of Technology, Atlanta, GA 30332.

Segal M. 1974, "The Operator Scheduling Problem: A Network Approach," *Operations Research*, Vol. 22, No. 4, pp. 808–823.

Shields, D., F. Glover, and R. Glover. 1987, "A Generalized Network with Material Routing for Mineral Supply Analysis," *Geostatistics*, Vol. 3, pp. 303–379.

Shier, D. 1982, "Testing for Homogeneity using Minimum Spanning Trees," *The UMAP Journal*, Vol. 3, No. 3, pp. 273–283.

Srinivasan A., M. Carey, and T. Morton. 1991, "Resource Pricing and Aggregate Scheduling in Manufacturing Systems," WP No. 88-89-58, Graduate School of Industrial Administration, Carnegie Mellon University, Pittsburgh, PA 15213.

Taha, H. 1971, *Operations Research*, Macmillan, New York.

Vladimirou, H. and J. Mulvey. 1991, "Parallel and Distributed Computing for Stochastic Network Programming," SOR-90-11, Department of Civil Engineering and Operations Research, Princeton University, Princeton, NJ 08544.

Williams, H. 1974, "Experiments in the Formulation of Integer Programming Problems," *Mathematical Programming Study 2*, pp. 180–197.

Wolters, J. 1979, "Minimizing the Number of Aircraft for a Transportation Network," *European Journal of Operational Research*, Vol. 3, pp. 394–402.

Zenios, S. 1991, "Network Based Models for Air-Traffic Control," *European Journal of Operational Research*, Vol. 50, pp. 166–178.

Zenios, S. 1990, "Integrating Network Optimization Capabilities into a High-Level Modeling Language," *ACM Transactions on Mathematical Software*, Vol. 16, No. 2, p. 113–142.

Zenios, S., A. Drud, and J. Mulvey. 1989, "Balancing Large Social Accounting Matrices with Nonlinear Network Programming," *Networks*, Vol. 19, pp. 569–585.

INDEX